Metazoan Life
without Oxygen

The cover photograph shows a natural cast of the discoidal, segmented 'worm' Dickinsonia *from Australia.*

Metazoan Life without Oxygen

EDITED BY

Christopher Bryant
Professor of Zoology
Australian National University

CHAPMAN AND HALL
LONDON · NEW YORK · TOKYO · MELBOURNE · MADRAS

UK	Chapman and Hall, 2–6 Boundary Row, London SE1 8HN
USA	Chapman and Hall, 29 West 35th Street, New York NY10001
JAPAN	Chapman and Hall Japan, Thomson Publishing Japan, Hirakawacho Nemoto Building, 7F, 1-7-11 Hirakawa-cho, Chiyoda-ku, Tokyo 102
AUSTRALIA	Chapman and Hall Australia, Thomas Nelson Australia, 102 Dodds Street, South Melbourne, Victoria 3205
INDIA	Chapman and Hall India, R. Seshadri, 32 Second Main Road, CIT East, Madras 600 035

First edition 1991

© 1991 Christopher Bryant

Printed in Great Britain by
St. Edmundsbury Press, Bury St Edmunds, Suffolk

ISBN 0 412 33360 0

All rights reserved. No part of this publication may be reproduced or transmitted, in any form or by any means, electronic, mechanical, photocopying, recording or otherwise, or stored in any retrieval system of any nature, without the written permission of the copyright holder and the publisher, application for which shall be made to the publisher.

The publisher makes no representation, express or implied, with regard to the accuracy of the information contained in this book and cannot accept any legal responsibility or liability for any errors or omissions that may be made.

British Library Cataloguing in Publication Data
Metazoan life without oxygen.
   1. Organisms. Oxygen
   I. Bryant, Christopher *1936*–
574.19214

ISBN 0–412–33360–0

Library of Congress Cataloging-in-Publication Data
Available

Contents

	Contributors	vii
	Preface	ix
	Introduction	x
1	The Physical Chemistry of Oxygen *Dereham L. Scott*	1
2	The Role of Oxygen Free Radicals in Biology and Evolution *Hosni M. Hassan* and *Joan R. Schiavone*	19
3	The Early Environment *Richard J. F. Jenkins*	38
4	Oxygen and the Early Evolution of the Metazoa *Bruce Runnegar*	65
5	Fumarate Reductase and the Evolution of Electron Transport Systems *Carolyn A. Behm*	88
6	Metazoan Adaptations to Hydrogen Sulphide *Russell D. Vetter*, *Mark A. Powell* and *George N. Somero*	109
7	Interstitial Meiofauna *Warwick L. Nicholas*	129
8	Parasitic Helminths *John Barrett*	146
9	Annelids *Udo Schöttler* and *Eva-Maria Bennet*	165
10	Molluscs *Albertus de Zwaan*	186
11	Arthropods *E. Zebe*	218

12	The Metabolic Arrest and Channel Arrest Concepts of Defence against Hypoxia in Vertebrates *Peter W. Hochachka*	238
13	Ruminant Animals and the Exploitation of Anaerobiosis *R.A.F. Chevis*	250
14	Anoxibiosis in Living Metazoa: an Overview *Carl S. Hammen*	271
	Index	279

Contributors

John Barrett, Department of Biological Sciences, University College of Wales, Aberystwyth, Wales, UK.

Carolyn A. Behm, Department of Biochemistry, Australian National University, Canberra, Australia.

Eva-Maria Bennet, Department of Zoology, Australian National University, Canberra, Australia.

R.A.F. Chevis, New South Wales Department of Agriculture, Glenfield, New South Wales, Australia.

Albertus de Zwaan, Department of Experimental Zoology, State University of Utrecht, The Netherlands.

Carl S. Hammen, Department of Zoology, The University of Rhode Island, Kingston, RI 02881, USA.

Hosni M. Hassan, Departments of Food Science and Microbiology and the Toxicology Program, North Carolina State University, Raleigh, USA.

Peter W. Hochachka, Department of Zoology, University of British Columbia, Vancouver B.C., Canada V6T 2A9.

Richard J.F. Jenkins, Department of Geology and Geophysics, University of Adelaide, Adelaide, South Australia.

Warwick L. Nicholas, Department of Zoology, Australian National University, Canberra, ACT 2601.

Mark A. Powell, Hopkins Marine Station, Stanford University, Pacific Grove, CA 93950.

Bruce Runnegar, University of California, Los Angeles, California, USA.

Joan R. Schiavone, Departments of Food Science and Microbiology and the Toxicology Program, North Carolina State University, Raleigh, USA.

Udo Schöttler, Zoologisches Institut der Universitat, Hindenburgplatz 55, Münster, Germany.

Dereham L. Scott, Department of Chemistry, Australian National University, Canberra, Australia.

George N. Somero, Marine Biology Research Division, Scripps Institution of Oceanography, La Jolla, CA 92093.

Russell D. Vetter, Marine Biology Research Division, Scripps Institution of Oceanography, La Jolla, CA 92093.

E. Zebe, Westfälische Wilhelms-Universitat Münster Zoologisches Institut, Lehrstuhl fur Tierphysiologie, Münster, Germany.

Preface

The genesis of this volume lies more than fifteen years in the past, when a reading of the current literature had clearly demonstrated that there were common features about the ways that organisms from all phyla coped either with the absence of oxygen in their environments or interruption to its supply to their vital organs. In spite of the success of journals like *Comparative Biochemistry and Physiology* and Marcel Florkin's great efforts in charting progress in comparative biochemistry, almost the only major text of recent years has been *Biochemical Adaptation* by Hochachka and Somero, published by Princeton University Press in 1984. While *Metazoan Life without Oxygen* does not attempt the integrated account of the of the last named book, it does lay out for the reader detailed summaries, for each major group of animals, of their strategies for anaero- or anoxybiosis.

It also and, I believe, uniquely takes an evolutionary approach to the problem. It starts with an description of the physical chemistry of oxygen, followed by a dissertation on the perils - and opportunities - created for life by oxygen derived free radicals. It moves on to examine the geochronology of the accumulation of oxygen in the environment and to analyse the first explosive adaptive radiation of the Metazoa in the Ediacarian and early Cambrian. It then explores the biochemistry of sulphide dependent organisms and follows with a detailed account of the evolution of fumarate reductase, the enzyme system that makes anaerobiosis possible in many invertebrate phyla. After the survey of invertebrate phyla, there is a chapter concerned with the strategies adopted by vertebrates for anoxibiotic survival, and the penultimate chapter is about one of the great ironies of adaptive biochemistry - the dependence of many vertebrates on anaerobic processes. This could, of course, be the take-off chapter for yet another book about the biochemistry of symbiosis and parasitism!

I owe all the authors who contributed to *Metazoan Life without Oxygen* a great debt of gratitude for their patience with a rather tardy Editor, whose only excuse is that the turmoil of the tertiary education system in Australia over the last three years proved a major distraction.

Introduction

Oxygen is a major constituent of the atmosphere. It is, perhaps, banal to begin a book of this nature with such a truism, but it is because oxygen is apparently omnipresent that organisms have evolved many strategies for doing without it. Oxygen is almost perfect for its role as terminal oxidant - its solubility in water is in just the 'right' range and its other chemical properties are such that its fitness as a major contributor to the energetic efficiency of life on Earth is recognised in its adoption by almost all phyletic groups. This is explored in detail in Chapter 1.

It was not always so. It is probable that physical reactions in the upper atmosphere ensured the continuous presence of oxygen in low concentrations near the surface of the Earth, but its great accumulation, beginning about three and a half billion years ago, was almost certainly due to the appearance and widespread proliferation of organisms with a primitive form of photosynthesis. These organisms were not oxygen respirers - they were anoxybiotic in the truest sense of the world.

Indeed, they were the first polluters of the pristine environment. Oxygen pollution of the environment brought with it a great many problems for early organisms. Because oxygen is kinetically sluggish its reduction requires a multistep process, leading to intermediates -superoxide, hydrogen peroxide and the hydroxyl radical - that are far more aggressive than molecular oxygen (Chapter 2). These reactions must be contained by any budding protobiont that uses oxygen as a terminal electron acceptor. Even traces of oxygen could be toxic to an anoxybiotic respirer, so defences against it were required. One of the earliest was probably superoxide dismutase, whose action interrupts the chain of reductions and prevents the appearance of toxic, highly reactive products.

There are many lines of evidence that suggest that the atmosphere and surface waters of the Earth became oxidative before animal life appeared. A recent estimate, based on chemical profiles of ancient strata, suggests that the partial pressure of free oxygen was about 0.2% of the present level during the early Proterozoic. It has been suggested that the earliest eukaryotes apeared about this time.

It is, however, a great leap from single celled eukaryotic organisms to multicelled protoMetazoa. Such clues that are available are to be found in Ediacarian strata which span the 120 million years before the base of the Cambrian. They contain an assemblage of (probably) metazoan fossils whose precise affinities remain problematical. It is safest to assume that they are the relics of organisms that had achieved a certain grade of organisation, certainly on a level with the Cnidaria and perhaps beyond. The point is, if organisms of those grades of organisation were present nearly 700 million years ago, then the origins of the Metazoa must go back much further, perhaps to one billion years before present (Chapters 3 and 4).

If ancestral organisms had achieved a platyhelminth grade of organisation at this time, it is interesting to speculate whether they were aerobic or anaerobic. The key elements in pathways of anaerobic metabolism encountered in extant Metazoa include an an anaerobic metabolic pathway between glucose (say) and phosphoenolpyruvate, a carbon dioxide fixation step for the synthesis of mitochondrial substrate, usually malate, and an electron transport system involving fumarate reductase. Fumarate reductase transfers electrons from reducing equivalents, via an electron transport chain to fumarate with the consequent formation of succinate, which may be excreted or metabolised further (Chapter 5).

Electron transport systems of the sort described have a number of important properties. Organisms that possess them are oxygen conformers; that is, their rates of oxygen uptake vary proportionally with the partial pressure of oxygen to which they are subjected. They show insensitivity to cyanide inhibition. They contain specialised cytochromes of the b group, and often an o-type alternative oxidase. *In vitro*, mitochondria accumulate hydrogen peroxide and finally, so-called classical electron transport involving cytochrome oxidase is usually present but in very low concentrations.

Were these pathways also present in the earliest Metazoa? If so were the organisms aerobic or anaerobic? Thes are important questions, for it must be remembered that, depending on the number of sites of ATP synthesis in the anaerobic pathway, aerobic oxidation is anything between five and twenty times more energy efficient than anaerobic. It is also possible for anaerobic pathways to compensate for their apparent low energy output but greatly increasing their rate of substrate utilisation.

It has long been suggested that cytochrome systems first arose as a mechanism for detoxifying oxygen, and Barrett has remarked that the o-oxidase has not yet been definitely implicated in oxidative phosphorylation and may therefore havea non-specific role to play. Cytochrome systems may therefore have started out on their evolutionary pathway as oxygen scavenging mechanisms and may have gradually acquired their role in ATP synthesis. The lower Metazoa may still employ cytochrome systems in both roles, perhaps reflecting the ancient grade of organisation first reached by the Ediacarian assemblage.

Perhaps the best place to obtain organisms for an investigation of this hypothesis would be the largely unexplored sulphide system. This occurs deep in the sediments that form the bottom of all major bodied of water. It houses an assemblage of primitive Metazoa, many of which seem to be close to the stem lines of the Platyhelminthes. Because the Earth is largely covered by water, this anoxic zone enormous both in extent and antiquity. Although there is some dispute as to whether the larger thiobiota are aerobic or anaerobic it seems almost cerrtain that the smaller ones are and that some of them use sulphide as an energy source. In this they are similar to the denizens of the sea bed surrounding deep sea hydrothermal vents (Chapters 6 and 7).

The ability to survive prolonged periods of anoxia seems to decrease with each step along the grades of organisation. More recent evolutionary arrivals thus appear to have less tolerance of anoxia than the more ancient ones. Like all generalisations this no doubt will break down under detailed scrutiny, but the evidence here suggests that there is indeed a series, with the interstitial meiofauna most tolerant, followed by platyhelminths and aschelminths, annelids, and molluscs. Higher vertebrates and arthropods appear to be the most susceptible to oxygen lack (Chapters 8 to 12).

Anaerobiosis, of course, is not confined to the sulphide system and deep-sea hydrothermal vents. The insides of an organism often provide excellent anoxic environments. In one case these environments are exploited by organisms that offer little if any advantage to the host; in another, that provided by ruminant animals, there is an exquisite exploitation of the anaerobic metabolism of its symbionts by the host, to the extent that ruminant life would be impossible without them (Chapter 13).

This introduction has been kept as short as is consistent with the object of laying before the reader a map of the territory to be covered. The last chapter (Chapter 14) might also be the first, an overview of metazoan life without oxygen. It raises some important questions about the study of adaptive biochemistry and some important challenges. Readers will be able to decide for themselves how well the other authors have addressed the issues raised there.

Chapter 1

The Physical Chemistry of Oxygen

Dereham L. Scott

Introduction

In the two centuries since Lavoisier rechristened dephlogisticated air as oxygen, the 'acid former', in the mistaken belief that oxygen was the essential principle of acids, and suggested that oxygen was involved in biological energy production - even hinting at its possible toxicity - perhaps 10^{14} tonnes of atmospheric dioxygen have cycled through the biosphere. While the mass of literature devoted to oxygen may be several orders of magnitude less, it has, in a manner reminiscent of an autocatalytic chemical process, grown to such an extent that a comprehensive bibliography would overflow the space allocated to this chapter. Mankind has, as Frimer (1983) says, indeed 'invested a great deal of time and resources in attempting to understand the exact role this life-supporting molecule plays in auto-oxidation, photo-oxidation and metabolic processes'. Recent years have seen the publication of a correspondingly large number of reviews, covering in great detail every aspect of what Dole (1965) elegantly described as the natural history of oxygen. Since these reviews typically run to hundreds of references, this introduction, covering some of those chemical aspects appropriate to the main text, draws mainly on reviews rather than the original papers. The reviews listed here provide a broad coverage of the field (Gilbert, 1981; Greenwood and Earnshaw, 1983; Hayaishi, 1974; Oberley, 1982; Royal Society of Chemistry, London, 1981; Patai, 1983; Taube, 1965).

It is common knowledge that dioxygen makes up about one-fifth of the Earth's atmosphere. Perhaps less-widely appreciated is the fact that 63% of the atoms in the crust are oxygen atoms, making up 94% of the total volume. But oxygen's fame does not rest on ubiquity alone. It is reasonable to argue that, if one were faced with the improbable task of illustrating the breadth and diversity of chemistry in terms of the behaviour of a single element and its compounds, oxygen should be considered for this exemplary role. True, it does not catenate to the same extent as carbon nor does it exhibit the same range of oxidation states as nitrogen but, because oxygen is second only to fluorine in its electronegativity, it tends to dominate the chemistry of its compounds, imparting a polarity which plays a large part in determining their physical and chemical properties. In this sense, oxygen-containing compounds may be said to belong to an oxygen system of compounds, with water as the parent solvent. Dioxygen is a molecule of some inspiration, while that strange substance water, with its unique suitability for the evolution and maintenance of life on this planet, constitutes a chemical story on its own. Oxides occupy a dominant position in solid state chemistry. The oxides of the non-metals and the oxy-acids, in great variety, are of considerable chemical and economic importance. Well may we shout 'Sing O for oxygen'.

Fitness of Oxygen

Because man derives reassurance from a belief in innate order ('to everything there is a season'), Henderson's (1913) concept of the fitness of the environment is intrinsically satisfying. Notwithstanding that this book is about organisms that manage very well without dioxygen - and, indeed, may be most unhappy in its presence - there can be no doubt that dioxygen is admirably fitted for its role as a terminal oxidant in aerobic metabolisms. As a major constituent of the atmosphere and through its adequate solubility, dioxygen is readily accessible to organisms in the requisite quantities. Its redox reactions with certain organic materials ('food') are sufficiently energetic to ensure efficient utilisation of this energy source. Recent re-evaluation of some of the relevant equilibrium constants has in no way invalidated George's (1965) exhaustive thermodynamic support for its fitness. However, the vigour of those same reactions would seem to pose a threat to the integrity of the whole organism, for organic material is thermodynamically unstable, tending towards oxidation to carbon dioxide and water. Only the fortunate kinetic sluggishness of dioxygen - even a potentially explosive mixture of hydrogen and oxygen remains unchanged indefinitely without external stimulation - prevents the spontaneous combustion of the biosphere. Yet this sluggishness in itself creates a further problem, in that it necessitates multi-stage reduction of dioxygen in the cell through a series of intermediates which are far more aggressive. There is no doubt that dioxygen can be toxic and that a price must be paid for its effectiveness.

These aspects have been widely discussed. Fridovich (1979) says 'like Janus, oxygen has two faces, one benign and the other malignant. Molecular oxygen is toxic to virtually all life forms and this toxicity becomes obvious on exposure to concentrations significantly greater than the ambient fifth of an atmosphere. Our margin of safety is evidently a narrow one. We possess defences against oxygen toxicity which are sufficient to meet the ordinary demands, but which can easily be overwhelmed.' On the appearance of dioxygen in the atmosphere, Hill (1978) comments 'the creation of a ubiquitous gaseous oxidant must have led to a selection for those organisms which could use molecular oxygen...as an electron acceptor. They simultaneously had to control and couple more oxidations and presumably protect themselves against side reactions. Those which could not became extinct or could, and can, only survive in the greatly decreased number of anaerobic environments...there exists a significant kinetic restraint without which the evolution of the Earth, both geologically and biologically, would have been very, very different...The paradox presented by the two opposing features of the role of dioxygen in biological processes has proved, and will continue to prove, a most seductive problem for all manner of men.'

As evidence for its toxicity, oxygen has been implicated in a number of deleterious physiological conditions, including ageing - and one can scarcely dispute Rosin's (1956) observation that 'ageing is invariably fatal'. A fascinating example of this toxicity, this time being turned to advantage, is the phagocytic action of macrophages, which absorb dioxygen and generate superoxide and peroxide as offensive weapons against bacteria. The reduction products of dioxygen are all oxidants, Bronsted bases and nucleophiles and are all more aggressive than the parent itself. Yet, despite a huge research effort, the exact basis for this toxicity has not been defined.

Because it was known that the hydroxyl and hydroperoxy radicals were amongst the primary products of the radiolysis of water and that radiation damage was enhanced by increasing the tension of dioxygen in tissue being irradiated - by a mechanism not fully understood even now - it was natural that one of the earlier approaches attributed oxygen toxicity to highly-reactive free radicals. Indeed, Hill (1978) has proposed that the presence of superoxide dismutases in some obligate anaerobes is a legacy from a past when their ancestors had to cope with higher levels of radiation. If toxicity is to be equated with reactivity, then the hydroxyl radical - and its derivatives - offer the greatest hazard, and since there are mechanisms for generating this from superoxide and peroxide, the case is argued (Hill, 1978) that the superoxide dismutases, and the catalases and peroxidases are front-line defences to destroy these precursors.

For a time, other reduction intermediates in turn became the favoured suspects. However, for the toxicity of any of these species to be realised *in vivo* a number of conditions must be satisfied and it is now being recognised, given the complexity of biochemical systems, that a single species operating through a single mechanism is an improbable principal culprit. Fee (1981, 1982) suggests that the best hope for understanding this complex phenomenon lies in an earlier proposal that the sledgehammer effects of hyperbaric dioxygen might be better explained in terms of a general metabolic overload, disrupting a number of delicate enzyme equilibria.

The History of Atmospheric Dioxygen

Although our knowledge of the Earth's primaeval atmosphere must be limited to a choice of educated guesses, it is generally agreed (Gilbert, 1981) that it was secondary in origin, replacing the primary volatiles lost from the weak gravitational field of the accreting planet by out-gassing of the crust and mantle, and reducing in character, with very little free dioxygen. There was accessible to the early atmosphere sufficient reduced lithospheric material, such as iron(II), to consume any dioxygen originally present or formed by photolysis of water. Further, the simple organic molecules which were the precursors of living organisms, synthesised by the prebiotic processes simulated by Miller, Fox and others, could not have formed nor survived under those conditions in the presence of dioxygen.

The banded iron formations of Greenland (about 3.8×10^9 years old), containing oxidised and reduced iron in alternate layers, reflect a greater abundance of dioxygen, suggesting some photosynthetic activity at this time. By 2×10^9 years ago, the replacement of the banded formations by the oxidised red beds gives evidence of the presence of significant levels of free atmospheric dioxygen. During this 1.8×10^9 year interval, the primitive anaerobes were building defences against dioxygen and the first eukaryotes probably appeared about 2×10^9 years ago. The protective ozone layer could have formed at an oxygen pressure about one-tenth of the present level. The level of dioxygen began a sharp upturn about 0.6×10^9 years ago, exceeding the Pasteur point for many anaerobes of about 10 Torr and allowing the explosion of metazoans with circulatory systems at the base of the Cambrian. The present level may have been reached about 0.4×10^9 years ago, but there may have been several marked oscillations since. Tappan (1974) has speculated that the biological catastrophes of 70 and 259 million years ago were associated with a large decrease in the oxygen level, while others have suggested that increases in level or destruction of the ozone layer by some cosmic event were responsible. Of course, even with dioxygen at its present level, there have always been sizeable local anaerobic environments, such as those presumably responsible for coal, oil and gas deposits in the past and those of today responsible for this book.

The present steady state represents a balance of a number of processes producing (photosynthesis, burial of reduced carbon, photolysis of water) and consuming (respiration, weathering) atmospheric dioxygen, which are interlocked with other variables, such as temperature and the level of carbon dioxide, a change in any one of which could cause future fluctuations. It is estimated that the weathering process alone consumes dioxygen equivalent to that in the atmosphere every four million years, so that the presence of appreciable amounts of oxygen in a planet's atmosphere is a good indication of the existence of an oxygen-producing biosphere, maintaining the level. Biofeedback between photosynthesis and respiration maintains the short-term balance, while the contributions from weathering and burial of organic material are longer-term. At the present rate, dehydration of the Earth through photolysis of water and loss of hydrogen to space would require 5.5×10^{12} years, a period about one thousand times as long as the estimated lifetime of the stable Sun. Gilbert (1981) concludes that the present steady state should continue for several million years.

Isotopes

Of the ten known isotopes, only ^{16}O, ^{17}O and ^{18}O are stable. The discovery of the heavier nuclides about sixty years ago and the recognition of some natural variation in isotopic abundance - the approximate percentages are 99.76, 0.04 and 0.20 - caused some concern in the chemical world, since oxygen was then being used as the atomic weight standard, and led to the search for a new standard, with ^{12}C as the ultimate choice. Extensive physical fractionation can produce almost pure ^{17}O and ^{18}O; the former, with a nuclear spin of 5/2 is useful for NMR and ESR studies (although low abundance and quadrupole broadening can be restrictive), while the latter is widely used as a stable tracer with mass spectrometry. Of the radioactive isotopes, ^{15}O is the longest-lived, but its half-life of just over two minutes makes it less than convenient as a tracer; nevertheless, a few biological studies have utilised this nuclide.

Of considerable theoretical interest and potential utility in studying the mechanisms of reactions are the equilibrium and kinetic isotope effects accompanying chemical processes and inducing isotopic fractionation, analogous to the physical differentiation utilised in effecting concentration of the heavier nuclides. The fractionation, arising because the zero-point energy of an X-O bond is less for the heavier species, with the consequence that the effective bond strengths and reaction rates are different, is slight for oxygen. The maximum theoretical effect for $^{16}O/^{18}O$ is only about twenty percent; in practice, the realised effects are much smaller, but, even so, they are readily resolvable by modern gas chromatography and mass spectrometry equipment and are thus potentially of much utility in studying biological mechanisms. The fact that the effect for $^{16}O/^{18}O$ is about twice as great as for $^{16}O/^{17}O$ has been utilised in characterising natural fractionations.

If the dioxygen in the atmosphere is essentially the product of photosynthesis, with a turnover time of two to three thousand years, it might be expected that photosynthetic and atmospheric dioxygen should have the same isotopic composition. In fact, the atmosphere contains a slightly higher proportion of ^{18}O because of the marginal preference for the photosynthetic oxidation of $H_2{}^{16}O$. Clearly, this isotopic differentiation can only be maintained by contrary isotope effects; the principal balancing factor is almost certainly the preferential consumption of $^{16}O_2$ by the processes of respiration, organic decomposition and weathering.

A number of other natural fractionations have been studied in detail. For example, atmospheric carbon dioxide and carbonate rocks are enriched in ^{18}O through equilibria such as

$$C^{16}O_3{}^{2-} + 3H_2{}^{18}O \leftrightarrow C^{18}O_3{}^{2-} + 3H_2{}^{16}O$$

with equilibrium constants differing slightly from unity. This particular fractionation is of the order of a few percent and its temperature-dependence has been utilised in the paleothermometry. Variations in isotopic ratios as small as 0.0001, corresponding to temperature differences of less than a degree, have been measured, but their interpretation rests on a number of assumptions and should be treated with some caution.

Some Oxygen Species

Although it shows very little tendency to catenate - the O-O single bond is weak and chains of three or more oxygen atoms are endothermic and inherently unstable - oxygen forms a remarkable array of simple molecules and ions, illustrated in Figure 1.1. Not all of these have a biological role and in the brief survey that follows only the principal species appear.

In considering the relationships between these species, and their relative stabilities under given conditions, it is helpful to picture them as rungs on a chemical energy ladder. In a close analogue of an ordinary ladder, for which the work expended in raising one's gravitational potential energy may be recovered, at least partially, by falling down, so the species with high chemical potential tend to fall down to more stable forms. One useful measure of the potential driving these chemical changes, analogous to the gravitational potential, is the Gibbs function (Gibbs free energy). The change in this function (ΔG) that accompanies any movement up or down the ladder (interconversion of species) is a measure of the work either available from a spontaneous (down-falling) process or from that required to climb upwards.

In the same way that gravitational potential energies are not absolute, being expressed relative to some datum such as sea level, so we can only measure changes in the Gibbs function and it is necessary to define some arbitrary reference point. Conventionally, the Gibbs function of dioxygen under specified conditions (the standard state) is assigned as zero. Species such as ozone and superoxide lie above dioxygen on the ladder, while water and hydrogen peroxide, with negative Gibbs functions, lie below. All other species are oxidising with respect to water, with the hydroxyl radical potentially the most powerful of all.

Because many of the interconversions are redox processes involving a change in the oxidation state of oxygen, it is informative to stagger the rungs laterally according to the oxidation number of oxygen to produce a two-dimensional plot called a Frost Diagram. Such processes are often formally represented as electrochemical, or electrode, half-reactions, for example, the dioxygen - water couple

$$O_2 + 4H^+ + 4e^- \leftrightarrow 2H_2O$$

for which we may, in principle, measure an electrode potential, E, directly related to the relevant change in Gibbs function ($-nFE = \Delta G$). Hence, the quantity nE, where n is the change in oxidation state, is also often plotted as abscissa (1nE unit is about equal to 100kJ mol^{-1}). In constructing and interpreting these diagrams, we must bear

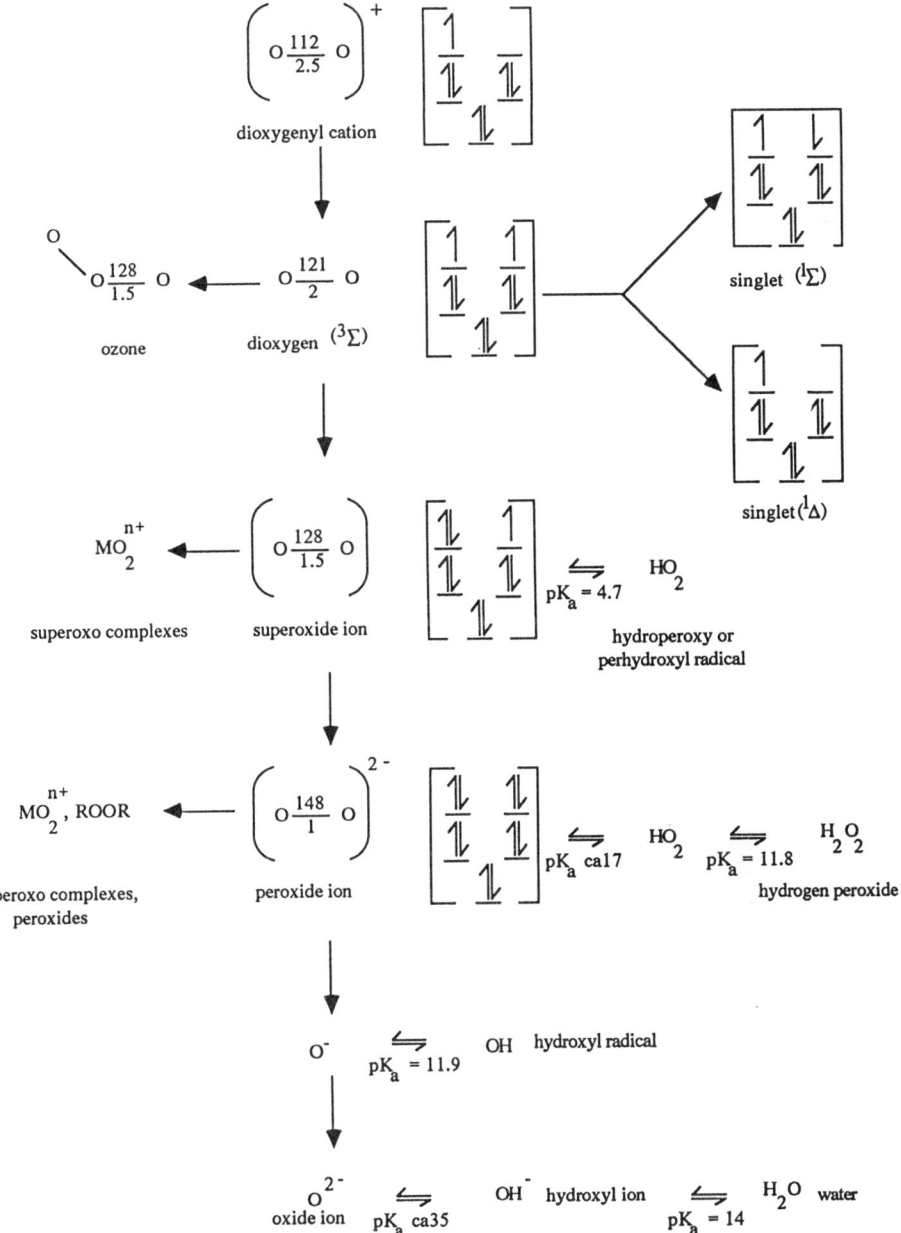

(where appropriate, bond lengths in pm and bond orders are shown above and below the bond lines, and partial molecular orbital diagrams are given inside square brackets).

Figure 1.1 The oxygen family tree.

8   *The Physical Chemistry of Oxygen*

in mind that many equilibria are pH-dependent, that species may be involved in complex formation with the proton or other cations and that the Gibbs function of a given species will depend upon the choice of standard states. Finally, we must remember that a thermodynamically spontaneous reaction may be kinetically hindered, and proceed at a useful rate only when catalysed.

Such a diagram (Figure 1.2) graphically summarises a great deal of information. The steeper the line joining the members of any couple, the greater the thermodynamic potential for reaction. Any species which lies above the line joining those on either side of it must be unstable towards disproportionation with a decrease in the Gibbs function; thus, superoxide and hydrogen peroxide are inherently unstable. In order to examine the potential for reaction between oxygen couples and those of other elements, we can superimpose diagrams. The susceptibility towards oxidation of organic material is clearly shown by the inclusion of the carbon diagram (Figure 1.2, inset). Williams and Da Silva (1978) have utilised these diagrams in a comprehensive review of the reactions of oxygen in biological systems.

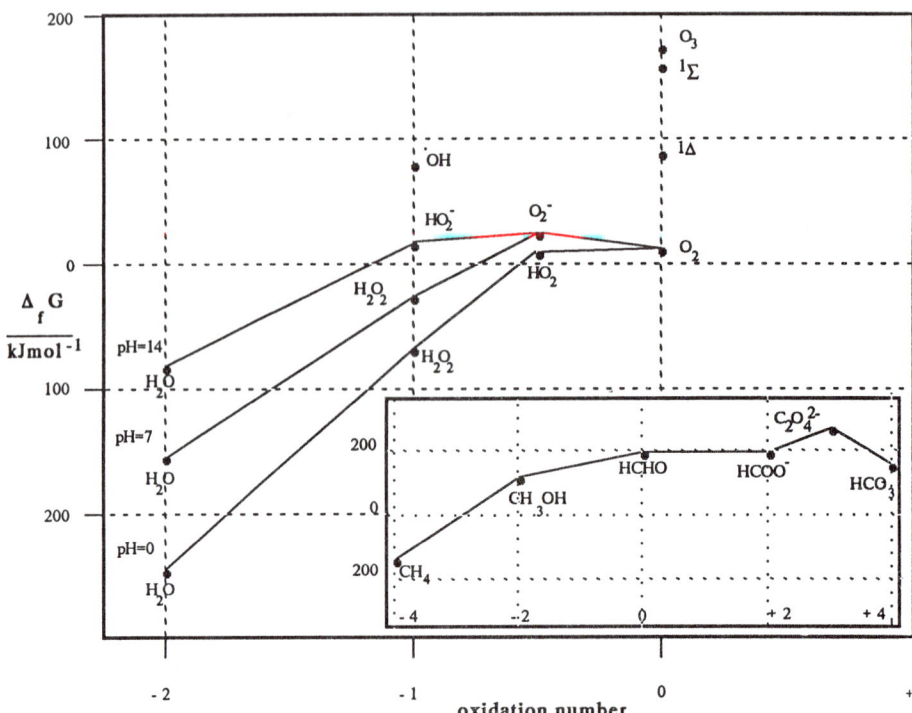

Figure 1.2 Sketch of Frost diagram for oxygen. Inserted, for comparison, is the diagram for carbon. (Redrawn from Hill, 1978 and Williams and Da Silva, 1978.)

Dioxygen

Given that the nature of the bonding in the dioxygen molecule provides the basis for the apparent paradox that living organisms survive in a bath of the same substance they use as the primary oxidant in the metabolism of other organic material, it is no great surprise that the ground electronic state of this molecule is a little unusual. The simple partial molecular orbital diagram provides a description adequate for our present purposes. The highest occupied levels are the degenerate antibonding orbitals, each containing one electron; the electronic spins are parallel, so the ground state in a triplet dioxygen is paramagnetic and the liquid is blue. The bond order is formally two but, at 495 kJ mol^{-1}, the bond is rather weak as double bonds go, and reactions in which it is replaced by other linkages, such as C=O (745 kJ mol^{-1}) or C-O + C-H (825 kJ mol^{-1}), are energetically favourable.

Nevertheless, the energy required to remove one of the unpaired electrons is surprisingly high at 1160 kJ mol^{-1} and the O_2^+ ion is generated only by powerful oxidising agents, such as PtF_6 as in Bartlett's celebrated accident. The first two excited states are a singlet with the two antibonding electrons spin-paired in the same orbital (95 kJ mol^{-1} above the ground state) and a singlet with these electrons spin-paired in separate orbitals (158 kJ mol^{-1} above the ground state). Singlet oxygen is generated in photochemical biological processes but fears that this highly-reactive species might be formed by the disproportionation of superoxide and hydrogen peroxide seem to be unfounded.

The diradical character of dioxygen is demonstrated by autooxidation reactions with organic compounds through radical chain mechanisms. These reactions are both useful - as in the formation of paint films, although further oxidation eventually leads to degradation of the protective layer - and destructive, as in the rancidification of edible oils, the polymerisation of petroleum products and the perishing of rubber. Anti-oxidants are commonly added to prolong the lives of these materials (Scott, 1985).

It is reasonable to expect that the combination of the unpaired electrons and the relative weakness of the bond would make dioxygen a highly-reactive and energetic reagent but this is only true under certain conditions. The triplet character imposes a kinetic barrier, which, so long as it is maintained, gives oxygen its sluggishness as an oxidant necessary for its coexistence with organic material (Taube, 1965). True, it is not so kinetically backward as perchlorate, for example, but, in any case, the basis for the kinetic constraint is different. Bond formation between triplet and singlet molecules necessitates a spin flip so that spin-pairing in the new singlet bonds can occur and this typically takes at least a million times longer than the lifetime of a collision complex. One-electron reductions then appear to be the preferred mode of reaction, but the affinity of dioxygen for another electron is low (the electron-attachment energy is only -41 kJ mol^{-1}) and the formation of the O_2^- radical is unfavourable in aqueous solution, even under optimum conditions.

The spin restriction can be overcome, leading to fast and energetic oxidation, in any one of the following ways:

(i) reaction with a free radical, as in the chain reaction with hydrogen, is spin-allowed
(ii) dioxygen may be energetically excited, for example by irradiation, to a singlet state or even to dissociation
(iii) bonding of triplet dioxygen to a metal which also has unpaired electrons can occur without violating the spin conservation rule, and the bound dioxygen is much more labile.

In biological systems, one-electron reduction is not the rule, because four-electron reductions to water and two-electron reductions to hydrogen peroxide are made possible through synchronous electron transfers to dioxygen bound to an activating metal site, but it nevertheless occurs in certain processes, generating the sequence of potentially toxic species which have been subject of much study.

The solubility of dioxygen is an important consideration in determining its availability to organisms. A saturated solution in water at 20° is about 0.0014 M in dioxygen, but this solubility is reduced by the presence of dissolved salts, so that seawater dissolves about 20% less. On the other hand, blood dissolves about thirty times as much. The solubility is considerably greater in many organic solvents, while the fluorocarbons are such good solvents for dioxygen that they have attracted attention as blood substitutes and even as 'liquid atmospheres' from which air-breathing animals might obtain oxygen.

Ozone

Although an improbable species in living systems, ozone is nevertheless exceedingly important to life. Produced in the upper atmosphere by the photochemical activation of dioxygen, it forms a thin layer at a height of around 25 km that effectively screens out the hard UV radiation (<300 nm) that would otherwise render life on the surface impossible. The energy transfer associated with its formation plays an important role in the temperature regulation of the lower atmosphere. Any risk of the large-scale disruption of this filter by the injection of pollutants into the atmosphere must thus be taken very seriously indeed.

As one of the most powerful oxidising agents known, it is toxic to life at low levels. Fortunately, it is detectable at about 0.01 ppm by its smell - hence its name - and the suggested permissible level for continuous exposure is 0.1 ppm. During daylight hours, it is naturally present in the lower atmosphere at about 0.04 ppm, but levels as high as 1 ppm have been detected during extreme episodes of photochemical smog; at this level, the permissible exposure time is only ten minutes.

Superoxide

After a humble beginning as one of the oxides of the heavier alkali metals and some menial employment as an air purifier in recycling closed atmospheric systems - thus

$$4KO_2 + 2CO_2 \rightarrow 2K_2CO_3 + 3O_2$$

- superoxide has latterly been accorded the attention appropriate to - or perhaps in slight excess - the first-born. The discovery that certain enzymes present in all aerobes, and even in some anaerobes, appear to have the sole function of destroying superoxide species set off an investigation of staggering magnitude founded on the belief that at least one of the major agents of oxygen toxicity had been identified (Michelson, McCord and Fridovich, 1977; Bannister and Bannister, 1980). There is now some agreement that superoxide, while perhaps something of a terrorist, is not, after all, the leader of the gang; certainly it is not the super reagent that its name may wrongly be taken to imply (Sawyer and Valentine, 1981). Although it seems to have all the necessary attributes - it is a Bronsted base, a nucleophile, a free radical, a reductant and a potentially powerful oxidant, its properties are very much a function of its environment, and its direct implication in any destructive action in living cells has been hard to substantiate. Among the important reviews of the properties of superoxide are those of Fee and Valentine (1977), Lee-Ruff (1977) and Oberley (1982).

Addition of one electron to the antibonding orbitals of dioxygen removes the degeneracy of these orbitals, with an energy splitting dependent upon the environment, and reduces the bond order to 1.5, with a consequent lengthening of the bond to 128-133 pm. As the conjugate base of the acid HO_2 (pK_a = 4.8), its reactions are strongly pH-dependent and its reactivity in aqueous solution is markedly modified by its strong solvation. Its basic properties, both in protic and aprotic media, have some resemblance to those of the fluoride ion, but its redox characteristics are quite different. In protic media it is unstable with respect to disproportionation to dioxygen and peroxide, but the rate of disproportionation is pH dependent, passing through a maximum at pH = pK_a, decreasing by a factor of ten for each pH unit above this point. One reason for this increase in stability is the electrostatic repulsion between the negatively-charged superoxide ions. The biological dismutases overcome this repulsion by forming positively-charged complexes between the metal centre and a superoxide ion; for example

$$Cu^{2+} + O_2^- \rightarrow CuO_2^+$$

$$CuO_2^+ + O_2^- + 2H^+ \rightarrow Cu^{2+} + H_2O_2 + O_2$$

Because of this charge repulsion and the unfavourable energetics of bare O_2^{2-}, superoxide is stable towards disproportionation in aprotic media, with a change in the standard Gibbs function in excess of 115 kJ mol^{-1}, compared with a value of around 75 kJ mol^{-1} in aqueous solution.

Any of the other modes of reaction of superoxide - as a one-electron oxidant or reductant, a nucleophile or a radical - must compete in aqueous solution with disproportionation. Its nucleophilicity is also markedly diminished by extensive solvation. For these reasons, superoxide is, in itself, by no means the highly dangerous species it was initially suspected to be; despite a huge and intensive search, it appears that no biologically deleterious reaction has been definitively established. Eternal youth is not, it appears, a bottle of superoxide dismutase pills.

Hydrogen peroxide

Its familiar applications as a bleaching agent and a disinfectant suggest that hydrogen peroxide and its derivatives may not be the most innocuous of the normal components of living cells. The addition of two electrons to the dioxygen molecule reduces the bond order to one. The single bond is long and weak because of intramolecular electronic repulsion with the consequence that hydrogen peroxide is highly unstable towards disproportionation. But again the activation energy barriers are high and the pure liquid or its aqueous solutions may be maintained for long periods in the absence of catalysts. In acidic solution, hydrogen peroxide is a moderate reductant and a very strong oxidant, but this redox behaviour is markedly pH dependent.

Labelling studies show that when hydrogen peroxide is itself oxidised, the O-O bond is not broken since the evolved dioxygen comes from the peroxide. While this indicates simple removal of electrons by the oxidant, many reactions of peroxides are not straightforward electron transfers but involve free radicals. The metal-catalysed reactions are of this type, as illustrated by the Fenton reaction

$$H_2O_2 + Fe(II) \rightarrow \cdot OH + OH^- + Fe(III)$$

Hydrogen peroxide is handled in the cell by being bound to metal centres which catalyse its disproportionation or control the reactions in which it is reduced to water.

Organic peroxides may be formed by the metal-catalysed free-radical autoxidation of organic materials by air. These species can be highly unstable - witness the numerous explosions of peroxidised ethers - and some find uses as free-radical initiators of polymerisation in adhesives and coatings.

Hydroxyl radical

Of the variety of ways in which this radical can be generated in solution, the just-mentioned Fenton reaction is here an appropriate useful source. It is a powerful and aggressive oxidant and free radical, which reacts so fast and so indiscriminately that it is difficult to conceive of any effective specific defence against it. However, this intense reactivity does not necessarily mean that it is the hydroxyl radical that inflicts the most serious damage; its longer-lived products have more time to diffuse to the sensitive sites. For a long time biologists were reluctant to embrace the hydroxyl radical as significant biochemical species and its role *in vivo* is still a matter of debate. Now, as Willson (1979) says, when much is known about the chemistry, 'fifty years after their formulation, hydroxyl radicals are at last becoming socially acceptable amongst biochemists' but much of their proposed biological involvement is a matter of inference. They have been implicated in the initiation and progress of a number of pathological conditions (Willson, 1979) and attempts have been made to ameliorate these conditions through treatment with radical scavengers. It has been suggested that the toxicity of ozone stems from the generation of these radicals.

All of the four main types of reaction of the hydroxyl radical

(i) charge transfer (redox): $X^- + \cdot OH \rightarrow X\cdot + OH^-$
(ii) hydrogen abstraction, the common reaction with saturated organic compounds:
$$RH + \cdot OH \rightarrow R\cdot + H_2O$$
(iii) addition: $RH + \cdot OH \rightarrow HROH$
$$2\cdot OH \rightarrow H_2O_2$$
(iv) displacement: $RI + \cdot OH \rightarrow ROH + I\cdot$

are potentially damaging to living organic material, but, as radiobiologists have discovered after decades of intensive study, it is difficult on the one hand to establish a definitive link between a particular effect and a given species, and, on the other hand, to modify that link once it is established.

Oxygen Carriers

The interactions between metal ions and oxygen in solution have been of practical interest to chemists for over a hundred years. For example, the classical preparations of cobalt(III) complexes involved aerial oxidation of the cobalt(II) compounds, and the oxidation of metal ions in solution under a range of conditions has been the subject of much fruitful study (Fallab, 1967). Almost every biological use of dioxygen involves the initial binding to a metal centre. The relatively recent isolation of dioxygen-containing intermediates has resulted in a change in emphasis from preparative and kinetic aspects to the elucidation of structure and bonding - of considerable theoretical interest - and to the use of these compounds as models for simulating and understanding the action of the biological oxygen-binders in their key roles of transporting, storing and activating dioxygen.

These biological molecules are examples of exquisite molecular engineering, in which a number of factors cooperate in the tailoring and maintenance of the oxygen-binding site in the state most fitted to its function (Spiro, 1980; Chang and Dolphin, 1978; Pratt, 1975). In deoxy-haemoglobin, five-coordinate, high-spin iron(II) is stabilised, with protection of the sixth site against ligands other than dioxygen; if it were necessary for dioxygen to displace another ligand, equilibration would be much less rapid and effective in terms of oxygen uptake. Oxygenation is accompanied by a transition to low-spin iron(II), with a reduction of about one-fifth in the effective radius of the iron ion, which can then drop into the plane of the porphyrin ring, with consequent minor, but important, adjustments to the neighbouring structure. The oxygenated structure is diamagnetic, implying a spin-pairing of six electrons - a remarkable change for any chemical process. There is a cooperative interaction between the four iron sites which facilitates the uptake of dioxygen at high partial pressure and its rapid release when the oxygen tension falls. The electronic interaction between the metal and dioxygen must be such as to ensure retention and activation of the dioxygen, without promoting irreversible auto-oxidation of the binding site.

Analysis of the oxidation of cobalt(II) complexes provides a starting point which immediately identifies two classes of metal-dioxygen adducts, depending on whether the bound dioxygen is behaving as a superoxo or a peroxo ligand.

$$Co^{II}(NH_3)_6^{2+} \xrightarrow{air} [(H_3N)_5Co^{III}-O_2-Co^{III}(NH_3)_5]^{4+}$$

$$\xrightarrow{H_2O_2} [(H_3N)_5Co^{III}-O_2-Co^{III}(NH_3)_5]^{5+}$$

$$Co^{II}(CN_6)^{4-} \xrightarrow{air} [(NC)_5Co^{III}-O_2-Co^{III}(CN)_5]^{6-}$$

$$\xrightarrow{Br_2/OH^-} [(NC)_5Co^{III}-O_2-Co^{III}(NC)_5]^{5-}$$

In both cases, the first product is diamagnetic, yields dioxygen on acidification, and the geometry of Co-O-O-Co and the O-O bond lengths are similar to those for hydrogen peroxide. These are peroxo complexes. The second products are paramagnetic, with geometries and bond lengths typical of superoxo complexes.

A great impetus for these studies was by the discovery that solutions of *trans*-[IrClCO(PPh$_3$)$_2$] (Vaska's compound) reversibly absorb dioxygen; the orange colour of the oxygenated solution was returned to the original yellow by the passage of dinitrogen. This system and related ones have been much used in studies of oxygen-carrying capability, which is clearly critically dependent upon charge distribution and stereochemistry. Replacement of chlorine by iodine in Vaska's compound renders oxygen uptake irreversible; presumably, the lower electronegativity of iodine leads to a higher electron density on the iridium, strengthening the M-O$_2$ bond, with a consequent increase in the O-O distance from 133 to 151 pm.

Quite a range of such adducts between dioxygen and metals ions in low oxidation states - not all reversible - are now known, with O$_2$:M ratios of 1:1 and 1:2. Essential to reversibility is a formation constant of sufficient magnitude to ensure the integrity of the complex, but with a metal-oxygen interaction less energetic than that leading to a redox reaction. For the equilibrium

$$M + O_2 \leftrightarrow MO_2$$

$Q = a_{MO_2}/(a_M \cdot p_{O_2})$, and we can characterise an oxygen carrier by $p_{O_2},1/2$, the partial pressure of dioxygen at which half the metal centres are oxygenated. For equilibrium at this pressure

$$\ln p_{O_2},1/2 = -\ln K = \Delta G°/RT = \Delta H°/RT - \Delta S°/R$$

The loss of translational freedom associated with solution of dioxygen and bringing together of the M/O$_2$ pair in solution so coordination can occur accounts for an entropy decrease of about 220 J K^{-1}mol^{-1}, necessitating an enthalpy change of around -65 kJ mol^{-1} at room temperature. Although a wide range would appear feasible, it is found that the enthalpy and entropy changes for the formation of carriers are remarkably constant and close to the values given above.

Although the exact nature of the bonding in these adducts has not been elucidated, it is clear that there is a significant transfer of electrons from the metal centre to the antibonding orbitals of dioxygen, to the extent that the dioxygen is partially reduced, with lengthening and weakening of the O-O bond. On the basis of the length and stretching frequency of the O-O bond, the coordinated dioxygen approaches O$_2^-$ or O$_2^{-}$, and these two classes, superoxo and peroxo, may be subdivided according to the mode of attachment of the dioxygen (Figure 1.3)

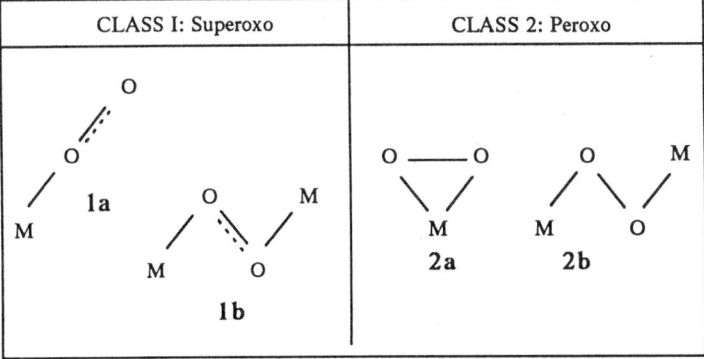

Figure 1.3 Classes of metal-oxygen bonding.

Compounds of type IIa are most common. Although types Ib and IIb appear similar, they differ significantly, as we have seen for the cobalt complexes above, in bonding and stereochemistry. In order to emphasise the degree of electron transfer, biochemists often formally represent the two classes as $M(III)O_2$ and $M(IV)O_2$. Summerville, Jones, Hoffman and Basolo (1979) have discussed the assignment of oxidation states for such complexes in terms of the valence bond and molecular orbital descriptions of the $M-O_2$ bond.

That the dioxygen is significantly activated in these coordinated states is evidenced by its reaction with species, such as carbon dioxide, that are not readily attacked by the free element. This activation may be attributed to any, or all, of the following factors

(i) through coordination to the metal, the dioxygen has been relieved of the spin restriction on its reactivity
(ii) a suitable stereochemistry for reaction may be generated
(iii) the partial reduction of dioxygen increases the electron density, conferring on it some of the aggressive character of the reduced oxygen species.

Since dioxygen is the cheapest oxidant available, there is some considerable interest in ways of activating it for carrying out bulk oxidations.

Conclusions

Living organisms need energy to survive and function. In principle, the more efficient the energy-producing metabolism, the fitter and more effective the organism. Two questions immediately come to mind - why are redox reactions employed for energy production and why is oxygen the paramount primary oxidant? The answer to the first is simply that, of the processes that are compatible with life, redox reactions give the best energy yield. In principle, the exergonic coupling of any two redox pairs could be utilised, with the provision that, if some of the energy is to be stored, rather than being used directly, the reduction in the Gibbs function must encompass the endergonic synthesis of the energy carrier; the ADP/ATP conversion requires about 35 kJ mol^{-1}.

In their review of energy production in anaerobic organisms, Decker, Jungermann and Thauer (1970) say 'The metabolic processes that occur in nature are characterized by the diversity of the substrates that can be used for the production of energy. This is particularly true of the electron donors, but also to a considerable extent of the acceptors; they can be either organic or inorganic.' They provide extensive lists of formal half-reactions for naturally occurring donors and acceptors, from which may be compiled a huge number of metabolic possibilities, the diversity of which is illustrated by some of their examples

$$2 \text{ crotonate}^- + 2H_2O \to \text{butyrate}^- + 2 \text{ acetate}^- + H^+$$
$$\text{glycine} + NADH + H^+ \to \text{acetate}^- + NH_4^+ + NAD^+$$
$$4 \text{ methanol} \to 3CH_4 + HCO_3^- + H^+ + H_2O$$
$$2 \text{ lactate}^- + SO_4^{2-} \to 2 \text{ acetate}^- + 2HCO_3^- + H_2S$$
$$\text{fumarate}^{2-} + FMNH_2 \to \text{succinate}^{2-} + FMN$$

Each of these is energetically-capable of synthesising at least one mole of ATP. The last is discussed in a later chapter.

George (1965) has dealt comprehensively with the second question, establishing that dioxygen is eminently suited to its task. But, in order to gain the benefit of the most efficient oxidation, must we inevitably have to meet the cost of some undesirable side effects? Organisms adapt to changes in their environment either by developing defences to the changes or by adaptation to the utilization of the new conditions for their benefit. Such is the complexity of living systems that, in either case, the adaptation is not without its consequences, which may include a new threat to the organism. For each advantage, there is a price to be paid, and, inherent in the concept of evolution, is the implication that organisms are always one step behind the optimum. Oxygenation of the atmosphere provided the biggest challenge that life on this planet has yet experienced; adaptation to it is not yet complete.

References

Bannister, W.H. and Bannister, J.V. (eds) (1980) *Biological and Clinical Aspects of Superoxide and Superoxide Dismutase* (Elsevier/North-Holland, Amsterdam)

Chang, C.K. and Dolphin, D. (1978) 'Oxidation and oxygen activation by heme proteins' in E.E. van Tamelin (ed), *Bioinorganic Chemistry*, Vol. 14 (Academic Press, New York)

Decker, K., Jungermann, K. and Thauer, R.K. (1970) 'Energy production in anaerobic organisms.' *Angewandte Chemie International Edition* 9(2), 138-158

Dole, M. (1965) 'The natural history of oxygen.' *Journal of General Physiology* 49(1) Pt 2, 5-27

Fallab, S. (1967) 'Reactions with molecular oxygen.' *Angewandte Chemie International Edition*, 6(6), 496-507

Fee, J.A. (1981) 'A comment on the hypothesis that oxygen toxicity is mediated by superoxide' in *Oxygen and Life* Spec. Publ., 39, Royal Society of Chemistry, London.

Fee, J.A. (1982) 'Is superoxide important in oxygen poisoning?' *Trends in Biochemical Sciences* 7, 84-86

Fee, J.A. and Valentine, J.S. (1977) 'Chemical and physical properties of superoxide' in A.M. Michelson, J.M. McCord and I. Fridovich (eds), *Superoxide and Superoxide Dismutases* (Academic Press, London)

Fridovich, I. (1979) 'Chairman's Introduction.' *Oxygen Free Radicals and Tissue Damage*, Ciba Foundation Symposium 65 (new series) (Exerpta Medica, Amsterdam)

Frimer, A.A. (1983) in S. Patai (ed), *The Chemistry of the Peroxides* (John Wiley, New York)

George, P. (1965) 'The fitness of oxygen' in T.S. King, H.S. Mason and M. Morrison

(eds), *Oxidases and Related Redox Systems* (John Wiley, New York)
Gilbert, D.L. (ed) (1981) *Oxygen and Living Processes: An Interdisciplinary Approach* (Springer Verlag, New York)
Greenwood, N.N. and Earnshaw, A. (1983) *Chemistry of the Elements* (Pergamon Press)
Hayaishi, O. (ed) (1974) *Molecular Oxygen in Biology: Topics in Molecular Oxygen Research,* (North Holland, Amsterdam)
Henderson, L.J. (1913) *The Fitness of the Environment* (Macmillan, New York) (Beacon Paperbacks, Boston, 1958)
Hill, H.A.O. (1978) 'The Superoxide Ion and the Toxicity of Molecular Oxygen' in R.J.P. Williams and J.R.R.F. Da Silva (eds), *New Trends in Bio-inorganic Chemistry* (Academic Press, London)
Lee-Ruff, E. (1977) 'The organic chemistry of superoxide.' *Chemical Society Reviews.* 6, 195-214
Michelson, A.M., McCord, J.M. and Fridovich, I. (eds) (1977) *Superoxide and Superoxide Dismutases* (Academic Press, London)
Oberley, L.W. (ed) (1982) *Superoxide Dismutase* , 2 Vols. (CRC Press, Florida)
Patai, S. (ed) (1983) *The Chemistry of Peroxides* (John Wiley, New York)
Pratt, J.M. (1975) in C.A. McAuliffe (ed), *Techniques and Topics in Bioinorganic Chemistry* (Macmillan, London)
Rosin, J. (1956) 'Oxygen ad absurdum.' *Chemical and Engineering News* 3 4(6), 546
Royal Society of Chemistry (1981) Oxygen and Life, Spec. Publ., 39. London
Sawyer, D.T. and Valentine, J.S. (1981) 'How super is superoxide?' *Accounts of Chemical Research* 1 4, 393
Scott, G. (1985) 'Antioxidants *in vitro* and *in vivo.*' *Chemistry in Britain,* 648-653
Spiro, T. (1980) *Metal Ion Activation of Dioxygen* (Wiley-Interscience, New York)
Summerville, D.A., Jones, R.D., Hoffman, B.M. and Basolo, F. (1979) 'Assigning oxidation states to some metal dioxygen complexes of biological interest.' *Journal of Chemical Education* 5 6, 157-162
Tappan, H. (1974) 'Molecular oxygen and evolution' in O. Hayaishi (ed) *Molecular Oxygen in Biology: Topics in Molecular Oxygen Research,* (North Holland, Amsterdam)
Taube, H. (1965) *Oxygen: Chemistry Structure and Excited States* (Little, Brown, Boston)
Williams, R.J.P. and Da Silva, J.R.R.F. (1978) 'High redox potential chemicals in biological systems' in R.J.P. Williams and J.R.R.F. Da Silva (eds), *New Trends in Bio-inorganic Chemistry* (Academic Press, London)
Willson, R.L. (1979) 'Hydroxyl radicals and biological damage *in vitro*; what relevance *in vivo* ?' in *Oxygen Free Radicals and Tissue Damage*, Ciba Foundation Symposium 65 (new series) (Exerpta Medica, Amsterdam)

Chapter 2

The Role of Oxygen Free Radicals in Biology and Evolution

Hosni M. Hassan and Joan R. Schiavone

Introduction

The Earth is the only planet in our solar system that contains molecular oxygen in its atmosphere and supports aerobic life. This has not always been the case, for the Earth's atmosphere was primarily reducing and essentially free of oxygen when the Earth was formed about 4.5 billion years ago. The basic elements required for making the first living cell (that is, amino acids, nucleic acids, fatty acids and so on) were made from the reduced compounds via free radical reaction mechanisms initiated by ionising radiation from the sun.

How the first form of life came about is speculative, but evidence from microfossils indicates that about one billion years elapsed before the simplest form of life arose. The first organisms were, most likely, anaerobic heterotrophs living in the deep oceans, where the opacity of water to UV light served to protect the cells against radiation damage. The continual presence of ionising radiation promoted the evolution and selection of cells that evolved defence and repair mechanisms designed to minimise the deleterious effect of ionising radiation.

Ionising radiation is also believed to have caused the formation of a small concentration of oxygen via photolytic dissociation of water in the outer atmosphere. However, the bulk of oxygen appeared only after the evolution of photosynthetic bacteria and plants, approximately 2.5-3.0 billion years ago. The concentration of oxygen in the atmosphere rose very slowly because of the presence of abundant supply of reduced elements that reacted with oxygen. Nevertheless, the appearance of oxygen in the environment presented both advantage and challenge to the primitive cells. Thus, microorganisms soon developed aerobic respiratory mechanisms to use the oxidative potential of oxygen and generate more energy than that provided by anaerobic fermentation; and the challenge was met by the development of various antioxidant mechanisms.

Oxygen itself is not toxic. Oxygen is tetravalently reduced to water via the cytochrome oxidase system, yet 1-2% of the oxygen undergoes univalent reduction and produces several reactive species (Taube, 1965; Czapski, 1971), namely superoxide radical (O_2^-), hydrogen peroxide (H_2O_2) and hydroxyl radical (OH·). Singlet oxygen and ozone are two other forms of reactive oxygen. Oxygen radicals possess unshared electrons in their outer shells, and for this reason they are even more reactive than molecular oxygen itself. Oxygen radicals may interact with a variety of organic molecules including proteins, nucleic acids, lipids and polysaccharides to render them defective and thus compromise the integrity of the organism.

Thus from the very beginning, the evolution of organisms was dictated by the selection and development of defence and repair mechanisms to protect against dangerous free radical reactions (Harmon, 1984). It is generally believed that oxygen radicals instituted the evolution of a whole line of antioxidant defence mechanisms including antioxidant enzymes such as the superoxide dismutases (SODs), catalases (CATs) and peroxidases, and antioxidant organic molecules such as vitamin C, vitamin E and β-carotenes. The mutagenic properties of oxygen radicals necessitated the need for DNA repair mechanisms to maintain the hereditary integrity of the cell. Thus, the toxicity of partially reduced oxygen intermediates served as a strong selective pressure in influencing the evolution of practically all life forms. It has even been suggested that the cytochrome oxidase system evolved primarily as a detoxification process to reduce oxygen to water instead of to oxyradicals, and its role in producing chemical energy is of secondary importance (Bulkley, 1983).

The topic of oxygen radicals in biology and evolution is so vast, that in this chapter we will concentrate on the potential sources and dangers of oxygen radicals and how these radicals are involved in the evolution of several antioxidant defence and DNA repair mechanisms. Emphasis will be placed on the superoxide radical and its role in biology and the evolution of the superoxide dismutases.

Sources of Oxygen Radicals

In order to address the role of oxygen free radicals in biology and evolution, one must ask what are the potential sources of oxyradicals? Many environmental factors have been implicated in the production of superoxide radicals *in vivo*. Ionising radiation and light-initiated free radical reactions can produce superoxide radicals directly or indirectly. *In vitro* studies indicate that oxygen radicals can be produced by photolysis (McCord and Fridovich, 1973), radiolysis (Rabani and Stein, 1962) and electrolysis (Forman and Fridovich, 1972) of aqueous solutions. One-electron transfers from transition metal ions to organic species are also capable of forming oxygen radicals. With the onset of photosynthesis, several different avenues were opened up that have the potential for generating oxyradicals. To date, it is known that superoxide radicals are biologically produced via enzymatic reactions, electron transport processes and the autoxidation of cellular components. Free radicals also arise from non-enzymatic reactions of oxygen with organic compounds.

Several enzymes, including xanthine oxidase (McCord and Fridovich, 1968), aldehyde oxidase (Rajagopalan, Fridovich and Handler, 1962; Rajagopalan and Handler, 1964), dihydroorotate dehydrogenase (Aleman and Handler, 1967), and flavoprotein dehydrogenases (Massey, Strickland, Mayhew, Howell, Engel, Matthews, Schuman and Sullivan, 1969) generate oxygen radicals. The autoxidation of pyrogallol (Marklund and Marklund, 1974), hydroquinones (Misra and Fridovich, 1972a), catecholamines (Ballou, Palmer and Massey, 1969; Misra and Fridovich, 1972b), reduced ferrodoxins and haemoproteins (Misra and Fridovich, 1971, 1972c; Nakamura and Kimura, 1972), thiols (Misra, 1974) and tetrahydropterins (Nishikimi, 1975) all generate superoxide radicals. Oxygen radicals are generated as intermediates during the catalytic action of several enzymes including intestinal tryptophan dioxygenase (Hirata and Hayaishi, 1971), galactose oxidase (Hamilton, Adolf, deJersey, Dubois, Dyrakcz and Libby, 1978), indoleamine dioxygenase (Hirata, Ohnishi and Hayaishi, 1977) and 2-nitropropane dioxygenase (Kido, Soda and Asada, 1978). Oxygenated haemoglobin liberates O_2^- as it is converted to methemoglobin (Misra and Fridovich, 1972c; Lynch, Thomas and Lee, 1977). Activated macrophages produce superoxide radicals during phagocytosis (Baehner, Boxer and Ingraham, 1982).

Another very important source of superoxide radicals, especially in today's society, is the uptake and/or metabolism of foreign compounds, that is, xenobiotics. These compounds often differ in their modes of action for producing superoxide radicals. Some xenobiotics, such as nitrogen oxides, are themselves free radicals (Pryor, 1984), while others, such as ozone, are not free radicals but react with other organic molecules to produce radical species (Pryor, Prier and Church, 1981, 1983). Several anti-cancer agents containing quinoid groups or bound metals often generate oxygen radicals when enzymatically reduced (Mason, 1982), while other quinoid compounds like paraquat and adriamycin undergo redox cycling and generate superoxide radicals in the presence of molecular oxygen. In addition to the xenobiotics mentioned above, many other environmental factors are themselves radicals or produce radicals when metabolised. Some of these include dietary substances, air pollutants, hyperbaric oxygen, pesticides, herbicides, tobacco smoke, anaesthetics and a variety of solvents (Freeman, 1984).

Dangers of Oxygen Radicals

Oxygen radicals are capable of causing many types of cellular damage because of their extreme reactivity with biomolecules. As to the exact dangers of oxygen free radicals, we are only just beginning to understand the full effects of these radicals on biological systems. We mentioned previously that oxyradicals can interact with proteins, lipids, polysaccharides and nucleic acids. Several molecules such as polyunsaturated fatty acids (PUFA), ascorbate, tocopherol, steroids and other unsaturated organic compounds like glutathione and thiol- containing enzymes are easily oxidised by superoxide radicals (Pryor 1981).

Recent studies suggest that several oxidants are involved in the chronic lung disease, emphysema. Oxidants in tobacco smoke such as nitrogen dioxide inactivate the principal human anti-protease, α1-proteinase inhibitor (α1PI) (Carp and Janoff, 1978; Cohen and James, 1982). This enzyme is further inactivated by superoxide and hydrogen peroxide that accumulate in smoker's lungs due to increased phagocyte activity (Carp and Janoff, 1978, 1980; Clarke, Stone, Hag, Calore and Franzblau, 1981; Cohen and James, 1982). The inactivation of α1PI is thought to make alveolar connective tissue more susceptible to leukocyte proteases, and for this reason oxygen radicals may play a role in the pathogenesis of emphysema (Carp, Janoff, Abrams, Weinbaum, Drew, Weissbach and Brot, 1983). Several researchers (Clarke and Lambertsen, 1971; Johnson, Fantone, Kaplan and Ward, 1981; Martin, Gadek, Hunninghake and Crystal, 1981) have suggested that oxygen radicals increase microvascular leakage of plasma proteins, pulmonary tissue fibrosis, and damage done to alveolar epithelia.

Oxygen radicals can lyse erythrocytes (Kellog and Fridovich, 1977) and damage granulocytes (Salin and McCord, 1975) leading to several diseased states. The autoxidation of oxyhaemoglobin within erythrocytes produces superoxide radicals that may damage red blood cell membrane thus possibly leading to the development of haemolytic anaemia. Babior, Kipnes and Curnutte (1973) demonstrated that polymorphonuclear leukocytes release superoxide radicals into their immediate environment as part of the phagocytic process. In the case of the autoimmune disease chronic granulomatous (Weitzman and Stossel, 1981), neutrophils lack the ability to generate superoxide radicals and hydrogen peroxide because of a defect in the NAD(P)H oxidase (superoxide generating system) which causes a decrease in the organism's ability to combat disease.

Oxygen radicals cause damage to organelle and cell membranes through the peroxidation of unsaturated fatty acids in the phospholipid bilayer. The disruption of lysosomes could cause havoc intracellularly because oxyradicals as well as hydrolytic enzymes would be released. Oxygen radicals may play a role in lipid and protein peroxidation in the eye, leading to cataract development (Varma, 1981) and retrolental fibroplasia (Hittner, Godio, Rudolf, Adams, Garcia-Prats, Friedman, Kautz and Monaco, 1981; Kretzer, Hittner, Johnson, Mehta and Godio, 1982).

Oxygen radicals have also been implicated as one of the causative agents in the heart disease, atherosclerosis. Atherosclerosis is a common form of arteriosclerosis in which yellowish plaques of cholesterol, lipid material and lipophages are formed in medium to large sized arteries. Harmon (1957) proposed a hypothesis explaining the steps which lead to this diseased state. He postulated that serum lipoproteins and arterial wall lipids undergo oxidative-polymerisation reactions and form high molecular weight complexes that anchor to the arterial wall. These complexes increase platelet aggregation and initiate as well as sustain an inflammatory reaction in the vessel wall. This condition leads to arterial wall damage as well as blockage of the arteries.

There is mounting evidence that superoxide radicals play a role in myocardial injury associated with ischaemia and reperfusion (Bulkley, 1983). Similarly, oxyradicals are thought to be involved in reperfusion injury to the kidney and skin, and indirect evidence indicates their involvement in circulatory shock (Parks, Bulkley and Granger, 1983a). Oxygen radicals possibly cause tissue injury in several digestive tract disorders, radiation injury and hepatic cirrhosis (Parks, Bulkley and Granger, 1983b). McCord (1983) proposed that superoxide plays a definite role in neutrophil-mediated inflammatory response. It is also plausible to speculate that oxygen radicals are responsible at least in part for brain damage associated with circumstances where an individual is deprived of oxygen for an acute period of time as in the case of stroke.

Superoxide anions can cause depolymerisation of acid polysaccharides (McCord, 1974) such as hyaluronic acid. Damage is especially apparent in connective tissue, bacterial cell walls, eyeball synovial fluid and in the human circulatory system. Cohen (1984) indicated that oxygen radicals are involved in alloxan-induced diabetes. Alloxan and dialuric acid undergo redox cycling and several reactive oxygen species are produced in the process. The pancreatic beta cells are the target of these oxygen radicals, and damage to these cells induces the diabetic state.

Oxygen radicals have been shown to cause strand scission in DNA (Brawn and Fridovich, 1980, 1981), and they may be involved in the pathogenesis of one of the leading diseases known to claim the most lives in the Western world, namely cancer (Totter, 1980). Harmon (1956) suggested that oxyradicals may act as tumour initiators and promoters. Oxygen radicals produced by human neutrophils give a positive Ames test indicating that these radicals are possibly mutagenic and/or carcinogenic agents (Weitzman and Stossel, 1981). Trioses and gloxyal derivatives have been found to be mutagenic in *Salmonella typhimurium* TA100, and free radicals generated by the autoxidation of these compounds are responsible for their mutagenicity (Yamaguchi and Nakagawa, 1983). Hassan and Moody (1982) and Moody and Hassan (1982) demonstrated that paraquat, a widely used herbicide and an effective intracellular generator of oxygen free radicals, is mutagenic in *S. typhimurium,* and they attributed the mutagencity of paraquat to its ability to exacerbate the intracellular production of superoxide radicals. Hyperbaric oxygen is also known to be mutagenic to *E. coli* (Fenn, Gershman, Gilbert, Terwilliger and Cothran, 1957; Gifford, 1968). In the autosomal recessive disease, Fanconi's anaemia, researchers (Joenje, Arwert, Erikson, deKoning and Oostra, 1981) have shown a positive correlation between increased oxygen tension and chromosomal aberrations. They believe that defects in the organism's defence and/or repair mechanisms against oxygen toxicity are responsible for the chromosomal aberrations.

Evolution of Antioxidant Defence Mechanisms

We hope that we have demonstrated that oxyradicals pose a very dangerous threat to living organisms. It is believed that oxyradicals have served and continue to serve as a strong selective pressure for the survival of different species. In response to this stress, the antioxidant enzymes, superoxide dismutase and the hydroperoxidases, evolved to protect organisms against oxidative damage. Throughout evolution, organisms that possessed the means to protect themselves against oxidative damage had a selective advantage over those that lacked antioxidant defence mechanisms.

Today, aerobic organisms possess at least one of the following antioxidant enzymes: superoxide dismutases, catalases or peroxidases. McCord and Fridovich (1969) first described SOD and its catalytic function to dismute superoxide radicals to hydrogen peroxide (reaction 1). The catalases and peroxidases reduce hydrogen peroxide to water (reactions 2 and 3). As one can see, if one or both enzymes are present in a biological system, the formation of the hydroxyl radical (reaction 4) is inhibited by removing the substrates of the reaction. It is very important to prevent the formation of the hydroxyl radical because it is by far the most potent and reactive oxidant known and is capable of producing very broad and non-specific oxidative damage.

superoxide dismutase: $\quad O_2^- + O_2^- + 2H^+ \rightarrow H_2O_2 + O_2 \quad (1)$

catalase: $\quad 2H_2O_2 \rightarrow H_2O + O_2 \quad (2)$

peroxidase: $\quad H_2O_2 + AH_2 \rightarrow 2H_2O + A^{2-} \quad (3)$

$\quad O_2^- + H_2O_2 \rightarrow OH\cdot + OH^- + O_2 \quad (4)$

There is ample evidence to support the theory that these antioxidant enzymes protect cells against oxidative damage. Several studies with microorganisms have shown that cells grown in the presence of redox active compounds (Hassan and Fridovich, 1977a, 1979a), elevated pO_2 (Hassan and Fridovich, 1977b), oxidisable substrates (Hassan and Fridovich, 1977c), and ozone (Whiteside and Hassan, 1987) exhibit growth inhibition and elevated levels of antioxidant enzymes. Hassan and Fridovich (1977b) illustrated that SOD and CAT are induced in *E. coli* when anaerobically grown cells are shifted to air. Hassan and Moody (1982) demonstrated that SOD exerts a protective effect against paraquat-mediated oxygen toxicity and mutagenicity in *S. typhimurium*. Mutants of *E. coli* deficient in SOD and CAT or CAT alone have been shown to be extremely sensitive to oxygen (Hassan and Fridovich, 1979b). More recently, Farr, D'Ari and Touati (1986) observed an enhanced mutation rate in SOD mutants of *E. coli* that was attributable to the accumulation of oxyradicals in the cells. Furthermore, a decline in the mutation rate was noted when a plasmid overproducing SOD was introduced into the mutants, thus confirming the biological role of SOD in protecting cells against oxyradicals. In *E. coli* (Richter and Loewen, 1981) and *S. typhimurium* (Finn and Condon, 1975), catalase was induced upon addition of H_2O_2 to the culture medium. Several thermophilic bacteria exhibit increased catalase levels when H_2O_2 or paraquat are added to the culture medium (Allgood and Perry, 1985). A peroxide-resistant mutant of *Proteus mirabilis* is much more resistant to H_2O_2 than the wild type because it contains elevated catalase levels (Jouve, Tessier and Pelmont, 1983). Loewen (1984) isolated several catalase-deficient mutants of *E. coli* and found that these mutants were 50-60 times more sensitive to killing by H_2O_2 than their wild type parents.

Shlafer, Kane and Kirsh (1982) demonstrated that injury resulting from myocardial ischaemia and reperfusion was significantly reduced upon administration of free radical scavengers such as SOD and CAT. Allopurinol, a competitive inhibitor of xanthine oxidase, effectively protected against haemorrhagic shock in dogs (Crowell, Jones and Smith, 1969), and the administration of hydroxyl radical scavengers to cats in cardiogenic shock reduced cardiovascular damage (Galvin and Lefer, 1978). Ischaemia reperfusion-induced damage in feline small intestine is minimised with the addition of allopurinol (Parks, Bulkley, Granger, Hamilton and McCord, 1982) or SOD (Granger, Rutili and McCord, 1981) to the tissue, and allopurinol protected against ischaemia related injury in the kidneys (Hansson, Gustafsson, Jonsson, Lundstrom, Pettersson, Schersten and Waldenstrom, 1982). Recently two other xanthine oxidase inhibitors, folic acid and pterin aldehyde, were shown to reduce ischaemia-induced vascular permeability in the feline intestine, further proving that xanthine oxidase is a major source of oxygen radicals in ischaemia-reperfusion conditions (Granger, McCord, Parks and Hollwarth, 1986). Allopurinol and/or oxygen radical scavenging agents may soon be applied in limb reimplantation surgery and organ preservation (Parks, Bulkley and Granger, 1983a). One problem with transplant surgery is that the organs in question rapidly deteriorate once removed from the donor, however, the use of antioxidants may provide protection and increase shelf-life.

Several naturally occurring antioxidants such as tocopherols (vitamin E), ascorbate, carotenoid pigments, glutathione, and phenolic compounds serve as a secondary line of defence against oxidative damage. Recent evidence suggests that vitamin E protects against lipid peroxidation in the eye, and low fat diets supplemented with antioxidants such as vitamin E may decrease the incidence of cardiovascular disease (Harmon, 1984). It is clear that oxygen and its radicals brought about the evolution of a whole line of antioxidant defence mechanisms, and without these defences, life as we know it would probably have never evolved.

Evolution of the Superoxide Dismutases

In this section we will concentrate on the evolution of the superoxide dismutases. This is not to say that the other antioxidants enzymes are not important, but with the exception of the genus *Lactobacillus,* SODs are ubiquitous throughout the five kingdoms. The lactobacilli seem to be the exception to the rule because they accumulate large amounts of Mn^{2+} ions which scavenge O_2^-, thus eliminating the need for SOD (Archibald and Fridovich, 1981). The appearance of catalase and peroxidases is varied throughout these kingdoms so the evolution of these enzymes will not be discussed here, but one should keep in mind that it is the combination of all the antioxidant enzymes and other antioxidants that provide the protection against oxygen toxicity necessary for aerobic survival.

To date, two families of superoxide dismutase (SOD) have been detailed, the iron- and manganese-containing (Fe/MnSODs) and the copper-zinc-containing (CuZnSOD) superoxide dismutases. Three SODs are generally found in prokaryotic organisms: MnSOD, HySOD and FeSOD. The MnSOD and FeSOD contain Mn and Fe in their active centres, respectively, whereas the HySOD contains both the Mn and Fe ions at the active centre of the enzyme. In eukaryotic systems, one will generally find the MnSOD in organelles and CuZnSOD in the cytosol. CuZnSODs are ubiquitous throughout the animal, plant and fungal kingdoms.

There are a few exceptions to this rule however. *Photobacterium leiognathi,* a symbiotic organism associated with a pony fish, contains the typically eukaryotic CuZnSOD (Puget and Michelson, 1974). Martin and Fridovich (1981) explained this phenomenon by proposing a gene transfer from one organism (pony fish) to the other (bacterium). This hypothesis has been ruled out however, for the CuZnSOD of *P. leiognathi* shows very little sequence homology to the eukaryotic CuZnSOD (Steffens, Bannister, Bannister, Floke, Gunzler, Kim and Otting, 1983) just as the CuZnSOD found in *Caulobacter crescentus* has little sequence homology to its eukaryotic cousin (Steinman, 1982). Steinman (1985) recently characterised two pseudomonads, *P. diminuta* and *P. maltophilia,* that contain CuZnSOD as well as MnSOD, and interestingly enough, the CuZnSOD of these strains is not cross-reactive with monoclonal antibody prepared against *P. leiognathi* CuZnSOD.

On the other hand, Salin and Bridges (1982) isolated FeSOD from the water lily, *Nuphar luteum,* and they also found the FeSOD in three isolated plant families out of 43 surveyed (Bridges and Salin, 1981). The FeSOD, MnSOD and the CuZnSOD have been isolated from the mustard plant, *Brassica campestris* (Salin and Bridges, 1980). Duke and Salin (1985) recently isolated an FeSOD from the eukaryotic *Ginkgo biloba.* This particular enzyme is localised only in the stroma of chloroplasts, and it shows sequence homology to the SODs from *Nuphar, Brassica* and *E. coli.* Photosynthetic bacteria, prokaryotic algae and eukaryotic algae all lack CuZnSOD and contain only the Mn- or FeSOD (Asada, Kanematsu and Uchida, 1977). Other eukaryotes that lack CuZnSOD include two anaerobic flagellates (Lindmark and Miller, 1974), the mushroom, *Pleurotus* (Lavelle, Durosky and Michelson, 1974) and several *Euglena* species (Asada, Kanematsu, Okaka and Hayakawa, 1980). Thus, one cannot simply say that the Fe/MnSODs are restricted to prokaryotes and the CuZnSOD to eukaryotes.

The FeSOD is the proposed ancestor of Fe- and MnSODs in present day prokaryotes because it is present in several primitive anaerobes including sulphate-reducing bacteria and clostridia. Ferrous iron in the primitive oceans possibly served as an oxygen acceptor (McElroy and Selinger, 1962). One may envision the evolution of the FeSOD beginning with the simple metal ion capable of undergoing redox changes, into the present day FeSOD with the metal ion at the active centre. Inspection of the *E. coli* genomic map (Taylor and Trotter, 1972) indicates that several biochemically related genes tend to lie 90° or 180° apart. Zipkas and Riley (1975) proposed that during the evolution of the *E. coli* genome it underwent two duplications so that four copies of each ancestral gene were possible. In *E. coli,* the MnSOD and FeSOD have been mapped at 87.5 minutes (Touati, 1983) and 36.5 minutes (Nettleton, Bull, Baldwin and Fee, 1984), respectively. It is possible, therefore, that the MnSOD arose from the FeSOD via gene duplication. Through time the MnSOD may have undergone subtle structural changes that favoured the Mn ion over the Fe ion in the active centre, or the appearance of abundant molecular oxygen may have initiated the change in the metal's active site.

One may ask what are the differences between the FeSODs and MnSODs, and what factors contribute to their distribution throughout the prokaryotic kingdom? Amino-terminal protein sequences of these two enzymes show a great deal of homology (Steinman, 1978), and these sequences seem to be highly conserved throughout the various phyla. The MnSOD and FeSOD are structural homologues (Stallings, Pattridge, Strong and Ludwig, 1984), and their catalytic properties are approximately equal. In some cases, as in the bacterium, *Propionibacterium shermanii,* both the MnSOD and FeSOD are synthesised from an identical protein moiety, dependent upon the metal supplied (Meier, Barra, Bossa, Calabrese and Rotilio, 1982).

There are no clear-cut answers to explain the distribution of these enzymes in prokaryotes, but several generalisations can be made. FeSODs are found in obligate anaerobes, facultative anaerobes, and a few aerobic organisms. In all cases, the FeSOD is expressed in its active form when these organisms are grown aerobically and anaerobically. The MnSOD is found in some facultative anaerobes and aerobic organisms. The MnSOD is actively expressed under aerobic and anaerobic conditions when it is the sole SOD in the organism, but in the case of *E. coli,* where FeSOD and MnSOD are both found, only the FeSOD is produced anaerobically. This suggests that *E. coli* has developed an elaborate mechanism of controlling the expression of MnSOD. Moody and Hassan (1984) proposed that MnSOD biosynthesis in *E. coli* is regulated by a repressor protein that inhibits the expression of the MnSOD gene under anaerobic conditions but not in the presence of air. Another model was proposed by Fridovich and coworkers (Pugh, DiGuiseppi and Fridovich, 1984; Pugh and Fridovich, 1985) in which the FeSOD and MnSOD are autogenously regulated by the availability of the appropriate metal cofactor and the presence of superoxide radicals. Fortunately both models are testable and the elucidation of the regulatory mechanism of MnSOD may soon prove to be more complex than originally thought.

The CuZnSOD evolved independently of the Fe/MnSODs sometime after the evolution of molecular oxygen. From the analysis of protein structure, one notices that the CuZnSOD is evolutionarily distinct from the Mn- and FeSODs (Tainer, Getzoff, Beem, Richardson and Richardson, 1982; Stallings, Powers, Pattridge, Fee and Ludwig, 1983; Stallings et al., 1984). The amino-terminal regions of the MnSOD and FeSOD are not homologous to the amino-terminal region of the CuZnSOD (Bridgen, Harris and Northrop, 1975; Harris and Steinman, 1977) and further sequencing of the MnSOD shows an absence of homology with the erythrocyte CuZnSOD throughout the entire polypeptide chain (Steinman, 1978).

One must ask at what point in Earth's history did the SODs begin to evolve? The appearance of FeSOD in obligate anaerobes has been a puzzle for years. No one knows for sure the exact role of FeSOD in these obligate anaerobes, but needless to say, this enzyme would not be found so consistently throughout these anaerobic systems unless it served a constructive purpose. Perhaps the answer to this puzzle may be found in the answer to the following question: At what point did oxygen become toxic to living organisms? Hatchikian, LeGall and Bell (1977) suggest that strict anaerobes were contained in anaerobic niches, and that transient exposures to oxygen resulted in the evolution of antioxidant enzymes. Lumsden and Hall (1975) propose that SOD evolved in anaerobes prior to oxygenic photosynthesis to protect them from molecular oxygen produced in the oceans via the Urey effect. However, it has been suggested that oxygen concentrations produced by the UV photolysis of water were kept to minimum by several sequestering mechanisms. McElroy and Selinger (1962) proposed that the light-emitting chemical reactions involved in bioluminescence evolved as a detoxifying process for the sequestering of oxygen and that ferrous iron possibly acted as an electron acceptor in the water. It is difficult to estimate what the oxygen concentrations were during this period of the Earth's history, and even if we did know, who is to say what concentrations would pose a threat to living systems? What may be insignificant levels to man may not be so insignificant from the microorganism's perspective.

Whether strict anaerobes were transiently exposed to oxygen or oxygen posed a small but constant stress to these organisms, superoxide dismutase probably begin to evolve before the appearance of abundant molecular oxygen. This must have been the case, for the appearance of an oxygen evolving system in the pre-Cambrian photosynthetic organisms may well have been lethal to the dominant fermentative anaerobes as well as to the photosynthesisers themselves. It is also possible that only a small percentage of the microorganisms that had the means to protect themselves against the deleterious effects of oxygen survived.

Further evidence to support the idea that FeSOD began to evolve prior to the appearance of abundant molecular oxygen is the fact that FeSODs and MnSODs are insensitive to cyanide whereas CuZnSOD is cyanide sensitive. The abiotic theory for the origin of life suggests that simple organic molecules such as cyanide, formate and formaldehyde resulted from bond rearrangements of water and methane, ammonia and hydrogen gases present in the reducing atmosphere catalysed by a variety of energy sources such as UV light, ionising radiation and lightning discharge. An anoxic atmosphere would have allowed the accumulation of cyanocompounds because they are fairly stable when not oxidised. If FeSOD began to evolve in the presence of cyanocompounds, it makes sense that it would be resistant to cyanide. With the evolution of the photosynthesisers, the cyanocompounds were rapdily oxidised and no longer served as a selective pressure in the evolution of superoxide dismutases. Therefore it seems logical that CnZnSOD would not need to be insensitive to cyanide because it is thought to have evolved after the appearance of molecular oxygen. This has been shown to be the case. Indeed, the cyanide sensitivity test is used to distinguish between CuZnSODs and Fe/MnSODs.

We would now like to propose a model for the evolution of the superoxide dismutases. The FeSODs were the first SODs to evolve and did so in the obligate anaerobes sometime prior to the evolution of the oxygenic photosynthesisers. The MnSOD arose via gene duplication from the FeSOD at least in the enteric bacteria, or in other organisms the MnSOD became dominant after molecular oxygen accumulated and favoured the Mn ion over the Fe ion in the active site of the enzyme. It is interesting to note that FeSODs are inhibitable by hydrogen peroxide, the product of the dismutation reaction, while the MnSODs are not sensitive to H_2O_2. Therefore, it is possible that MnSODs have evolved for the aerobic lifestyle where the dismutation of O_2^- and the generation of H_2O_2 is consistently higher than that encountered anaerobically. In any case, the ancestral enzyme was conserved throughout evolution as witnessed by the similarities in amino acid sequence, structure, catalytic properties and cyanide insensitivity in the Mn- and FeSODs. It is also interesting to note that only the MnSOD is found in the organelles of eukaryotes. This suggests that two types of prokaryotes existed, and the ones containing the MnSOD were favoured to undergo an endosymbiotic relationship with the eubionts leading to present day eukaryotes.

The acquisition of CuZnSOD has been previously used as a measure to determine the phylogenetic relationships of various organisms. It is generally believed that the CuZnSOD appeared as the more complex photosynthetic organisms began to evolve, i.e. ferns, mosses, and higher plants, but its appearance in an increasing number of prokaryotes raises doubts as to its exact descent and evolution. We propose that the CuZnSOD evolved independently of the Fe/MnSODs because of differences in amino acid sequence, protein structure and catalytic properties. It must have evolved after oxygenic photosynthesis began because of its sensitivity to cyanide. This is not to say that all organisms which evolved after the blue green algae contain CuZnSOD. We are simply suggesting that two independent protein families convergently evolved. Some factor(s), possibly increased oxygen pressure or the move to a terrestrial environment, favoured the CuZnSOD over the Fe/MnSODs. The absence of CuZnSOD in some of the lower eukaryotic photosynthesisers may be due to the possibility that these organisms evolved from an anaerobic prokaryotic ancestor instead of arising as a new phylogenetic line after oxygen appeared. The appearance of CuZnSOD in prokaryotes may be explained by the fact that with the accumulation of molecular oxygen, new niches were opened up for prokaryotes to occupy, and the evolution of many new species may well have favoured CuZnSOD over the typically prokaryotic SODs. The dissimilarities between prokaryotic CuZnSODs, eukaryotic CuZnSODs and prokaryotes Fe/MnSODs suggests that there were several evolutionary lines for the SODs and not just the typical two.

Evolution of DNA Repair Mechanisms

The evolution of antioxidant mechanisms would have been pointless without the evolution of DNA repair mechanisms. It may be relatively simple for the cell to replace proteins, lipids and other cellular components damaged by oxygen radicals that slipped by the antioxidant mechanisms, but radical damage to nucleic acids jeopardises the hereditary integrity of the cell. The cell cannot simply replace the damaged DNA because the replicative process requires an intact correct template. Thus DNA repair mechanisms evolved to ensure that the genetic material of the cell was conserved and passed along to future generations. We refer the reader to several excellent reviews on DNA repair mechanisms (Grossman, 1981; Ganesan, Cooper, Hanawalt and Smith, 1982; Lindahl, 1982; Walker, 1984). Many chemical and physical agents can cause DNA damage. Ultraviolet irradiation, alkylating agents and intercalating agents are capable of damaging cellular DNA. More recently, oxygen free radicals were added to the list of agents that can cause DNA strand scission and mutation (Brawn and Fridovich, 1981; Moody and Hassan, 1982). Damage to cellular DNA has been shown to induce the SOS system in *E. coli* which consists of a set of unlinked genes that are under the control of the recA and lexA gene products (Walker, 1984). In this system damage to DNA activates the proteolytic function of the RecA protein which cleaves the LexA repressor protein and causes the depression of the SOS genes (Quillardet, Huisman, D'Ari and Hofnung, 1982, 1984) whose functions are essential for the ultimate survival of the cell. This unique inducible DNA repair system (SOS) has been shown to be inducible by numerous DNA-damaging agents including oxyradicals (Little and Mount, 1982; Brawn and Fridovich, 1985). The fact that both the manganese-containing superoxide dismutase (Hassan and Fridovich, 1980) and the SOS system (Brawn and Fridovich, 1985) are induced in *E. coli* in response to elevated levels of superoxide radicals raised the possibility that the gene encoding the MnSOD is part of the SOS system. Hancock and Hassan (1985) investigated the induction of MnSOD in several strains of *E. coli* with different mutations in the recA and lexA genes and concluded that the induction of MnSOD is independent of the SOS system. It is clear that the induction of both MnSOD and the SOS system by increased levels of oxyradicals represent a coordinated response by the cells for improving their chances for survival. Thus, the induction of MnSOD will prevent the accumulation of toxic superoxide radicals, whereas the induction of the SOS DNA repair system will repair the damage caused by the oxygen free radicals that escape the first line of defence, superoxide dismutase.

This work was supported in part by DCB-8910153 from the National Science Foundation. This is paper number 10677 of the Journal Series of the North Carolina Agriculture Research Service, Raleigh, NC 27695- 7601. The use of trade names in this publication does not imply endorsement by the North Carolina Research Service of the products named, nor criticism of similar ones not mentioned.

References

Aleman, V. and Handler, P. (1967) 'Dihydroorotate Dehydrogenase I. General Properties', *Journal of Biological Chemistry* 242, 4087-4096
Allgood, G.S. and Perry, J.J. (1985) 'Paraquat Toxicity and Effect of Hydrogen Peroxide on Thermophilic Bacteria', *Journal of Free Radicals in Biology Medicine* 1, 233-237
Archibald, F.S. and Fridovich, I. (1981) 'Manganese and Defences Against Oxygen Toxicity in *Lactobacillus planterum*', *Journal of Bacteriology* 145, 442-451
Asada, K., Kanematsu, S., Okaka, S., and Hayakawa, T. (1980) 'Phylogenetic Distribution of Three Types of Superoxide Dismutase in Organisms and in Cells' Organelles', in J.V. Bannister and H.A.O.Hill (eds.), *Chemical and Biological Aspects of Superoxide and Superoxide Dismutases* (Elsevier, New York), pp. 136-153
Asada, K., Kanematsu, S., and Uchida, K. (1977) 'Photosynthetic Organisms: Absence of the Cuprozinc Enzyme in Eukaryotic Algae', *Archives of Biochemistry and Biophysics* 179, 243-256
Babior, B.M., Kipnes, R.S., and Curnutte, J.T. (1973) 'Biological Defence Mechanisms. The Production by Leukocytes of Superoxide, a Potential Bacteriocidal Agent', *Journal of Clinical Investigation* 52, 741-744
Baehner, R.L., Boxer, L.A., and Ingraham, L.M. (1982) 'Reduced Oxygen By-Products and White Blood Cells', in W.A. Pryor (ed.), *Free Radicals in Biology* (Academic Press, New York), 5, 91-127
Ballou, D., Palmer, G., and Massay, V. (1969) 'Direct Demonstration of Superoxide Anion Production During the Autoxidation of Reduced Flavin and its Catalytic Decomposition by Erythrocuprein', *Biochemical and Biophysical Research Communications* 36, 898-904
Brawn, K. and Fridovich, I. (1980) 'Superoxide Radicals and Superoxide Dismutases: Threat and Defence', *Acta Physiologica Scandinavia* 492, 9-18
Brawn, K. and Fridovich, I. (1981) 'DNA strand Scission by Enzymatically Generated Oxygen Radicals', *Archives of Biochemistry and Biophysics* 206, 414-419
Brawn, K. and Fridovich, I. (1985) 'Increased Superoxide Radical Production Evokes Inducible DNA Repair in *Escherichia coli*', *Journal of Biological Chemistry* 260, 922-925
Bridges, S.M. and Salin, M.L. (1981) 'Distribution of Iron-Containing Superoxide Dismutase in Vascular Plants', *Plant Physiology* 68, 275-278
Bridgen, J., Harris, J.I., and Northrop, F. (1975) 'Evolutionary Relationships in Superoxide Dismutase', *FEBS Letters* 49, 392-395
Bulkley, G.B. (1983) 'The Role of Free Radicals in Human Disease Processes', *Surgery* 94, 407-411
Carp, H. and Janoff, A. (1978) 'Possible Mechanisms of Emphysema in Smokers: *In vitro* Suppression of Serum Elastase-Inhibitory Capacity by Fresh Cigarette Smoke and its Prevention by Antioxidants', *American Reviews of Respiratory Diseases* 118, 617-621
Carp, H. and Janoff, A. (1980) 'Potential Mediator of Inflammation', *Journal of Clinical Investigation* 66, 987-995

Carp, H., Janoff, A., Abrams, W., Weinbaum, G., Drew, R.T., Weissbach, H., and Brot, N. (1983) 'Human methionine sulfoxide-Peptide Reductase, an Enzyme Capable of Reactivating Oxidized alpha-1-Proteinase Inhibitor in vitro', *American Reviews of Respiratory Diseases* 127, 301-305

Clarke, J.M. and Lambertsen, C.J. (1971) 'Pulmonary Oxygen Toxicity: A Review', *Pharmacological Reviews* 23, 38-117

Clarke, R.A., Stone, P.J., Hag, E.A., Calore, J.D., and Franzblau, C. (1981) 'Myeloperoxidase-Catalyzed Inactivation of α1-Protease Inhibition by Human Neutrophils', *Journal of Biological Chemistry* 256, 3348-3353

Cohen, A.B. and James, H.L. (1982) 'Reduction of Elastase Inhibitory Capacity of alpha-1-Antitrypsin by Peroxides in Cigarette Smoke', *American Reviews of Respiratory Diseases* 126, 25-30

Cohen, G. (1984) 'Oxy-Radical Production in Alloxan-Induced Diabetes: An Example of an in vivo Metal-Catalyzed Haber-Weiss Reaction', in D. Armstrong, R.S. Sohal, R.G. Cutler and T.F. Slater (eds.), *Free Radicals in Molecular Biology, Aging, and Disease* (Raven Press, New York), pp. 307-316

Crowell, J.W., Jones, C.E., and Smith, E.E. (1969) 'Effect of Allopurinol on Hemorrhagic Shock', *American Journal of Physiology* 216, 744-748

Czapski, G. (1971) 'Radiation Chemistry of Oxygenated Aqueous Solutions', *Annual Reviews of Physical Chemistry* 22, 171-208

Duke, M.V. and Salin, M.L. (1985) 'Purification and Characterization of an Iron-Containing Superoxide Dismutase from the Eucaryote, Ginkgo biloba', *Archives of Biochemistry and Biophysics* 243, 305-314

Farr, S.B., D'Ari, R., and Touati, D. (1986) 'Oxygen-Dependent Mutagenesis in *Escherichia coli* lacking Superoxide Dismutase', *Proceedings of the National Academy of Science U.S.A.* 83, 8268-8277

Fenn, W.A., Gershman, R., Gilbert, D.L., Terwilliger, D.E., and Cothran, F.W. (1957) 'Mutation Rate of *E. coli* Exposed to High Pressure Oxygen for Several Hours Increased over the Spontaneous Rate', *Proceedings of the National Academy of Science U.S.A.* 43, 1027-1032

Finn, G.J. and Condon, S. (1975) 'Regulation of Catalase Synthesis in *Salmonella typhimurium*', *Journal of Bacteriology* 123, 570-579

Forman, H.J. and Fridovich, I. (1972) 'Electrolytic Univalent Reduction of Oxygen in Aqueous Solution Demonstrated by SOD', *Science* 175, 339

Freeman, B. (1984) 'Biological Sites and Mechanisms of Free Radical Production', in D. Armstrong, R.S. Sohal, R.G. Cutler and T.F. Slater (eds.), *Free Radicals in Molecular Biology, Aging and Disease* (Raven Press, New York), pp. 43-52

Galvin, M.J. and Lefer, A.M. (1978) 'Salutory Effects of Cysteine on Cardiogenic Shock in Cats', *American Journal of Physiology* 235, H657-663

Ganesan, A.K., Cooper, P.K., Hanawalt, P.C., and Smith, C.A. (1982) 'Biochemical Mechanisms and Genetic Control of DNA Repair', *Progress in Mutation Research* 4, 313-323

Gifford, G.D. (1968) 'Mutation of an Auxotrophic Strain of *E. coli* by High Pressure Oxygen', *Biochemical and Biophysical Research Communications* 33, 294-298

Granger, D.N., MCord, J.M., Parks, D.A., and Hollwarth, M.E. (1986) 'Xanthine Oxidase Inhibitors Attenuate Ischemia-Induced Vascular Permeability Changes in the Cat Intestine', *Gastroenterology* 90, 80-84

Granger, D.M., Rutili, G., and McCord, J.M. (1981) 'Superoxide Radicals in Feline Intestinal Ischemia', *Gastroenterology* 81, 22-29

Grossman, L. (1981) 'Enzymes Involved in the Repair of Damaged DNA', *Archives of Biochemistry and Biophysics* 211, 511-522

Hamilton, G.A., Adolf, P.K., deJersey, J., Dubois, G.C., Dyrkacz, G.R., and Libby, R.D. (1978) 'Trivalent Copper, Superoxide, and Galactose Oxidase', *Journal of the American Chemical Society* **100**, 1899-1912

Hancock, L.C. and Hassan, H.M. (1985) 'Regulation of the Manganese-Containing Superoxide Dismutase is Independent of the Inducible DNA Repair System in *Escherichia coli*', *Journal of Biological Chemistry* **260**, 12954-12956

Hansson, R., Gustafsson, B., Jonsson, O., Lundstrom, S., Pettersson, S., Schersten, T., and Waldenstrom, J. (1982) 'Effect of Xanthine Oxidase Inhibition on Renal Circulation After Ischemia', *Transplant. Proceedings* **14**, 51-58

Harmon, D. (1956) 'Aging: A Theory Based on Free Radical and Radiation Chemistry', *Journal of Gerontology* **11**, 298-300

Harmon, D. (1957) 'Atherosclerosis: a hypothesis concerning the limiting steps in pathogenesis'. *Journal of Gerontology* **12**, 199-202

Harmon, D. (1984) 'Free Radicals and the Origination, Evolution, and Present Status of the Free Radical Theory of Aging', in D. Armstrong, R.S. Sohal, R.G. Cutler and T.F. Slater (eds.), *Free Radicals in Molecular Biology, Aging and Disease* (Raven Press, New York), pp. 1-12.

Harris, J.I. and Steinman, H.M. (1977) 'Amino Acid Sequence Homologies Among Superoxide Dismutases', in A.M. Michelson, J.M. McCord and I. Fridovich (eds.), *Superoxide and Superoxide Dismutase* (Academic Press, New York), pp. 225-230

Hassan, H.M. and Fridovich, I. (1977a) 'Regulation of the Synthesis of Superoxide Dismutase in *Escherichia coli* : Induction by Methyl Viologen', *Journal of Biological Chemistry* **252**, 7667-7672

Hassan, H.M. and Fridovich, I. (1977b) 'Enzymatic Defences Against the Toxicity of Oxygen and of Streptonigrin in *Escherichia coli*', *Journal of Bacteriology* **129**, 1574-1583

Hassan, H.M. and Fridovich, I. (1977c) 'Regulation of Superoxide Dismutase Synthesis in *Escherichia coli:* Glucose Effect.' *Journal of Bacteriology* **132**, 505-510

Hassan, H.M. and Fridovich, I. (1979a) 'Intracellular Production of Superoxide and Hydrogen Peroxide by Redox Active Compounds', *Archives of Biochemistry and Biophysics* **196**, 385-395

Hassan, H.M. and Fridovich, I. (1979b) 'Superoxide, Hydrogen Peroxide and Oxygen Tolerance of Oxygen-Sensitive Mutants of *Escherichia coli*'. *Review of Infectious Diseases* **1**, 357-367

Hassan, H.M. and Fridovich, I. (1980) 'Superoxide Dismutases: Detoxification of a Free Radical', in W.B. Jakoby (ed.), *Enzymatic Basis of Detoxification* (Academic Press, New York) 311-332

Hassan, H.M. and Moody, C.S. (1982) 'SOD Protects Against Paraquat-Mediated Dioxygen Toxicity and Mutagenicity: Studies in *Salmonella typhimurium*', *Canadian Journal of Physiology and Pharmacology* **60**, 1367-1373

Hatchikian, C.E., LeGall, J., and Bell, G.R. (1977) 'Significance of Superoxide Dismutase and Catalase Activities in the Strict Anaerobes, Sulfate Reducing Bacteria', in A.M. Michelson, J.M. McCord and I. Fridovich (eds.), *Superoxide and Superoxide Dismutases* (Academic Press, New York), pp. 159-172

Hirata, F. and Hayaishi, O. (1971) 'Possible Participation of Superoxide Anion in the Intestinal Tryptophan 2,3-Dioxygenase Reaction', *Journal of Biological Chemistry* **246**, 7825-7826

Hirata, F., Ohnishi, T. and Hayaishi, O. (1977) 'Indoleamine 2,3-Dioxygenase: Characterization and Properties of Enzyme-O_2-Complex', *Journal of Biological Chemistry*. **252**, 4637-4642

Hittner, H.M., Godio, L.B., Rudolf, A.J., Adams, J.M., Garcia-Prats, J.A., Friedman, Z., Kautz, J.A., and Monaco, W.A. (1981) 'Retrolental Fibroplasia: Efficacy of

Vitamin E in a Double-Blind Study in Preterm Infants', *New England Journal of Medicine* **305**, 1365-1371

Joenje, H., Arwert, F., Erikson, A.W., deKoning, H. and Oostra, A.B. (1981) 'Oxygen-Dependence of Chromosomal Aberrations in Fanconi's Anaemia', *Nature (London)* **290**, 142-143

Johnson, K.J., Fantone, J.C., Kaplan, J. and Ward, P.A. (1981) 'In vivo Damage of Rat Lungs by Oxygen Metabolites', *Journal of Clinical Investigation* **67**, 983-993

Jouve, H.M., Tessier, S., and Pelmont, S. (1983) 'Properties of a Catalase from a Peroxide-Resistant Mutant of *Proteus mirabilis*' *Canadian Journal of Biochemistry and Cell Biology* **61**, 1219-1226

Kellog, E.W.III and Fridovich, I. (1977) 'Liposome Oxidation and Erythrocyte Lysis by Enzymatically Generated Superoxide and Hydrogen Peroxide', *Journal of Biological Chemistry* **252**, 6721-6728

Kido, T., Soda, K., and Asada, K. (1978) 'Properties of 2-Nitropropane Dioxygenase of *Hansenula markii*: Formation and Participation of Superoxide', *Journal of Biological Chemistry* **253**, 226-232

Kretzer, F.L., Hittner, H.M., Johnson, A.T., Mehta, R.S. and Godio, L.B. (1982) 'Vitamin E and Retrolental Fibroplasia: Ultrastructural Support of Clinical Efficacy', *Annals of the New.York Academy of Science* **393**, 145-166

Lavelle, F., Durosky, P., and Michelson, A.M. (1974) 'Purification et Proprietes de la Superoxyde Dismutase du Champignon *Pleurotus olearius*' *Biochemie* **56**, 451-458

Lindahl, T. (1982) 'DNA Repair Enzymes', *Annual Reviews of Biochemistry* **51**, 61-87

Lindmark, D.G. and Miller, M. (1974) 'Superoxide Dismutase in the Anaerobic Flagellates, *Tritrichmonas foetus* and *Monocercomonas* sp.', *Journal of Biological Chemistry* **249**, 4634-4637

Little, J.W. and Mount, D.W. (1982) 'The SOS Regulatory System of *Escherichia coli*', *Cell* **29**, 11-22

Loewen, P.C. (1984) 'Isolation of Catalase-Deficient *Escherichia coli* Mutants and Genetic Mapping of katE, a Locus that Affects Catalase Activity', *Journal of Bacteriology* **157**, 622-626

Lumsden, J. and Hall, D.O. (1975) 'Superoxide Dismutase in Photosynthetic Organisms Provides an Evolutionary Hypothesis', *Nature (London)* **257**, 670-672

Lynch, R.E., Thomas, J.E., and Lee, G.R. (1977) 'Inhibition of Methemoglobin Formation From Purified Oxyhemoglobin by Superoxide Dismutase', *Biochemistry* **16**, 4563-4567

Marklund, S. and Marklund, G. (1974) 'Involvement of the Superoxide Anion Radical in the Autoxidation of Pyrogallol and a Convenient Assay for Superoxide Dismutase', *European Journal of Biochemistry* **47**, 469-474

Martin, J.P., Jr. and Fridovich, I. (1981) 'Evidence for a Natural Gene Transfer from the Ponyfish to its Bioluminescent Bacterial Symbiont *Photobacter leiognathi*', *Journal of Biological Chemistry* **256**, 6080-6089

Martin, W.J., III, Gadek, J.E., Hunninghake, G.W. and Crystal, R.G. (1981) 'Oxidant Injury of Lung Parenchymal Cells', *Journal of Clinical Investigation* **68**, 1277-1288

Mason, R.P. (1982) 'Free-Radical Intermediates in the Metabolism of Toxic Chemicals', in W.A. Pryor (ed.), *Free Radicals in Biology* (Academic Press, New York), **5**, 161-196

Massey, V., Strickland, S., Mayhew, S.G., Howell, L.G., Engel, P.C., Matthews, R.G., Schuman, M., and Sullivan, P.A. (1969) 'The Production of Superoxide Anion Radicals in the Reaction of Reduced Flavins With Molecular Oxygen', *Biochemical and Biophysical Research Communication* **36**, 891-897

McCord, J.M. (1974) 'Free Radicals and Inflammation: Protection of Synovial Fluid by Superoxide Dismutase', *Science* **185**, 529-531

McCord, J.M. (1983) 'The Superoxide Free Radical: Its Biochemistry and Pathophysiology', *Surgery* **94**, 412-414

McCord, J.M. and Fridovich, I. (1968) 'Reduction of Cytochrome c by Milk Xanthine Oxidase', *Journal of Biological Chemistry* **243**, 5753-5760

McCord, J.M. and Fridovich, I. (1969) 'Superoxide Dismutase: An Enzymatic Function for Erythrocuprein (Hemocuprein)', *Journal of Biological Chemistry* **244**, 6049-6055

McCord, J.M. and Fridovich, I. (1973) 'Production of O_2^- in Photolyzed Water Demonstrated Through the Use of SOD', *Photochemistry and Photobiology* **17**, 115-121

McElroy, W.D. and Selinger, H.H. (1962) 'Origin and Evolution of Bioluminescence' in M. Kasha and B. Pullman (eds.), *Horizons in Biochemistry* (Academic Press, New York), pp. 91-101

Meier, B., Barra, D., Bossa, F., Calabrese, L., and Rotilio, G. (1982) 'Synthesis of Either Fe- of Mn- SOD with an Apparently Identical Protein Moeity by an Anaerobic Bacterium Dependent on the Metal Supplied', *Journal of Biological Chemistry* **257**, 13977-13980

Misra, H.P. (1974) 'Generation of Superoxide Free Radicals During the Autoxidation of Thiols', *Journal of Biological Chemistry* **249**, 2151-2155

Misra, H.P. and Fridovich, I. (1971) 'The Generation of Superoxide Radical During the Autoxidation of Ferredoxins', *Journal of Biological Chemistry* **246**, 6886-6890

Misra, H.P. and Fridovich, I. (1972a) 'The Univalent Reduction of Oyxgen by Reduced Flavins and Quinones', *Journal of Biological Chemistry* **247**, 188-192

Misra, H.P. and Fridovich, I. (1972b) 'The Role of Superoxide Anion in the Autoxidation of Epinephrine and a Simple Assay for Superoxide Dismutase', *Journal of Biological Chemistry* **247**, 3170-3175

Misra, H.P. and Fridovich, I. (1972c) 'The Generation of Superoxide Radical During the Autoxidation of Hemoglobin', *Journal of Biological Chemistry* **247**, 6960-6962

Moody, C.S. and Hassan, H.M. (1982) 'Mutagenicity of Oxygen Free Radicals', *Proceedings of the National Academy of Science U.S.A.* **79**, 2855-2859

Moody, C.S. and Hassan, H.M. (1984) 'Anaerobic Biosynthesis of Manganse-Containing Superoxide Dismutase in *Escherichia coli*', *Journal of Biological Chemistry* **25**, 12821-12825

Nakamura, S. and Kimura, T. (1972) 'Studies on Aggregated Multienzyme Systems: Stimulation of Oxygen Uptake by Ferredoxin-Nicotinamide Adenine Dinucleotide Phosphate Reductase-Ferredoxin Complex by Cytochrome c', *Journal of Biological Chemistry* **247**, 6462-6468

Nettleton, C.T., Bull, C., Baldwin, T.O., and Fee, J.A. (1984) 'Isolation of the *Escherichia coli* Iron Superoxide Dismutase Gene: Evidence That Intracellular Superoxide Concentration Does Not Regulate Oxygen-Dependent Synthesis of the Manganese Superoxide Dismutase', *Proceedings of the National Academy of Science U.S.A* **81**, 4970-4973

Nishikimi, M. (1975) 'The Generation of Superoxide Anion in the Reaction of Tetrahyropteridines With Molecular Oxygen', *Archives of Biochemistry and Biophysics* **166**, 273-279

Parks, D.A., Bulkley, G.B., and Granger, D.N. (1983a) 'Role of Oxygen Free Radicals in Shock, Ischemia, and Organ Preservation', *Surgery* **94**, 428-432

Parks, D.A., Bulkley, G.B., and Granger, D.N. (1983b) 'Role of Oxygen-Derived Free Radicals in Digestive Tract Disease', *Surgery* **94**, 415-422

Parks, D.A., Bulkley, G.B., Granger, D.N., Hamilton, S.R., and McCord, J.M. (1982) 'Ischemic Injury in the Cat Small Intestine: Role of Superoxide Radicals', *Gastroenterology* **82**, 9-15

Pryor, W.A. (1981) 'Radical-Induced Oxidations in Biology', in M.A.J. Rodgers and E.L. Powers (eds.), *Oxygen and Oxy-Radicals in Chemistry and Biology* (Academic Press, New York), pp. 133-139

Pryor, W.A. (1984) 'Free Radicals in Autoxidation and in Aging', in D. Armstrong, R.S. Sohal, R.G. Cutler and T.F. Slater (eds.), *Free Radicals in Molecular Biology, Aging, and Disease* (Raven Press, New York), pp. 13-41

Pryor, W.A., Prier, D.G., and Church, D.F. (1981) 'Radical Production From the Interaction of Ozone and PUFA as Demonstrated by Electron Spin Resonance Spin-Trapping Techniques', *Environmental Research* **24**, 42-52

Pryor, W.A., Prier, D.G., and Church, D.F. (1983) 'Detection of Free Radicals From Low-Temperature Ozone-Olefin Reactions by ESR Spin Trapping: Evidence That the Radical Precursor is a Trioxide', *Journal of the American Chemical Society* **105**, 2883-2888

Puget, K. and Michelson, A.M. (1974) 'Isolation of a New Copper-Containing Superoxide Dismutase Bacteriocuprein', *Biochemical and Biophysical Research Communications* **58**, 830-838

Pugh, S.Y.R., DiGuiseppi, J.L., and Fridovich, I. (1984) 'Induction of Superoxide Dismutase in *Escherichia coli* by Manganese and Iron', *Journal of Bacteriology* **160**, 137-142

Pugh, S.Y.R. and Fridovich, I. (1985) 'Induction of Superoxide Dismutase in *Escherichia coli* by Metal Chelators', *Journal of Bacteriology* **162**, 196-202

Quillardet, P., Huisman, O., D'Ari, R. and Hofnung, M. (1982) 'SOS Chromotest, a Direct Assay of Induction of an SOS Function in *Escherichia coli* K-12 to Measure Genotoxicity', *Proceedings of the National Academy of Science U.S.A.* **79**, 5971-5975

Quillardet, P., Huisman, O., D'Ari, R. and Hofnung, M. (1984) 'The SOS Chromotest: Use of a Gene Fusion to Measure the Genotoxicity of Chemicals', *Bulletin de l'Institut Pasteur* **82**, 75-82

Rabani, J. and Stein, G. (1962) 'The Radiation Chemistry of Aqueous Solutions of Cytochrome c', *Radiation Research* **17**, 327-340

Rajagopalan, K.V., Fridovich, I. and Handler, P. (1962) 'Hepatic Aldehyde Oxidase I.Purification and Properties', *Journal of Biological Chemistry* **237**, 922-928

Rajagopalan, K.V. and Handler, P. (1964) 'Hepatic Aldehyde Oxidase II.Differential Inhibition of Electron Transfer to Various Electron Acceptors', *Journal of Biological Chemistry* **239**, 2022-2026

Richter, H.E. and Loewen, P.C. (1981) 'Induction of Catalase in *Escherichia coli* by Ascorbic Acid Involves Hydrogen Peroxide', *Biochemical and Biophysical Research Communications* **100**, 1039-1046

Salin, M.L. and Bridges, S.M. (1980) 'Isolation and Characterisation of an Iron-Containing Superoxide Dismutase from *Brassica compestris*', in J.V. Bannister and H.A.O. Hill (eds.), *Chemical and Biological Aspects of Superoxide and Superoxide Dismutases* (Elsevier, New York), pp. 176-284

Salin, M.L. and Bridges, S.M. (1982) 'Isolation and Characterization of an Iron-Containing Superoxide Dismutase From Water Lily *Nuphar luteum*', *Plant Physiology* **69**, 161-165

Salin, M.L. and McCord, J.M. (1975) 'Free Radicals and Inflammation: Protection of Phagocytosing Leukocytes by Superoxide Dismutase', *Journal of Clinical Investigation* **56**, 1319-1323

Shlafer, M., Kane, P.F., and Kirsh, M.M. (1982) 'Superoxide Dismutase Plus Catalase Enhances the Efficacy of Hypothermic Cardioplegia to Protect the Globally Ischemic, Reperfused Heart', *Journal of Thoracic and Cardiovascular Surgery* **83**, 830-839

Stallings, W.C., Pattridge, K.A., Strong, R.K. and Ludwig, M.L. (1984) 'Manganese and Iron Superoxide Dismutase are Structural Homologs', *Journal of Biological Chemistry* **259**, 10695-10699

Stallings, W.C., Powers, T.B., Pattridge, K.A., Fee, J.A. and Ludwig, M.L. (1983) 'Iron Superoxide Dismutase From *Escherichia coli* at 3.1-Ang. Resolution: A Structure Unlike that of Copper, Zinc Protein at Both Monomer and Dimer Levels', *Proceedings of the National Academy of Science U.S.A.* **80**, 3884-3888

Steffens, G.J., Bannister, J.V., Bannister, W.H., Floke, L., Gunzler, W.A., Kim, S.M.A. and Otting, F. (1983) 'The Primary Structure of Copper Zinc Superoxide Dismutase From *Photobacterium leiognathi:* Evidence for a Separate Evolution of Copper Zinc Superoxide Dismutase in Bacteria', *Hoppe-Seyler's Zeitschrift für Physiologische Chemie* **364**, 675- 690

Steinman, H.M. (1978) 'The Amino Acid Sequence of Mangano SOD From *E. coli'*, *Journal of Biological Chemistry* **253**, 8707-8720

Steinman, H.M. (1982) 'Copper-Zinc Superoxide Dismutase From *Caulbacter crescentus,* CB15', *Journal of Biological Chemistry* **257**, 10283-10293

Steinman, H.M. (1985) 'Bacteriocuprein Superoxide Dismutases in Pseudomonads', *Journal of Bacteriology* **162**, 1255-1260

Tainer, J.A., Getzoff, E.D., Beem, K.M., Richardson, J.S. and Richardson, D.C. (1982) 'Determination and Analysis of the 2 Ang. Structure of Copper Zinc Superoxide Dismutase', *Journal of Molecular Biology* **160**, 181-217

Taube, H. (1965) 'Mechanisms of Oxidation With Oxygen', *Journal of General Physiology* **49**, (Suppl.) 29-52

Taylor, A.L. and Trotter, C.D. (1972) 'Linkage Map of *Escherichia coli* K12', *Bacteriological Reviews* **36**, 504-524

Totter, J.R. (1980) 'Spontaneous Cancer and its Possible Relationship to Oxygen Metabolism', *Proceedings of the National Academy of Science U.S.A.* **77**, 1763-1767

Touati, D. (1983) 'Cloning and Mapping of the Manganese Superoxide Dismutase Gene (sodA) of *Escherichia coli* K-12', *Journal of Bacteriology* **155**, 1078-1087

Varma, S.D. (1981) 'Superoxide and Lens of the Eye: A New Theory of Cataractogenesis', *International Journal of Quantum Chemistry* **20**, 479-484

Walker, G.C. (1984) 'Mutagenesis and Inducible Responses to Deoxyribonucleic Acid Damage in *Escherichia coli'*, *Microbiological Reviews* **48**, 60-93

Weitzman, S.A. and Stossel, T.P. (1981) 'Mutation Caused by Human Phagocytes', *Science* **212**, 546-547

Whiteside, C. and Hassan, H.M. (1987) 'Induction and Inactivation of Catalase and Superoxide Dismutase of *Escherichia coli* by Ozone', *Archives of Biochemistry and Biophysics* **257**, 464-471

Yamaguchi, T. and Nakagawa, K. (1983) 'Mutagenicity of and Formation of Oxygen Radicals by Trioses and Glyoxal Derivatives', *Agricultural and Biological Chemistry* **47**, 2961-2965

Zipkas, D. and Riley, M. (1975) 'Proposal for Mechanism of Evolution of the *E. coli* Genome', *Proceedings of the National Academy of Science U.S.A.* **72**, 1354-1359

Chapter 3

The Early Environment

Richard J.F. Jenkins

Introduction

Palaeobiologists have long been intrigued by the abrupt appearance of metazoan skeletal remains at the base of the Cambrian. Recognition of soft-bodied Ediacaran assemblages over the past 40 years emphasised the geologically sudden advent of large invertebrates and greatly fuelled speculation concerning the early evolution of the Metazoa and the timing of their origin. Was the evolution of animal life in some way triggered by major changes in the environment, perhaps the great glaciations which occurred during the late Precambrian, or possibly an increase in the concentration of oxygen in the biosphere? Literature pertaining to such questions is extensive (Runnegar, 1982b; Cloud, 1984; Glaessner, 1984). Modern advances in the understanding of the history and controls of the biosphere coupled with closer fixing of the timing of the advent of major sections of the biota, based on new palaeontological and geochronological studies as well as molecular evolution, cast fresh light on these problems.

Older Precambrian Environments

Characteristics of the Early Biosphere: Tectonic and Biological Influence

The more than 10,000 publications relating to the evolution and history of the Earth's atmosphere (Lewis and Prinn, 1984) signal wide interest in this topic. It has been commonly considered that the early atmosphere was lacking in free oxygen. The finding that young (T-Tauri) stars emit levels of ultraviolet radiation initially thousands, and subsequently hundreds of times that of the present Sun, indicates the likely generation of a relatively high flux of oxygen through photolysis of water vapour and CO_2 in the upper atmosphere (Canuto, Levine, Augustsson and Imhoff, 1983). The oxygen so produced would react with methane or residual traces of ammonia and also with labile volcanic effusives. Current notions of rapid differentiation of the Earth's mantle suggest that the concentration of oxygen probably remained low during early stages of formation of the lithosphere (Holland, 1984).

Sutton (1968, 1971) recognised that the Earth's sialic crust records a history of major tectonic processes occurring at intervals of 800-1000 Ma; these 'Chelogenic Cycles' signify large additions to the cratons and herald the development of new tectogenes that underwent complex processes of filling, orogeny and lateral migration during the subsequent part of each cycle. The earliest part of the history of the Earth is unrecorded save for rare examples of its sediments metamorphosed during subsequent cycles. The earliest sialic rocks recognised date back 3.8-3.5 Ga (here Cycle I). Further major phases of cratonisation occured at 3.0-2.6 Ga (Cycle II), 2.0-1.6 Ga (Cycle III), and 1.2-0.85 Ga ('Grenville Cycle' or Cycle IV; Figure 3.1).

Observed changes in the geological record signalling possible alteration of chemical and physical processes in the biosphere broadly correspond with particular chelogenic cycles. Indeed, it would be surprising if this were not so since the processes of cratonisation involve alteration of huge amounts of materials through related anatexis and volcanism. Uplift after cratonic fusion would have released a flood of chemical components through the weathering cycle. For example, silicates containing Ca^{2+} and Mg^{2+} ions react with CO_2 during weathering, leading subsequently to precipitation of carbonates and removal of carbon from the exigenic environment (Berner and Barron, 1984).

The profound influences on the chemistry of the oceans and atmosphere engendered by biogenic processes are well known; life has probably existed on Earth from a very early time. Carbon associated with 3.8 Ga old metasedimentary rocks of the Isua region, Greenland, already shows fractionation towards the light isotope, suggesting carbon-based, autotrophic metabolism (Schidlowski, Hayes and Kaplan, 1983).

During Cycle I, a thin crust supported large stratovolcanoes that generated great quantities of relatively calcic lavas. The common occurrence of graphitic black shales presumably indicates burial of organic matter. Silicified fossil remains of sheath-enclosed unicells occurring in bedded cherts within 3.5-3.3 Ga strata of the Warrawoona Group in the Pilbara, Western Australia, have been compared to modern cyanobacteria capable of oxygenic photosynthesis (Schopf and Packer, 1987). Rare stromatolites from the middle and later parts of this cycle show layering suggesting a phototropic response in the organisms that constructed them, also a likely indication of forms of photosynthesis. An appreciable flux of oxygen was probably generated through processes of photodissociation (e.g. Canuto *et al.*, 1983).

Assuming a large reservoir of reduced iron present in solution in the ocean, Kasting and Walker (1981) calculate that the atmosphere entered a sustained oxidative mode when the surface equilibrium concentration of free oxygen reached approximately 4×10^8 PAL (present atmospheric level). Banded grey and brick-red

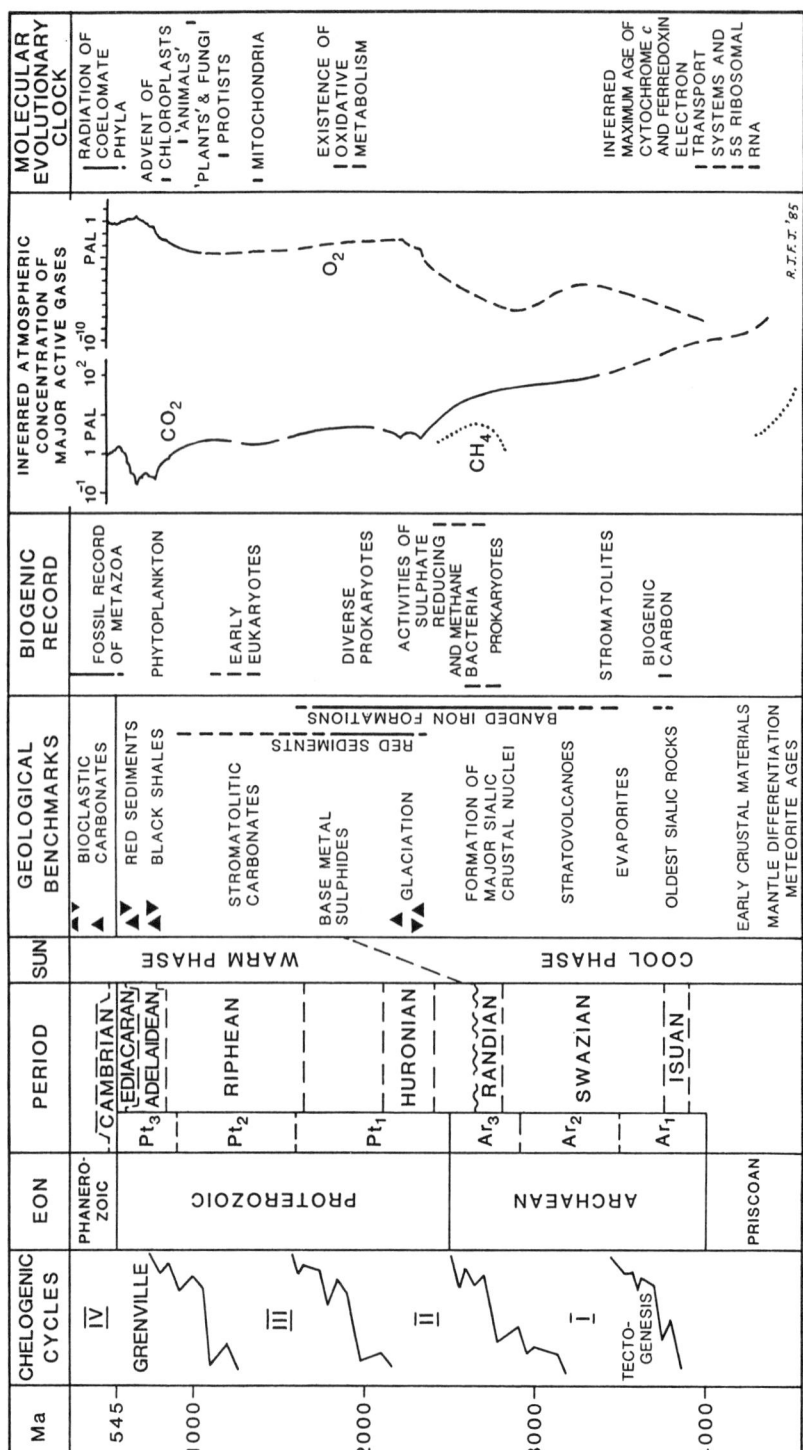

Figure 3.1 The Precambrian: summary of time scales, major geological phenomena, inferred changes in active atmospheric gases and molecular evolutionary events.

cherts which include small, shiny euhedra of unweathered, authigenic pyrite occur in the Warrawoona Group (J.A. Cooper, pers. comm.), and there are also indications of sulphates in this sequence (Holland, 1984).

Weathering of uplifted parts of vast granodioritic terrains formed during the chelogenic phase of Cycle II probably led to significant lowering of the partial pressure of CO_2. Fossil remains of prokaryotes, mainly cyanobacteria, become relatively abundant in cherts dating from this cycle. Occasional very low $\delta^{13}C$ values of -35 to -50°/oo for kerogens from rocks 2.8 to 2.5 Ga old may reflect activities of anaerobic methane bacteria, and fractionation of $\delta^{34}S$ attributed to bacterial reduction of sulphate has been detected back to about 2.7 Ga ago (Schidlowski et al., 1983).

Oxygen released by algal photosynthesis likely played a part in the precipitation of the banded iron formations, a common lithotype of Cycle I and perhaps the most distinctive facies association of Cycle II. Volatiles from volcanic sources probably reacted with traces of oxygen in the atmosphere to form H_2SO_3 and H_2SO_4 and acid rains so generated possibly helped rapidly to leach iron from the regolith. The complex mineralogy of the iron formations evidently reflects their primary precipitation as iron-rich carbonates and silicate complexes and is consistent with a partial pressure of CO_2 between c. 10 and 100 times greater than at present (Mel'nik, 1982).

Early Proterozoic Oxygenation

Widespread red beds indicative of an oxidative atmosphere appear during the later part of Cycle II (early Proterozoic), especially within the time interval 2.3 to 1.8 Ga (Walker and Hays, 1981). Glacial deposits in the Huronian sequence of Canada are underlain by sediments that appear to have been deposited under reducing conditions and are overlain by oxidised sediments, suggesting that the process of glaciation was in some way associated with an increase in free oxygen. Glacigenic rocks of broadly similar age occur in several regions (Harland, 1983). Possibly these glaciations were related to lowering of the partial pressure of CO_2 and lessening of the greenhouse effect brought about both by the weathering of newly formed continental cratons and by the enhanced productivity of organismic photosynthesis as additional nutrients entered the sea (cf. Kasting, 1987). Also about the same time the Sun may have begun to brighten into its mature phase. Red beds persist through the succeeding Cycle III.

Clearly, the atmosphere and surface waters of the Earth became oxidative long before the generally suggested time of appearance of animal life. A key question is the actual concentration of free oxygen attained during the early and middle Proterozoic. Based on various lines of evidence such as the chemical profile of ancient fossil soils or occurrences of detrital uraninite ores, and his postulation that the major banded iron formations probably formed in open marine basins, Holland

(1984) estimates that the partial pressure of free oxygen was c. 0.002 PAL during the early Proterozoic. Newer analyses of fossil soils led Pinto and Holland (1988) to suggest that their preferred range for the oxygen pressure in the atmosphere between c. 2·5 and 1·8 Ga is 5×10^{-4} to 1×10^{-3} atm (0·0025 to 0·005 PAL).

Kasting and Donahue (1981) calculate that the formation of a stratospheric ozone shield sufficiently dense to protect surface life from the lethal ultraviolet radiation of the Sun required an O_2 mixing ratio of 0.05 PAL, or assuming a steeper solar-zenith angle, about 0.1 PAL. These concentrations are marginally greater than estimated values for the early Proterozoic, suggesting that at this time the surface of the Earth may still have been subject to considerable ultraviolet radiation (see also Kasting, 1987). Barren and dusty red continents and frozen polar icecaps may have presented an aspect similar to that of the planet Mars.

Study of the homology of cytochrome c in the photosynthetic or respiratory electron transport chains of autotrophic bacteria, cyanobacteria and eukaryotes supports the notion that oxidative respiration evolved through modification of the energy transference system of photosynthetic prokaryotes (Dickerson, 1980). As will be documented below, the molecular evolutionary clock for cytochrome c indicates that cytochrome c_6 in cyanobacteria and protists diverged from cytochrome c_2 of facultatively aerobic photosynthetic bacteria earlier than about 1·85 Ga, suggesting that an evolved form of oxidative metabolism is linked to the early Proterozoic oxygenation of the atmosphere. Scholarly arguments for the probable evolution of efficient aerobic respirers during the early Proterozoic are presented by Knoll (1983, 1985). The concentration of oxygen necessary to support organismic aerobiosis is about 0.005 PAL or greater, and presumably such a level has been maintained since the start of the chelogenic phase of Cycle III.

During the post-chelogenic part of Cycle III (middle Proterozoic), transgressive seas flooded broad, shallow epeiric basins, and widespread deposition of carbonates occurred. Banks of stromatolites constructed by flourishing algal microfloras characterised wide swathes within these seaways. Platform deposits reflecting such environments are extensively developed in Eurasia and Siberia, and constitute the older Riphean of Soviet authors. There is little geochronological data that would place significant glaciations during this time. Though stromatolites may form today in widely variable climates, the negative evidence for the scarcity of glaciation suggests that during the older Riphean the climate was generally warm and equable. Under such climates, epeiric basins commonly become stratified and develop anoxic bottom waters (Goodfellow and Jonasson, 1984); persistence of such conditions over a long period may lead to the vast bulk of the ocean beneath the thermocline becoming anoxic (Jenkyns, 1980). A continuing trend towards enrichment of $\delta^{34}S$ in sulphides relative to sulphates precipitated by evaporation of seawater (Schidlowski et al., 1983) testifies to the global activities of sulphate reducing bacteria flourishing under these conditions. Bacterial decay also probably made the main body of the ocean slightly acidic, so that hydrated oxides of iron delivered in colloidal state by rivers became redissolved, generating a substantial reservoir of iron in solution.

If processes of oxygen formation were to cease on the present Earth, free oxygen would be removed by natural reductants in a time as short as 3 my (e.g. Carver, 1981). An equilibrium relationship exists between the release of oxygen due to photosynthesis and the oxygen fugacity of the main body of oceanic waters (Bralower and Thierstein, 1984). As seas of the middle Proterozoic became more stratified and waters below the thermocline more anoxic, the equilibrium concentration of free oxygen can be expected to have diminished.

There is an important clue suggesting that oxygen tensions were generally low for the greater part of the duration of the Proterozoic. This is the common association between concretionary cherts and Precambrian carbonates, especially carbonate-rich stromatolites. Such cherts preserve most of the fossil microbiotas that have been discovered. The process of silification which preserved these very fragile organisms evidently precluded decay, and thus presumably took place over a matter of hours. Jones and Fitzgerald (1986) indicate that silica becomes more soluble by several orders of magnitude at ambient temperatures and pressures in the presence of organic chelating agents released by decay of organisms, particularly the decomposition of algae. Once the silica is in solution a slight change in chemical conditions may cause its rapid precipitation; this process has been used to 'petrify' plant debris experimentally and is the likely manner in which the Precambrian fossil microfloras were preserved. The chelating agents involved are significantly stable only under anoxic conditions.

Carbonate beds containing buried organic materials may well be anoxic at shallow depth. However, not only do stromatolite heads form in the productive upper waters of aqueous environments, but they are usually rather porous and are likely well permeated by water in near equilibrium with the surroundings. For silica gels to impregnate the living algal mats on the surface of the stromatolites, the surrounding waters presumably contained little oxygen. Chert preserved microfloras occur widely in Proterozoic sequences until as late as c. 800 Ma ago.

Biotas of this kind are apparently unknown from the terminal Precambrian. Phanerozoic cherts are commonly interstratal; chert preserved assemblages of this eon (e.g. abundant plant remains) possibly formed in unusual circumstances involving eutrophication or rapid burial.

A 1·1 Ga old fossil soil near Sturgeon Falls, Michigan, indicates oxidative weathering and the presence of free oxygen, but the concentration necessary to account for the oxidation state of the palaeosol could have been 10^3 times lower than in the atmosphere today (Zbinden, Holland, Feakes and Dobos, 1988).

Environments of Early Grenville Cycle

The molecular clock for cytochrome c indicates that the lineages leading to the higher plants, the fungi and animals differentiated near the start of this cycle. Thus the Grenville Cycle rather precisely encapsulates the major radiation of higher

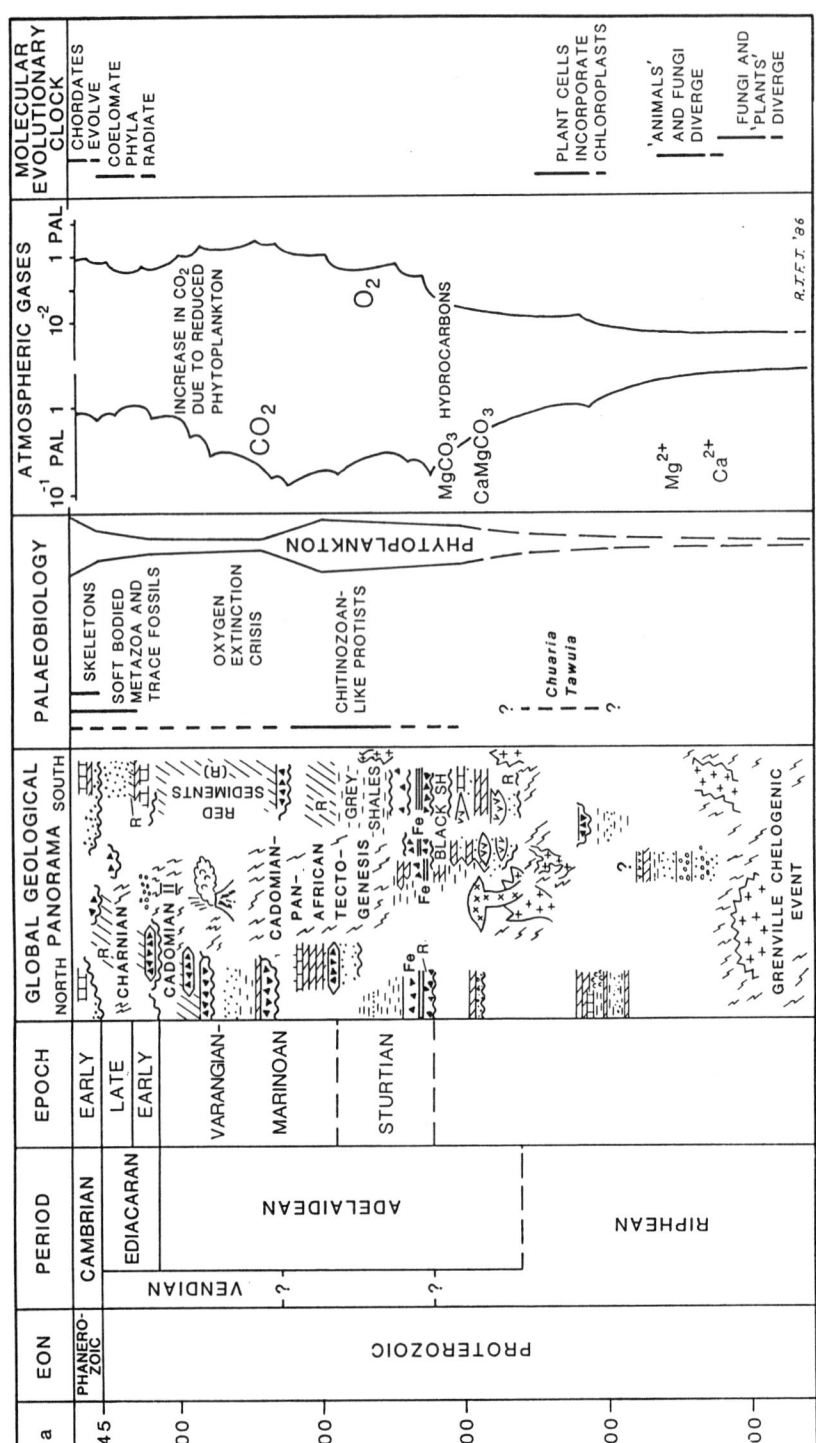

Figure 3.2 Geohistorical perspective of early part of the Grenville chelogenic cycle; inferred atmospheric changes and molecular evolutionary events.

eukaryotes (Figure 3.2). At about 900-800 Ma ago robust walled acritarchs representing probable eukaryotic plankters became numerous and studies of their distribution suggest that they constituted a major part of a relatively diverse oceanic phytoplankton (Vidal and Knoll, 1983). Black shales of about this age occurring in southwestern and southern-central Africa and southern Australia are rich in carbon, indicating burial of organic material. Analyses of carbonates of similar age (900-780 Ma) show anomalous concentrations of $\delta^{13}C$ of + $5^0/00$ or greater (Schidlowski *et al.*, 1983; Knoll *et al.*, 1986; Aharon *et al.*, 1987). This may be interpreted as indicating a strong partitioning of carbon isotopes due to extensive burial of light carbon in organic detritus, and is in general accord with the known carbon-rich shales. These several lines of investigation are consistent with the existence of a productive phytoplankton and high rate of burial of sapropel. The rate of production of oxygen is likely to have been high, but its rate of accumulation in the atmosphere may have been buffered by reactions with reductants present in the body of the ocean.

Adelaide Fold Belt

Relatively well studied late Precambrian sediments of the Adelaide Fold Belt, South Australia, comprise one of the better known sequences documenting various environmental characteristics of the early part of the Grenville Cycle (e.g. Preiss, 1987; Figure 3.3). New geochronological studies (Compston, Williams, Jenkins, Gostin and Haines, 1987; Preiss, 1987) provide a useful time framework.

The clastics and carbonates chiefly comprising the older parts of the sequence commonly show drab colours and include dark shales; however, red siltstones and sandstones are also reported. Magnesite-rich carbonates in the Burra Group include abundant black chert concretions containing microfloras with elements similar to those in the Bitter Springs Formation in central Australia (T.F. Fairchild, pers. comm.).

Two major glacial episodes are represented, an older Sturtian phase separated by up to 4700 m of strata from glacigenic sediments of Marinoan Age. The complex nomenclature applied to these glacigenic division (Coats, 1981) need not be of concern here.

In the northeastern Flinders Ranges (Figure 3.4) the Sturtian glacigenic sediments comprise the Yudnamutana Subgroup, which reaches 5000 m in thickness; the cumulative thickness of comparable sediments in the eastern Flinders Ranges exceeds 5600 m. Modern analysis of these rocks is incomplete, but there is increasing evidence of turbidite - 'greywacke' associations and related synsedimentary faulting of the basins (Preiss, 1985; 1987). Understanding of such ancient subaquatic glacial environments is given new perspective by comparisons with sediments on the shelves and slopes of present polar regions (e.g. Gravenor, von Brunn and Dreimans, 1984).

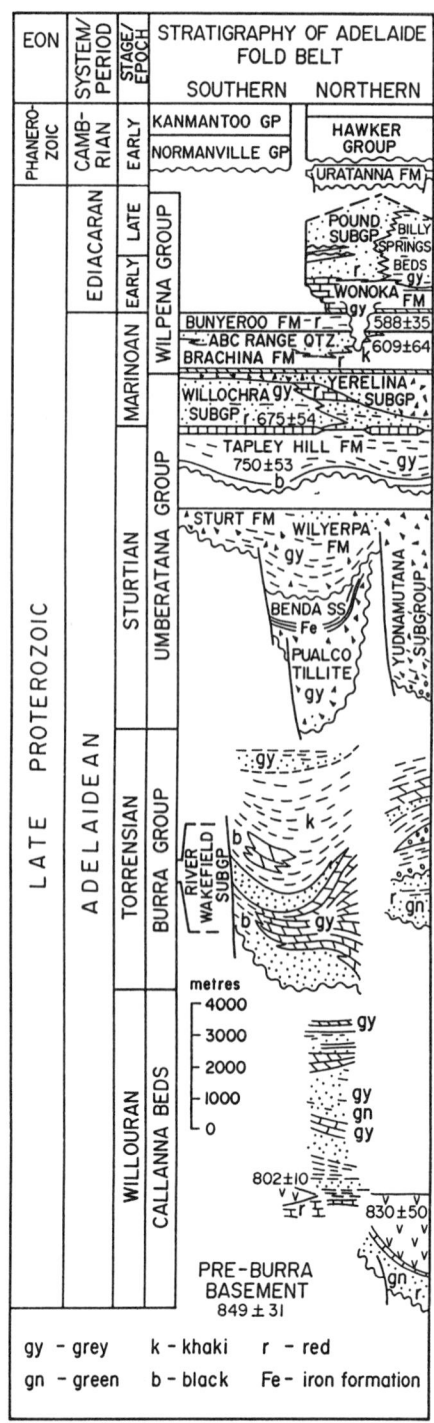

Figure 3.3 Simplified stratigraphy and summary of age data for late Precambrian and older Cambrian sequences of Adelaide Fold Belt, South Australia.

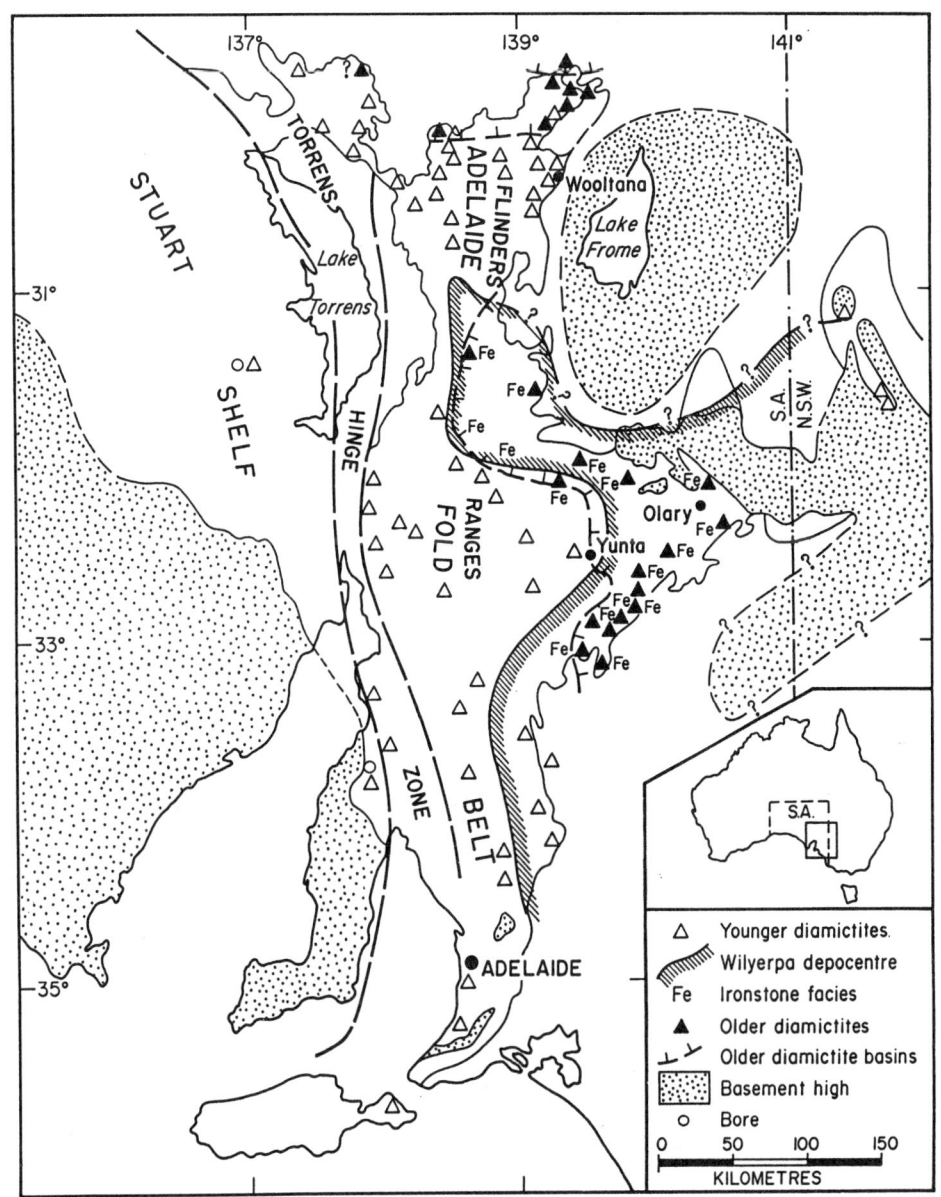

Figure 3.4 Palaeogeographical map for Sturtian of Adelaide Fold Belt about 790–750 Ma ago (after Figure 1 of Coats, 1981).

Glacial environments of the Sturtian in South Australia possibly resembled the Ross Sea of modern Antarctica; glaciers flowed off uplifted northeasterly situated highlands and a partly grounded ice shelf lay above the continental ramp and platform now incorporated in the main axis of the Flinders Ranges. The open ocean was probably towards the southeast. Several major glacial advances seem to have occurred. The timing of this great ice-age was probably somewhat later than 800 Ma ago.

Of special interest is the widespread occurrence of laminated ironstones in the upper part of the Pualco Tillite and in the Benda Siltstone. In the central Flinders Ranges they are termed the Holowilena Ironstone, and in the Yunta-Olary region, the Braemar Ironstone 'facies' (Coats, 1981). There is no outcrop link between the separate occurrences, but their stratigraphic placement is similar. Study of the Braemar Ironstone shows that the iron concentration consists principally of euhedral, martized magnetite varying between 10-120 micron in diameter and fine hematite flakes (1-20 micron) in a diagenetic association with quartz, chlorite and dolomite as well as clastic material (Whitten, 1970). Water rounded magnetite crystals in parts of the deposit demonstrate that this mineral was either primary or formed at the earliest stages of deposition. Comparable iron minerals have been described from the Holowilina Ironstone.

Varve-like banding in the ironstones (Figure 3.5) may reflect deposition in a quiet environment. The Benda Siltstone is a ribbon-layered, graphitic siltstone reaching

Figure 3.5 Varve-like banding and minor erosional features in ironstone bed of Braemar Ironstone 'fascies', Benda Silstone, south of Yunta, South Australia. Scale bar = 1 cm.

about 730 m in thickness and probably represents a euxinic facies. I consider that the ironstones formed on a subsiding continental slope or marginal continental apron at depths in excess of a few hundred metres to possibly several kilometres.

Most Sturtian diamictites are metamorphosed to some extent and show drab colours. In the northeastern Flinders Ranges unmetamorphosed equivalents of the Sturt Formation are pink or reddish in colour, suggesting deposition in oxygenated water (V.A. Gostin, pers. comm.).

Iron formation occurs in association with late Precambrian glacigenic rocks on several continents. Ironstones occur in the Numees Formation, Gariep Group, and below the Chuos Formation of the Otavi Group in southwestern Africa, and a magnetite-hematite bed is present in the Mwashia Group below the 'Grand Conglomerat' in Zambia and adjacent Zaire (e.g. Whitten, 1970). Glacigenic conglomerates at the base of the Vindhyan Supergroup of central India have a ferruginous(magnetite-haematite) matrix. In the northern Yukon of Canada several bands of iron formation occur with dropstone facies and glacigenic conglomerates forming part of the Rapitan Group of the lower Windermere Supergroup (Yeo, 1981). Information as to the ages of these occurrences and data on the composition of some of the ironstones are summarised in Table 3.1.

The Numees Formation of the Gariep Geosyncline remarkably resembles Sturtian diamictites in its association with carbonaceious argillites and dolomites (Martin, 1965, Figure 12). Notably, the Grand Conglomerat is also associated with carbonaceous shales (one sample contains 4.85% free carbon), and carbon-rich intervals are present in similar mixtites in the western Congo and Angola (3.5 to 8% free carbon). Glacigenic conglomerates of the Rapitan Group are uncomfortable above the carbonate-quartzite assemblage of the Mackenzie Mountains Supergroup and occur in a tectonic setting closely parallel to that of the older diamictites of the Sturtian.

The huge iron formation and manganese deposit of Mutum, Bolivia, and Morro do Urucum, Brazil, is conformable above diamictites and is commonly considered as being late Precambrian. However, its stratigraphic position may be above other supposedly glacigenic sediments of 'Eocambrian' age (Fairchild, Barbour and Haralyi, 1978), and millimetric tubular structures in the underlying Corumbia Group resemble *Cloudina* Germs, a form of Ediacaran aspect.

The wide occurrence of chemically precipitated ironstones in close association with late Precambrian glacigenic diamictites of basinal origin and broadly similar age signals a common process leading to their formation. The comparatively high tenor (iron content, Table 3.1) of these ores supports ideas that they were precipitated by the interaction of oxygen-rich water and dissolved iron (Martin, 1965; Whitten, 1970; Yeo, 1981). Martin suggested that oxygen was carried down by sinking cold waters. The carbon content of the adjacent shales indicates burial of abundant sapropel and probably reflects high levels of organic productivity in top waters of the ocean. Surely, the iron was originally contained in solution in the vast bulk of oceanic waters made anoxic during the long period of equable climate characterising

Table 3.1 Possible Sturtian ironstones associated with glacigenic deposits and comparison with older and (?) younger iron formations.

Locality and Horizon	General Age Costraints in Ma		Composition %	
			Fe	SiO$_2$
South Australia Braemar Ironstone, Yunta-Olary area	<800	>?750 ± 50	33-43	20.4-30
Southwestern Africa Chuos Formation, Damara Geosyncline	ca800 ± 20	>756 ± 30	37.2	22.0
Numees Formation, Gariep Geosyncline	<780 ± 10	>700	-	-
Zambia - Zaire Mwashia Group below Grand Conglomerat	?<870-820	>700	31.2	30.1
Central India Lower Vindhyan Supergroup	?Middle or ?Late Proterozoic		-	-
Yukon Canada Rapitan Group	ca776 ± 24 - 769 ± 27		46	25
World occurrences of mid Precambrian iron formations	2700 - 1850		30.0 (mean)	45.2 (mean)
Bolivia - Brazil Morro do Urucum	None ? Early Palaeozoic		51	23

the older Riphean.

Sinking of cold waters at high latitudes is currently transporting large amounts of oxygen into the deep oceans (Deacon, 1984). Wind induced upwelling introduces nutrients into surface waters and promotes high productivity of phytoplankton during summer, thus leading to concentration of dissolved oxygen generated through photosynthesis. Floating sea ice, such as occurs around Antarctica, tends to inhibit the degree of oxygenation. Sinking of open surface waters, as in the North Atlantic, generates richly oxygenated bottom waters containing only little less oxygen than the level in equilibrium with the atmosphere. Thus the process is likely to be maximally efficient during cool interglacial intervals when sea ice is restricted. Similar processes evidently occurred during the great glaciations of the past, and the replacement of reductive waters comprising the main body of the ocean by oxygen charged waters of surface origin is a likely explanation of the link between Precambrian glaciations and rapid increase in ambient oxygen concentrations (see also Cook and Shergold, 1984; Knoll *et al.*, 1986; Kasting, 1987; Wilde, 1987).

Interglacial Succession

Sturtian glacigenic rocks are succeeded by the Tapley Hill Formation, mainly thick (up to c. 3000 m) rhythmically banded, grey siltstones deposited below wave base. It is the most widely transgressive unit in the Adelaide Fold Belt. High $\delta^{34}S$ values averaging $+20^0/00$ in dispersed fine pyrite and sulphide veinlets present in parts of the unit may be attributed to bacterial sulphate reduction in a markedly stratified, restricted water mass (Lambert, Donnelly, Etminan and Rowlands, 1984). The development of such euxinic facies is characteristically associated with times of equable climate and major transgression marking interglacial episodes (Jenkyns, 1980; Goodfellow and Jonasson, 1984). These processes need not especially reflect levels of oxygenation in the top waters and atmosphere.

Ripple marked and sun-cracked, intertidal facies in the lower parts of the Marinoan are maroon or red coloured, due to oxidation. Rapid changes in water depth indicated in parts of this succession may reflect glacio-eustatic phenomena linked to formation of distant ice caps.

Marinoan - Varangian Glaciations

In the Flinders Ranges the Marinoan glaciation is recorded in glacio-fluvial or deltaic pebbly sandstones and diamictites of the Elatina Formation, and by the thicker (1400 m) glacio-marine diamictites and feldspathic sandstones of the Yerelina Subgroup in the northeast.

In sea cliffs between Marino and Hallett Cove, south of Adelaide, basinal sediments immediately below the glacigenic interval show drab colours indicative of reducing conditions. The glacigenic sediments of the Reynella Siltstone are mauve coloured, carbonate-rich, granule diamictites with rare dropstones. The coloration is due to finely dispersed hematite with maghemite and minor gegothite; the environment was paralic or fluviatile, with varved intervals and casts of ice wedges suggesting seasonal freezing.

Through much of the Adelaide Fold Belt the Marinoan glacigenic interval is capped by a thin marker dolomite or dolomitic intervals heralding transgressive shales and siltstones of the Brachina Formation. In the south, lower parts of this formation include mauve coloured turbidites formed below wave base and succeeding storm deposited sandstones interbedded with purple shales. These colours are again due to finely dispersed hematite and lesser amounts of other iron oxides and hydrated oxides; irregularly shaped hematite grains are abundant. Red and purple shales are intercalated with the sandstones of the ABC Range Quartzite. Distinctive red-purple, basinal shales of the overlying Bunyeroo Formation also include a dark coloured pyritic interval.

Sprigg (1984) links the colour of the rocks just described to the time of major oxygenation of the biosphere. While the red coloration of some of the sediments may partly reflect a lessening in abundance of sapropelic compounds that cause reduction during diagenesis, it is possible and perhaps more likely that the maroon hues of the turbidites and other basinal facies resulted from relatively high concentrations of dissolved oxygen.

Glacigenic deposits believed to fall broadly within the 90 Ma age spectrum of the Marinoan occur widely around the globe. In northern Europe they are assigned to the 'Varangian' and in European USSR, comprise part of the early 'Vendian' (Harland, 1983; Keller, 1984). At least two major phases of glaciation are represented; age information for most occurrences is equivocal. The glacigenic Smalfjord Formation of Finmark, northern Norway, is older that 654 ± 23 Ma, while the Mortensnes Tillite Formation is unconformable above the dated sequence; the Moelv Tillite of southern Norway is older than 610 ± 18 Ma (Hambrey, 1983). A diamictite in the Urals is older than 609-602 Ma (Keller, 1984). There is a common association of diamictites with dolostones and a frequent occurrence of red coloration in these deposits (Hambrey, 1983, 1984). The finely dispersed hematite colouring diamictites in Spitzbergen may be of diagenetic origin, possibly formed by processes of oxidation during burial of iron-magnesium minerals representing glacially comminuted remnants of igneous and metamorphic rocks (Hambrey, 1984). The Gaskiers Tillite of the Avalon Peninsula, Newfoundland, also passes up into red sediments; its age is less than c. 600 Ma, and late intrusives in an underlying volcanic complex are dated by Krogh, Strong, O'Brian and Papezik (1988) at $585 \cdot 9 + 3 \cdot 4, -2 \cdot 4$ Ma (U -Pb zircon age).

The conclusion seems inescapable that the several refrigerations associated with the Marinoan or Varangian led to increasingly large concentrations of oxygen being injected into the oceans. The Marinoan evidently records most of the span of Earth history during which the biosphere first became richly oxygenated, between about 675-585 Ma ago.

Ediacaran

Sediments of this age in the Flinders Ranges evidence a spectrum of environments, from basinal deeps shallowing to paralic or fluviatile sands (Haines, 1988; von der Borch, Christie-Blick and Grady, 1988). Early Ediacaran muds deposited by suspension settling on an outer shelf are oxidised red, indicating an appreciable concentration of free oxygen in adjacent ocean waters. Rapidly deposited storm-beds show reduced (green-grey) colours. The concentration of oxygen in the biosphere at this time may well have been similar to that at present.

Timing the Origin of the Metazoa

Evidence from 'Fossil Remains'

There are numerous claims of fossil indications of metazoans predating the large and clearly defined imprints and moulds of organisms and distinctive trace fossils so characteristic of the Ediacaran. Readily recognisable fossils of the latter kind probably relate to a time interval between about 590 Ma and 545-540 Ma (Jenkins, 1984; Conway Morris, 1988). Cloud (1968 and later studies) and various other American workers consider that most pre-Ediacaran markings attributed to being of metazoan origin can be better explained as resulting from inorganic processes. A new wave of literature has revived the notion that fossil indications of Metazoa extend back 1000 Ma ago or beyond (e.g. Glaessner, 1983, 1984; Walter, 1987).

A primary difficulty with many such claims is that the markings suggested to be of metazoan origin are found singly or in low numbers; this contrasts sharply with the plethora of fossils in the Ediacaran and Phanerozoic. If indeed some of the earliest metazoans attained significant size and robust musculation, why did they not rapidly diversify? Among the supposed pre-Ediacaran indications of metazoans cited by Glaessner (1983, 1984) are a 'faecal extrusion' from the Riphean of the Urals resembling collapsed hollow ooids (elephantine ooids), spindle shaped forms grading into polygonal mud-cracks occurring in western North Greenland, and burrows from probable Ediacaran sediments in central Australia. I have previously suggested that the marking *'Bunyerichnus'* is of inorganic origin. Ribbon-like, transversely-ribbed, carbonaceous structures ('1.4 mm wide') from a purported Precambrian stratum in China and similar forms from the Vindhyan Supergroup of India are undoubtedly true fossils, but are preserved in just the same manner as the presumed algal material associated with the Chinese examples. The reliability of shale datings on these sequences also poses questions in terms of the possible relationship between the ribbon structures and sabelliditids (Vidal and Moczydlowska, 1987).

Hofmann (1985) discusses the possibility that three-dimensionally preserved examples of *Tyrasotaenia?* in the Little Dal Group of the Mackenzie Mountains Supergroup, are ichnofossils made by smaller metazoans, but he leaves this interpretation as an open question. Striking striate and transversely ribbed markings discovered in the Elatina Formation in the southern, Flinders Ranges (Dyson, 1985) may be best interpreted as grooves and 'jigger' marks made by sharp edges of floating ice touching bottom (Jenkins, 1986).

Molecular Evolution

Of substantive interest is the finding by Runnegar (1982a, 1986) and several earlier authors that percentage differences in the amino acid sequences of globins and haemoglobins of various animals indicate that the initial radiation of the metazoan phyla occurred 1000-900 Ma ago. Differences in homologous proteins of organisms are commonly utilised as 'molecular evolutionary clocks'.

Mathematical principles underlying molecular clocks are complex and form the subject of considerable investigation. The 'minute hand' of such clocks is represented by the substitution of nucleotides in the three-base codons of DNA. It is now known that about half the nucleotide substitutions that accumulate during evolution are silent, and probably selectively neutral (Jukes, 1980b). Formulation of probability functions describing this process poses considerable difficulties.

Number theory underlying the reading of the 'hour hand', the substitution of amino acids in proteins, is less complicated. It is commonly considered that the substitution of nucleotides is stochastic, thus generating a 'Monte Carlo' effect in the substitution of amino acids. Such a model leads to a concept of multiple substitutions increasing in logarithmic fashion with time (e.g. Runnegar, 1982a, Table 2). However, different proteins show substitutions at different rates despite the notion that the substitutions are brought about by the same mechanism. Moreover, the frequency of occurrence of individual amino acids in large samples of completely sequenced proteins differs from that expected from random use of the nucleotide code in DNA (Jukes, 1980a). This, and other arguments concerning the neutral evolutionary acceptance of amino acid changes in proteins, suggests that substitutions of amino acids at a given site are remote in time.

The implications of this may be simulated by a simple model provided by drawing rows of dots in a rectilinear matrix (Figure 3.6A). The length of the lines signifies a given passage of time, and the rows represent the number of codons, C. Each dot indicates an amino acid substitution. In order to simulate the 'remote' aspect of substitutions the dots should be set at rather even intervals along the rows. The mean rate of substitution is the function K, usefully expressed as a fraction per year; T_K is the mean time for a cycle of substitutions in all codons.

The model suggests that for a lapsed time $t^1 << T_K$ the number of multiple substitutions is zero (or very low). As t^1 approaches T_K the number of multiple substitutions may approach a few percent. As t^1 increases beyond T_K the number of multiple substitutions increases substantially and the clock becomes unworkable (Figure 3.6B). The apparent rate of substitution of amino acids between the proteins of any *two* compared taxa will be 2K, because each evolutionary branch embraces an equal length of time since their separation (this approximation does not take account

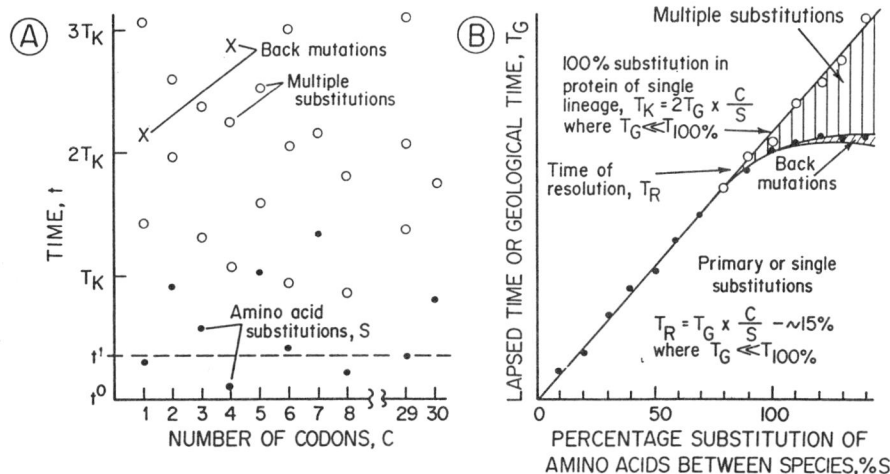

Figure 3.6 A model for the estimation of the accuracy of molecular clocks. A full description is provided in the text.

of coincidental or 'hidden' substitutions at equivalent sites in compared proteins).

The model indicates a means of estimating the time for which molecular clocks begin to lose capacity for resolving past events,

$$T_R = T_G \times C/_S - 15\%$$

where T_G is the estimated geological timing for an evolutionary event, and is large but significantly less than the time when the number of substitutions, S, between the taxa compared approaches 100%. On this basis the clock for fibrinopeptides give good resolution for evolutionary events back to ca140 Ma in the past, myoglobin and haemoglobin may resolve events back to ca580 Ma, and cytochrome c back to ca1.85 Ga. The apparent rate of amino acid substitution for a single codon in the myoglobin or haemoglobin of a *particular* lineage is 0.74×10^9 yr^{-1}, and for cytochrome c, 0.23×10^{-9} yr^{-1} or about one substitution per codon over the entire history of the Earth!

Figure 3.7 presents examples of molecular clocks based on globins and haemoglobin, and on cytochrome c. For the globin clock, the separation of vertebrate myoglobin and lamprey globin indicates a time 550-545 Ma ago, or at the close of the Ediacaran. This may be a reasonable estimate for the origin of the chordates; tubes of sabelliditids possibly belonging to the protochordates occur in terminal Precambrian and earliest Cambrian rocks. The separations between the globins of various invertebrates and vertebrates suggest that the coelomate phyla differentiated during the Ediacaran; however, these results be marginally too low as they are so near the limit of resolution of this clock (for a somewhat different result

56   *The Early Environment*

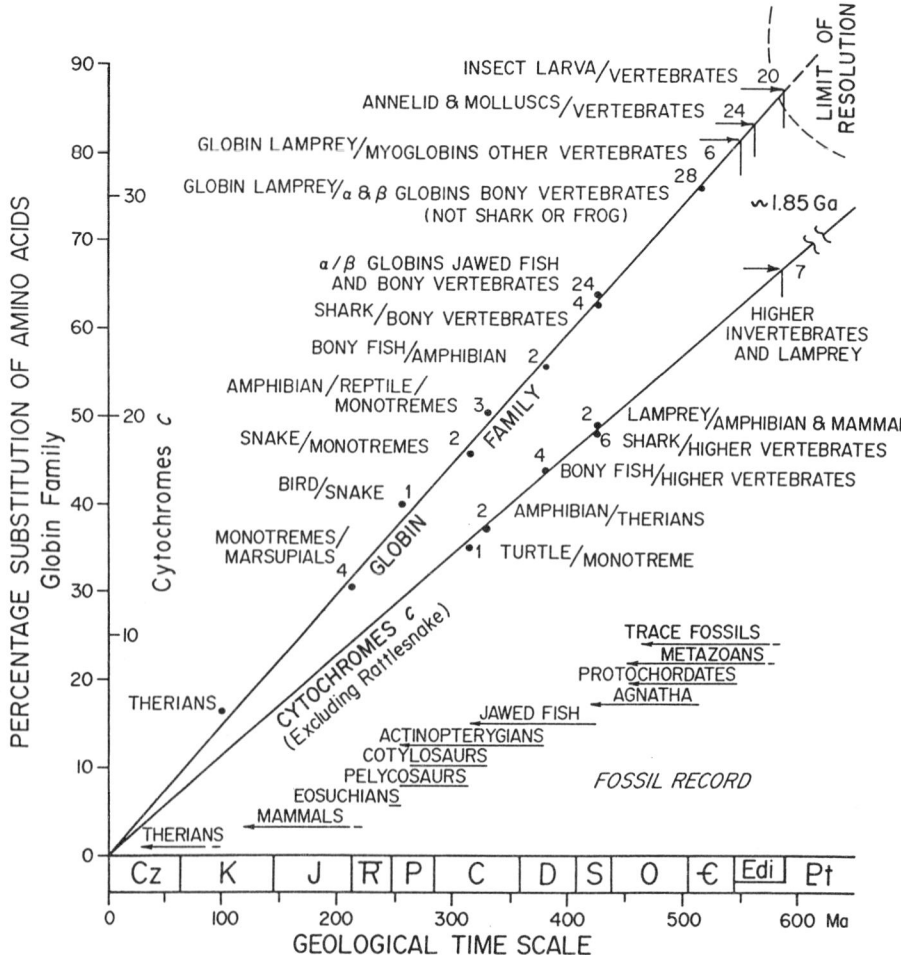

Figure 3.7 Examples of molecular evolutionary clocks, emphasising maximal differences of amino acids in proteins of conservative evolutionary clades or so-called 'living fossils'. Data principally from Dayhoff (1978) and Jukes (1980b); numbers of measurements averaged indicated adjacent to plotted points.

based on corrected mutation values and a timescale varying from that used herein see Goodman *et al.*, 1988).

Turning to the cytochrome *c* molecular clock, and with reference to Table 3.2 showing the percentage differences in amino acids of this protein in several vertebrate and invertebrate stocks, the coelomate phyla differentiated at about 58(5) Ma. This is in close agreement with palaeontological and geochronological information presently

Table 3.2 Cytochromes c of several vertebrates and invertebrates.

	1	2	3	4	5	6	7
1. Horse	104	16	19	21	21	23	25
2. Pacific lamprey	15	104	23	24	29	28	28
3. Echinoderm	18	22	103	21	21	24	23
4. Mollusc	20	23	20	104	24	30	26
5. Insect	20	28	22	23	107	24	28
6. Crustacean	22	27	23	29	23	104	27
7. Annelid	24	27	22	25	26	26	108

Note: upper right, numbers of different amino acids; diagonal row, number of compared residues; lower left, percentage difference in amino acids. Data from Dayhoff (1978) and Lydiatt *et al.* (1978).

available (e.g. Jenkins, 1984). Bergstrom (1986) has investigated the phyletic relationships implied by parsimony or maximal retentions of amino acids in cytochromes c.

The common role of homologous molecules of cytochrome c in the photosynthetic or respiratory electron transport chains of prokaryotes and in chloroplasts and mitochondria of all forms of eukaryotic life indicates that this molecular 'family' comprises ancient proteins which occupy pivotal positions in respect of energetic metabolism (Dayhoff and Barker, 1976; Dickerson, 1980). The percentage separation of amino acids in cytochrome c_6 of cyanobacteria and chloroplasts of protists and cytochrome c_2 of aerobic photosynthetic bacteria is about 84% and hence probably near the limit the cytochrome c clock can resolve back in time; thus this divergence likely occurred earlier than 1.85 Ga. Bacterial cytochrome c_2 and mitochondrial cytochrome c show a divergence of 65%, suggesting a time of separation c. 1400 Ma ago. Cytochromes c of several protist lineages and higher forms of life seem to have separated at c. 1200 Ma; higher plants and fungi diverged about 1000 Ma ago and fungi and animal life differentiated at c. 950 Ma. These results also broadly match lines of geological evidence. Robust walled acritarchs representing probable eukaryotic plankters occur as early as 1400 Ma and gradually become numerous after 900-800 Ma (Knoll, 1983; Vidal and Knoll, 1983). Enigmatic, coiled, megascopic 'algal remains' are reported in the 1300 Ma old Greyson Shale of the Belt Supergroup (Walter, Oehler and Oehler, 1976), while other problematic megascopic carbonaceous forms are widespread in sediments broadly 1000-700 Ma in age (Hofmann, 1985; Walter 1987).

Dickerson (1971) suggested that the rate of substitution of amino acids in proteins is likely a function of the structural configuration of the protein itself. Thus in essence, the theoretical time for 100% substitution of amino acids, T_K, may reflect the time the particular protein has been in existence. The value of T_K for cytochrome c is 4·4 Ga or 0·96 the age of the Earth. Is this a measure of the time when organisms first acquired an electron transport system? It is notable that the percentage separations of amino acids of ferredoxins in several comparisons between cyanobacteria and anaerobic sulphate-reducing and anaerobic photosynthetic bacteria are statistically indistinguishable from the corresponding values for the cytochrome c family, suggesting that these two groups of electron transport proteins are of comparable antiquity. Percentage differences of recognisable mutations identified from nucleotides in 5S ribosomal RNA of eukaryotes are also similar to values for separations of amino acids in the cytochrome c of equivalent organisms. 5S rRNA occurs in larger ribosomal subunits and is thought to function in the nonspecific binding of transfer RNA to the ribosome during protein synthesis; it is also extremely ancient, and may even predate the contemporary form of the genetic code (Schwartz and Dayhoff, 1978). The rate of molecular clocks for the cytochrome c family, ferredoxins and 5S rRNA may reflect fundamental steps in the biochemical evolution of life.

The Earliest Metazoans

Towe (1981) favoured the fungi (or fungus-like protists) as possibly being ancestral to the metazoans, pointing out that as a group the fungi have the ability to synthesise hydroxyproline, chitin and ferritin, and have replaced phototrophy with heterotrophic absorption. Hydroxyproline and hydroxylysine biosynthesis, actually found in most higher life, is necessary for production of the metazoan structural protein collagen. Towe's suggestion derives support from studies of cytochrome c (e.g. Bergstrom, 1986) Molecular clocks provide strong circumstantial evidence that the coelomate phyla evolved several hundred million years after the initial divergence of the 'animal' lineage from 'plants'.

Extending his review of the more probable classical ideas concerning the origin of the Metazoa, Glaessner (1983, 1984) also considered that the origins of animal life occurred some 300 Ma or more prior to the dramatic Ediacaran radiation. Protein assay now places the older speculative notions in sharper perspective, but different analyses give varied results concerning the near ancestry of the Metazoa. There is no way of determining if the proliferation of Chitinozoan-like, vase-shaped, protists after about 800 Ma (Knoll, 1983, 1985) was related to the evolution of the metazoan lineage. The respective origins of these divisions seem to be separated in time and possibly the evolution of both was favoured by a particular series of environmental changes.

Acquisition of photosynthetic endosymbionts by different eukaryotic lineages was almost certainly iterative, and probably took place at several intervals widely separated in geological history (Knoll, 1983, 1985; Walter, 1987; for a recent 5S rRNA study see Van den Eynde *et al.*, 1988).). The certain identification of eukaryotes from the rock record is difficult and, needless to say, any interpretation of the possible physiological function of purported organelles in ancient silicified cells is open to question (Walter, 1987). Thus, while the possible fossil record of eukaryotes extends back to 1.4 Ga and perhaps before, the earliest timing of the incorporation of photosynthetic plastids is enigmatic. Though rates of amino acid substitution in ferredoxins evidently differ between major clades, based on the broad approximation that the overall rate of evolution of this class of proteins is close to that of cytochrome *c*, a coccoid cyanobacterium was incorporated into cells of the primary lineage leading to higher plants to form chloroplasts at approximately 900-850 Ma. This event may well be reflected in the known radiation of a diverse eukaryotic phytoplankton 900-800 Ma ago (see also Perasso, Baroin, Qu, Bachellerie and Adoutte, 1989). The enhanced concentration of free oxygen generated by these photosynthetic plankters evidently permitted the necessary interdependence between heterotrophy and oxidative metabolism and also provided a large food resource. As argued above, the great glacial period during the older part of the Grenville Cycle also favoured a marked increase in the degree of oxygenation of the oceans.

Molecular studies confirm that the divergence of the coelomates was linked closely to the appearance of the spectacular Ediacaran fossil assemblages. Coelomates have alimentary tracts and pellets likely of faecal origin occur in near association with trace fossils in Early Ediacaran carbonates in the Flinders Ranges (Jenkins, 1984). The pre-adaptive potential of the coelomate body plan inevitably engendered an explosive radiation.

So called pre-Ediacaran 'fossil' indications of the Metazoa are highly dubious, and thus current attempts to identify the primal metazoan stocks are essentially speculative. Observations that the metabolism of some parasitic nematodes is able to synthesise collagen at low oxygen tensions (Towe, 1981) provide a possible lead, but this trait might well be a secondary adaptation. It is likely that pre-Ediacaran metazoans were too small to influence sediments mechanically and leave fossil imprints (Glaessner, 1972; Boaden, 1975).

During Marinoan-Varangian time episodic glaciation occurred in numerous regions. The common red coloration of the associated sediments probably reflects significant injection of oxygen into the oceans. The major extinction of plankters documented at ca650 Ma (Vidal and Knoll, 1983) may well have occurred due to the increasing concentration of oxygen poisoning many forms of the pre-existing phytoplankton adapted to survive at relatively low oxygen tensions. Oxidative processes in the sea progressively increased in intensity until hematite coloured even shelf sediments deposited below wave base, perhaps an indication of oxygen concentrations greater than in the present ocean.

Utilising globins evolved by earlier eukaryotes (T_K for globins is about 1350 Ma) the ancestors of the metazoans were able to scavenge the rather low concentrations of free oxygen then available. Pre-adapted to oxidative heterotrophy, they survived the Varangian oxygen crisis and began to exploit the new biochemical pathways offered by the surfeit of this gas. The near extinction of the plankton kept their numbers low. The novel evolution of multicellularity permitted complex body configurations, especially mechanisms for carrying aerated fluids to subdermal musculation. Subsequently, as global climates began to ameliorate about 580 Ma ago, bacterial activity in the transgressive epeiric seas of the Early Ediacaran diminished the excess concentration of free oxygen and phytoplankters again gradually started to proliferate. The metazoans also rapidly diversified as they exploited this renewed food resource. The great radiation of animal life had begun.

Summary

Astronomical, atmospheric and geological studies suggest that surficial environments of the Earth were weakly oxygenated for much of the Precambrian; reducing oceanic waters below the thermocline inhibited significant accumulation of free oxygen. At times of great glaciations, cooled surface waters aerated by organismic photosynthesis sank and oxidised the deep body of the oceans, thus allowing ambient oxygen concentrations to increase markedly. The incorporation of chloroplasts into eukaryotic plankters 900-850 Ma ago was followed by a major global refrigeration which significantly altered the oxygen balance. Another great glaciation 675-585 Ma ago evidently led to such an abundance of oxygen as partly to poison the biosphere. Recovery of biotas from the oxygen crisis is manifest by the appearance of the distinctive Ediacaran invertebrate assemblages, which represent the primary radiation of multicellular organisms adapted to exploit oxygenic regimes.

Acknowledgements

Dr Bruce Runnegar, University of California, Los Angeles, and Drs Victor A. Gostin and Brian McGowran, University of Adelaide, South Australia, are warmly thanked for ideas and valuable discussion. Ms M.S. Proferes drafted Figures 3, 4 and 6, 7. This research was supported by funding from Esso Australia Ltd.

References

Aharon, P., Schidlowski, M., and Singh, I.B. (1987) 'Chronostratigraphic Markers in the End-Precambrian Carbon Isotope Record of the Lesser Himalaya'. *Nature (London)* **327**, 699-702

Bergstrom, J. (1986) 'Metazoan Evolution. A New Model', *Zoologica Scripta* **15**, 189-200

Berner, R.A. and Barron, E.J. (1984) 'Factors Affecting Atmospheric CO_2 and Temperature Over the Past 100 Millions Years', *American Journal of Science* **284**, 1183-1192

Boaden, P.J.S. (1975) 'Anaerobiosis, Meiofauna and Early Metazoan Evolution', *Zoologica Scripta* **4**, 21-14

Bralower, T.J. and Thierstein, H.R. (1984) 'Low Productivity and Slow Deep-water Circulation in Mid-Cretaceous Oceans', *Geology* **12**, 614-618

Canuto, V.M., Levine, J.S., Augustsson, T.R. and Imhoff, C.L. (1983) 'Oxygen and Ozone in the Early Earth's Atmosphere', *Precambrian Research* **20**, 109-120

Carver, J.H. (1981) 'Prebiotic Atmospheric Oxygen Levels', *Nature (London)* **292**, 136-138

Cloud, P. (1968) 'Pre-Metazoan Evolution and the Origins of the Metazoa', in T. Drake (ed.), *Evolution and Environment* (Yale University Press, New Haven and London), 1-72

Cloud, P. (1984) 'The Cryptozoic Biosphere: Its Diversity and Geological Significance', Proceedings of the 27th International Geological Congress Moscow 4-14 August 1984, Vol. 5, *Precambrian Geology*, (VNU Science Press, Utrecht), 173-198

Coats, R.P. (1981) 'Late Proterozoic (Adelaidean) Tillites of the Adelaide Geosyncline', in M.J. Hambrey and W.B. Harland (eds.), *Earth's Pre-Pleistocene Glacial Record*, (Cambridge University Press, Cambridge), 537-548

Compston, W., Williams, I.S., Jenkins, R.J.F., Gostin, V.A. and Haines, P.W. (1987) 'Zircon Age Evidence for the Late Precambrian Acraman Ejecta-blanket' *Australian Journal of Earth Sciences* **34**, 435-445

Conway Morris, S. (1988) 'Radiometric Dating of the Preecambrian-Cambrian Boundry in the Avalon Zone', in E. Landing, G.M. Narbonne and P. Myrow (eds), *Trace Fossils, Small Shelly Fossils and the Precambrian-Cambrian Boundary* (New York State Museum Bulletin), 463, 53-58

Cook, P.J. and Shergold, J.H. (1984) 'Phosphorus, Phosphorites and Skeletal Evolution at the Precambrian-Cambrian Boundary', *Nature (London)* **308**, 231-236

Dayhoff, M.O. (1978) *Atlas of Protein Sequence and Structure*, Vol. 5, Suppl. 3, (National Biomedical Research Foundation, Washington D.C.)

Dayhoff, M.O. and Barker, W.C. (1976) 'Cytochromes' in M.O. Dayhoff (ed.), *Atlas of Protein Sequence and Structure*, Vol. 5, Suppl. 2 (National Biomedical Research Foundation, Washington D.C.), 25-49

Deacon, G.E.R. (1984) 'Oxygen in Antarctic Water', *Deep-Sea Research* **11**, 1369-1371

Dickerson, R.E. (1971) 'The Structure of Cytochrome *c* and the Rates of Molecular Evolution', *Journal of Molecular Evolution* **1**, 26-45

Dickerson, R.E. (1980) 'The Cytochromes *c*: An Exercise in Scientific Serendipity' in D.S. Sigman and M.A.B. Brazier (eds.), *The Evolution of Protein Structure and Function* (Academic Press, New York), 173-202

Dyson, I.A. (1985) 'Frond-like Fossils from the Base of the Late Precambrian Wilpena Group, South Australia', *Nature (London)* **318**, 283-285

Fairchild, T.R., Barbour, A.P. and Haralyi, N.L.E. (1978) 'Microfossils in the "Eopaleozoic" Jacadigo Group at Urucum, Mato Grosso, Southwest Brazil', *Boletino 1G. Instituto Geosciences, Universidado de Sao Paulo* **9**, 74-79

Glaessner, M.R. (1972) 'Precambrian Palaeozoology', in J.B. Jones and B. McGowran (eds.), *Stratigraphic Problems of the Later Precambrian and Early Cambrian*, Centre for Precambrian Research Special Paper No. 2, University of Adelaide, Adelaide, 43-52

Glaessner, M.F. (1983) 'The Emergence of Metazoa in the Early History of Life', *Precambrian Research* 20, 427-41

Glaessner, M.F. (1984) *The Dawn of Animal Life* (Cambridge University Press, Cambridge)

Goodfellow, W.D. and Jonasson, I.R. (1984) 'Oxygen Stagnation and Ventilation Defined by $\delta^{34}S$ Trends in Pyrite and Barite, Selwyn Basin, Yukon', *Geology* 12, 583-586

Goodman, M., Pedwaydon, J., Czelusniak, J., Susuki, T., Gotoh, T., Moens, L., Shishikura, F., Walz, D. and Vinogradov, S. (1988) 'An Evolutionary Tree for Invertebrate Globin Sequences', *Journal of Molecular Evolution* 27, 236-249

Gravenor, C.P., von Brunn, V. and Dreimans, A. (1984) 'Nature and Classification of Waterlain Glacigenic Sediments, Exemplified by Pleistocene, Late Paleozoic and Late Precambrian Deposits', *Earth-Science Reviews* 20, 106-166

Haines, P.W. (1988) 'Storm-Dominated Mixed Çarbonate/Siliciclastic Shelf Sequence Displaying Cycles of Hummocky Cross-stratification, Late Proterozoic Wonoka Formation, South Australia', *Sedimentary Geology* 58, 237-254

Hambrey, M.J. (1983) 'Correlation of Late Proterozoic Tillites in the North Atlantic Region and Europe', *Geological Magazine* 120, 209-232

Hambrey, M.J. (1984) 'Comment on Late Precambrian Palaeoclimates', *Geological Magazine* 121, 367-368

Harland, W.B. (1983) 'The Proterozoic Glacial Record', *Geological Society of America Memoir* 161, 279-288

Hofmann H.J. (1985) 'The Mid-Proterozoic Little Dal Macrobiota, Mackenzie Mountains, North-West Canada', *Palaeontology* 28, 331-354

Holland, H.D. (1984) *The Chemical Evolution of the Atmosphere and Oceans* (Princeton University Press, Princeton, New Jersey)

Jenkins, R.J.F. (1984) 'Ediacaran Events: Boundary Relationships and Correlation of Key Sections, Especially in Armorica', *Geological Magazine* 121, 635-643

Jenkins, R.J.F. (1986) 'Are Enigmatic Markings in Adelaidean of Flinders Ranges Fossil Ice-tracks?', *Nature (London)* 323, 472

Jenkyns, H.C. (1980) 'Cretaceous Anoxic Events: From Continents to Oceans', *Journal of the Geological Society London* 137, 171-188

Jones, J.B. and Fitzgerald, M.J. (1986) 'Silica Rich Layering at Blanche Point, South Australia', *Australian Journal of Earth Sciences* 33, 529-551

Jukes, T.H. (1980a) 'Neutral Changes Revisited' in D.S. Sigman and M.A.B. Brazier (eds), *The Evolution of Protein Structure and Function* (Academic Press, New York), 203-219

Jukes, T.H. (1980b) 'Silent Nucleotide Substitutions and the Molecular Evolutionary Clock', *Science* 210, 973-978

Kasting, J.F. (1987) 'Theoretical Constraints on Oxygen and Carbon Dioxide Concentrations in the Precambrian Atmosphere', *Precambrian Research* 34, 205-229

Kasting, J.F. and Donahue, T.M. (1981) 'Evolution of Oxygen and Ozone in Earth's Atmosphere' in J. Billingham (ed.), *Life in the Universe* (MIT Press, Cambridge, Massachusetts), 149-162

Kasting, J.F. and Walker, J.C.G. (1981) 'Limits on Oxygen Concentration in the Prebiological Atmosphere and the Rate of Abiotic Fixation of Nitrogen', *Journal of Geophysical Research* 86, 1147-1158

Keller, B.M. (1984) 'Upper Precambrian Systems', *International Geology Review* **26**, 247-261 [translated from Russian]

Knoll, A.H. (1983) 'Biological Interactions and Precambrian Eukaryotes' in M.J.S. Teresz and P.L.McCall (eds.). *Biotic Interactions in Recent and Fossil Benthonic Communities* (Plenum Press, New York), pp. 251-253

Knoll, A.H. (1985) 'Patterns of Evolution in the Archean and Proterozoic Eons', *Paleobiology* **11**, 53-64

Knoll, A.H., Hayes, J.M., Kaufman, A.J., Swett, K. and Lambert, I.B. (1986) 'Secular Variation in Carbon Isotope Ratios from Upper Proterozoic Successions of Svalbard and East Greenland' *Nature (London)* **321**, 832-838

Krogh, T.E., Strong, D.F.,O'Brian, S.J. and Papezik, V.S. (1988) 'Precise U-Pb zircon dates from the Åvalon Êerrane in Newfoundland' *Canadian Journal of Earth Sciences* **25**, 442-453

Lambert, I.B., Donnelly, T.H., Etminan, H. and Rowlands, N.J. (1984) 'Genesis of Late Proterozoic Copper Mineralization, Copper Claim, South Australia', *Economic Geology* **79**, 461-475

Lewis, J.S. and Prinn, R.G. (1984) *Planets and Their Atmospheres* (Academic Press, London)

Lydiatt, A., Peacock, D. and Boulter, D. (1978) 'Evolutionary Change in Invertebrate Cytochrome c' *Journal of Molecular Evolution* **11**, 35-45

Martin, H. (1965) *The Precambrian Geology of South West Africa and Namaqualand* (Precambrian Research Unit, University of Cape Town)

Mel'nik, Y.P. (1982) *Precambrian Banded Iron-Formations* (Elsevier, Amsterdam, Oxford, New York) [translation from Russian]

Perasso, R., Baroin, A., Qu, L.H., Bachellerie, J.P. and Ådoutte, A. (1989) 'Origin of the Algae', *Nature (London)* **339**, 142-144

Pinto, U.P. and Holland, H.D. (1988) 'Paleosols and the Evolution of the Atmosphere; Part 11' in J. Reinhardt and W.R. Sigleo (eds), *Paleosols and Weathering Through Geologic Time: Principles and Applications, Geological Society of America Special Paper* **216**, 21-34

Preiss, W.V. (1985) *Stratigraphy and Tectonics of the Worumba Anticline and Associated Intrusive Breccias, Geological Survey of South Australia Bulletin* **52**

Preiss, W.V. (1987) (compiler) *The Adelaide Geosyncline - Late Proterozoic, Stratigraphy, Sedimentation, Palaeontology and Tectonics. Geological Survey of South Australia Bulletin* **53**

Runnegar, B. (1982a) 'A Molecular-Clock Date for the Origin of the Animal Phyla', *Lethia* **15**, 199-205

Runnegar, B. (1982b) 'The Cambrian Explosion: Animals or Fossils?', *Journal of Geological Society of Australia* **29**, 395-411

Runnegar, B. (1986) 'Molecular Palaeontology' *Palaeontology* **29**, 1-24

Schidlowski, M., Hayes, J.M. and Kaplan, I.R. (1983) 'Isotopic Inferences of Ancient Biochemistries: Carbon, Sulphur, Hydrogen, and Nitrogen' in J.W. Schopf (ed.), *Earth's Earliest Biosphere* (Princeton University Press, Princeton), 149-186

Schopf, J.W. and Packer, B.M. (1987) 'Early Archean (3.3 Billion to 3.5 Billion-year-old) Microfossils from Warrawoona Group, Australia', *Science* **237**, 70-73

Schwartz, R.M. and Dayhoff, M.O. (1978) 'Origins of Prokaryotes, Eukaryotes, Mitochondria, and Chloroplasts', *Science* **199**, 395-403

Sprigg, R.C. (1984) *Arkaroola-Mount Painter in the Northern Flinders Ranges, S.A.: The Last Billion Years* (Arkaroola Pty Ltd, Adelaide)

Sutton, J. (1968) 'The Extension of the Geological Record into the PreCambrian', *Proceedings of the Geologists' Association* **78**, 493-543

Sutton, J. (1971) 'Some Developments in the Crust', *Geological Society of Australia Special Publication* **3**, 1-10

Towe, KM. (1981) 'Biochemical Keys to the Emergence of Complex Life' in J. Billingham (ed.), *Life in the Universe* (MIT Press, Cambridge, Massachusetts), 297-306

Van den Eynde, H., De Baere, R., De Roek, E., Van de Peer, Y., Vandenberghe, A., Willekens, P. and De Wachter, R. (1988) 'The 5S Ribosomal RNA Sequences of a Red Algal Rhodoplast and a Gymnosperm Chloroplast. Implications for the Evolution of Plastids and Cyanobacteria', *Journal of Molecular Evolution* **27**, 126-131

Vidal, G. and Knoll, A.H. (1983) 'Proterozoic Plankton', *Geological Society of America Memoir* **161**, 265-277

Vidal, G. and Moczydlowska, M. (1987) 'Further Reflections on Metazoan Evolution' *Precambrian Research* **36**, 345-348

von der Borch, C.C., Christie-Blick, N. and Grady, A.E. (1988) 'Depositional sequence analysis applied to upper Proterozoic Wilpena Group, Adelaide Geosyncline, South Australia', *Australian Journal of Earth Sciences* **35**, 59-72

Walker, J.C.G. and Hays, P.B. (1981) 'A Negative Feedback Mechanism For the Long-Term Stabilization of Earth's Surface Temperature', *Journal of Geophysical Research* **86**, 9776-9782

Walter, M.R. (1987) 'The Timing of Major Evolutionary Innovations from the Origin of Life to the Origins of the Metaphyta and Metazoa: the Geological Evidence' in K.S.W. Campbell and M.F. Day (eds) *Rates of Evolution* (Allen and Unwin, London), pp. 15-38

Walter, M.R., Oehler, J.H. and Oehler, D.Z. (1976) 'Megascopic Algae 1300 Million Years Old From the Belt Supergroup, Montana: A Reinterpretation of Walcott's Helminthoidichnites'. *Journal of Paleontology* **50**, 872-881

Whitten, G.C. (1970) *The Investigation and Exploitation of the Razorback Ridge Iron Deposit, Report of Investigations Geological Survey of South Australia* **33**

Wilde, P. (1987) 'Model of Progressive Ventilation of the Late Precambrian-Early Paleozoic Ocean' *American Journal of Science* **287**, 442-459

Yeo, G.M. (1981) 'The Late Proterozoic Rapitan Glaciation in the Northern Cordillera' in F.H.A. Campbell (ed.), *Proterozoic Basins of Canada*, Geological Survey of Canada Paper, 81-10, pp 25-46

Zbinden, E.A., Holland, H.D., Feakes, C.R. and Dobos, S.K. (1988) 'The Sturgeon Falls Paleosol and the Composition of the Atmosphere 1·1 Ga BP', *Precambrian Research* **42**, 141-163

Chapter 4

Oxygen and the Early Evolution of The Metazoa

Bruce Runnegar

Introduction

In 1959, J.R. Nursall suggested that the origin and subsequent rapid radiation of the Metazoa could be linked to the addition of sufficient free oxygen to the atmosphere and hydrosphere to allow for the evolution of oxidative (aerobic) metabolism. A similar hypothesis was developed by Berkner and Marshall (1965) using quantitative atmospheric models. They concluded that only very small amounts of molecular oxygen could be generated by abiotic processes (the photochemical dissociation of water vapour in the upper atmosphere) and that about a hundredth of the present concentration of atmospheric oxygen would be required to provide adequate protection from solar ultraviolet (UV) rays for organisms inhabiting the upper layers of the oceans. Because about the same concentration of oxygen would have permitted higher organisms to begin to replace anaerobic metabolic pathways with aerobic ones, the synchronous development of aerobic respiration and an ozone-based UV shield at the end of the Precambrian was thought to account for the subsequent explosive radiation of multicelled animals.

The Berkner-Marshall hypothesis fell into disfavour when different photochemical models of the early atmosphere pointed to the much earlier production of a biologically effective ozone shield (Margulis, Walker and Ramber, 1976; Carver, 1981) and geological evidence indicated relatively high concentrations of atmospheric oxygen throughout much of the Precambrian (Dimroth and Kimberley, 1976; Clemmey and Badham, 1982). Most of the evidence for oxygen concentrations in the Precambrian atmosphere and hydrosphere is obtained from the oxidation states of compounds or uranium and iron in sedimentary rocks (Walker, Klein, Schidlowski, Schopf, Stevenson and Walker, 1983; Holland, 1984). The consensus at present is that oxygen was being produced by both abiotic and biotic processes during the first half of the Precambrian but it failed to accumulate in large amounts in the atmosphere because too many reduced substances were present in the oceans (Towe, 1978; Walker et al., 1983; Holland, 1984). Once these sinks were exhausted, oxygen began to accumulate eventually to reach the present amount of about 21 per cent Present Atmospheric Level (PAL). The principal oxygen sink seems to have been ferrous iron so it is normally assumed that after the great iron ores of middle Precambrian age were formed, oxygen levels in the atmosphere and oceans began to climb towards modern values (Cloud, 1972). The presence of widespread red-coloured sediments and the absence of reduced compounds such as uraninite (UO_2) and pyrite (FeS_2) distinguish younger Precambrian sedimentary rocks from those of earlier times. This is taken to indicate the presence of an oxidising atmosphere but not necessarily one that was able to support aerobic respiration, although 'no plausible explanation has been offered why oxygen should not have achieved values similar to those of the present relatively soon after the disappearance of inorganic reduced species from the atmosphere and oceans' (Walker *et al*, 1983, p. 228).

In the absence of a realistic quantitative model for the rate of growth of the oxygen fraction of the atmosphere in the later Precambrian, the most direct evidence for the level of atmospheric oxygen is likely to be obtained from analyses of the probable physiology of late Precambrian and early Cambrian animals (Holland, 1984). Fossil organisms, if correctly interpreted, are much more sensitive environmental indicators than inorganic compounds or sedimentary structures. Thus a single sea-shell may provide unequivocal evidence for marine conditions not suspected on other criteria. However, the use of fossils for environmental interpretation becomes more difficult as the age of the fossils, and hence their evolutionary distance from the modern biota, increases. This is particularly true for the late Precambrian because the fossils are rare, difficult to interpret and perhaps are wholly unrelated to modern forms (Seilacher, 1984; Gould, 1984). It is therefore necessary to proceed cautiously by questioning the most basic conclusions regarding the age, preservation and biological affinities of the known late Precambrian animals. The purpose of this critical reappraisal is to identify those fossils that may, with some confidence, be related to the living animal phyla. It should then be possible to use knowledge of the biochemistry, physiology and ecology of the modern organisms to test aspects of the Nursall and Berkner-Marshall hypotheses. Using this approach it will be shown that all of the known Ediacarian animals could have lived in a low-oxygen world in which the atmosphere and oceans contained about a tenth of the present concentration of molecular oxygen. It was perhaps for this reason that the Ediacarian fossil faunas are dominated by the remains of soft-bodied, sheet-like animals that had high surface area to volume ratios. Such organisms are adapted for life in low-oxygen environments.

Precambrian History of the Metazoa: Direct Evidence

The sudden appearance of many kinds of animals with mineral skeletons about 550 million years (Ma) ago marks the end of the Precambrian and the beginning of the period of Earth history known as the Phanerozoic. Although a large number of reports of Precambrian animal fossils have been published, all of those older than about 600 Ma (Jenkins, 1984) have been discredited, and only those from the

Ediacaran Period (650-550 Ma; Jenkins, 1981; Cloud and Glaessner, 1982; Glaessner, 1984) provide any direct evidence for the existence of Precambrian metazoans. These Ediacarian fossils are the casts and moulds of sizeable (1 cm) to large (1 m) soft-bodied organisms.

The Ediacarian fauna is now known from many places on five continents (Figure 4.1). Some of the more distinctive genera such as the sheet-like 'worm' *Dickinsonia* (Figures 4.2-4.3) and the paddle-shaped 'sea-pen' *Charnia* (Figure 4.4) come from sites that could not have been closer than 90° of great-circle distance during the late Precambrian. At least one form (*Charnia*) is found in near-intertidal sediments in Australia and in deposits formed at the base of the continental slope in Newfoundland (Anderson and Conway Morris, 1982). These facts indicate that the observed component of the Ediacaran biota had broad biogeographic and environmental ranges. Therefore, these animals cannot be dismissed as aberrant organisms preserved in unusual environments; they must be regarded as typical of their time.

The most diverse and best-studied of the Ediacarian assemblages comes from the Flinders Ranges of South Australia (Glaessner, 1984). Most of this review will therefore focus on fossils from South Australia and only limited reference will be made to fossils from other sites.

Preservation of the Ediacarian Faunas

Seilacher (1984) has rightly drawn attention to the fact that the preservation of the Ediacarian animals is as exceptional as the fossils themselves. If - as has been frequently stated - many of the fossils are the remains of jellyfish (cnidarian medusae), why are similar fossil jellyfish not found throughout the Phanerozoic? In other words, why is this kind of preservation so uncommon in younger strata?

Figure 4.1 Distribution of the most important sites yielding fossils of Ediacarian age.

68   *Oxygen and the Early Evolution*

Ediacarian fossils are known from more than 20 well-separated localities in the Flinders Ranges (Jenkins, Ford and Gehling, 1983). All but a few poor specimens are from a relatively thin but widespread horizon (Ediacara Member) within the middle part of the Pound Quartzite, a vast rock unit that outcrops over an area of some 40,000 km^2 and reaches a thickness of about 2.7 km. (The 'frond-like fossils' reported by Dyson (1985) from well below the Pound Quartzite appear from the published photograph to be pseudofossils.)

It is inconceivable that the time of existence of the Ediaracrian biota coincided precisely with the time of deposition of the Ediacara Member so the nature of the depositional environment and the kind of sediments being deposited were largely responsible for the preservation of the fossils. Almost all of the fossils in the Pound Quartzite are positive (convex) or negative (concave) impressions on the lower surfaces of thin ripple- marked beds of sandstone (Wade, 1968). In general, the more complex animals are preserved as external moulds (negative impressions) as if the animals had been cast by a foundry sand as they lay on the sea floor (Glaessner, 1984, p. 49). In some cases contraction of the animal after burial has increased its

Figure 4.2 Natural casts of the discoidal, segmented 'worm' *Dickinsonia* from Australia and Russia (insert).

The two Australian specimens (13 cm and 8 cm in length) have about the same number of segments and may be regarded as the expanded and contracted states of a single individual. Photograph of Russian specimens (3 cm in length) by courtesy of M.A. Fedonkin, Academy of Sciences, Moscow.

Figure 4.3 Well-preserved Ediacarian fossils from South Australia.

A, artificial cast of *Spriggina* showing the animal (4 cm in length) as it would have appeared in life. B, natural cast of the ventral surface of a small specimen of *Dickinsonia* (2.5 cm in length). C, natural cast of a crushed cone (4 cm in length). D, natural cast of the holdfast and part of the stem of *Charniodiscus* (disk is 9 cm in diameter). E, artificial cast of folded specimen of *Dickinsonia* (10 cm in length).

Figure 4.4 The 'sea-pen' *Charnia* from England preserved as a natural cast about 18 cm in length.

Photograph by courtesy of R.J.F. Jenkins, University of Adelaide.

thickness and thus presumably improved its impression in the casting medium (Figure 4.2; Runnegar, 1982a). This kind of preservation will be limited to those soft-bodied organisms that cannot escape horizontally or vertically after being buried and which are tough enough to remain intact for some time. The fossilised leaves of deciduous trees are a good example. By contrast, a less-resistant soft-bodied organism such as jellyfish will collapse quickly and the sand that buried it will flow downwards into the cavity formed by the base of the body (Figure 4.6; Schafer, 1941). A good example of the best results of such a process is given by a very fine middle Jurassic trace fossil (Seilacher, 1953). It is a sandstone cast from the impressions left in an underlying muddy layer by the ventral surface of a starfish (the trace-fossil *Asteriacites quinquefolius*) and a number of individuals of a species of bivalve mollusc (the trace fossil *Lockeia amygdaloides*). Impressions of the tube feet of the starfish are clearly visible on the sandstone cast (Hantzschel, 1975, p. 77). The trace fossils were produced when the sand buried the starfish and the bivalves; all moved up through sand which then fell downwards to cast impressions left in the underlying mud. This example is entirely analogous to some of the Ediacarian fossils discussed below and it shows that structures as soft as the tube feet of starfish can be expected to be preserved by exceptional casting media. The point is that this kind of preservation is uncommon throughout the Phanerozoic despite the fact that starfish and bivalves have abounded in marine environments since at least Ordovician times (500 Ma ago).

Figure 4.5 An Ediacarian 'by-the-wind sailor', *Ovatoscutum*, from South Australia (artificial cast 5 cm in diameter).

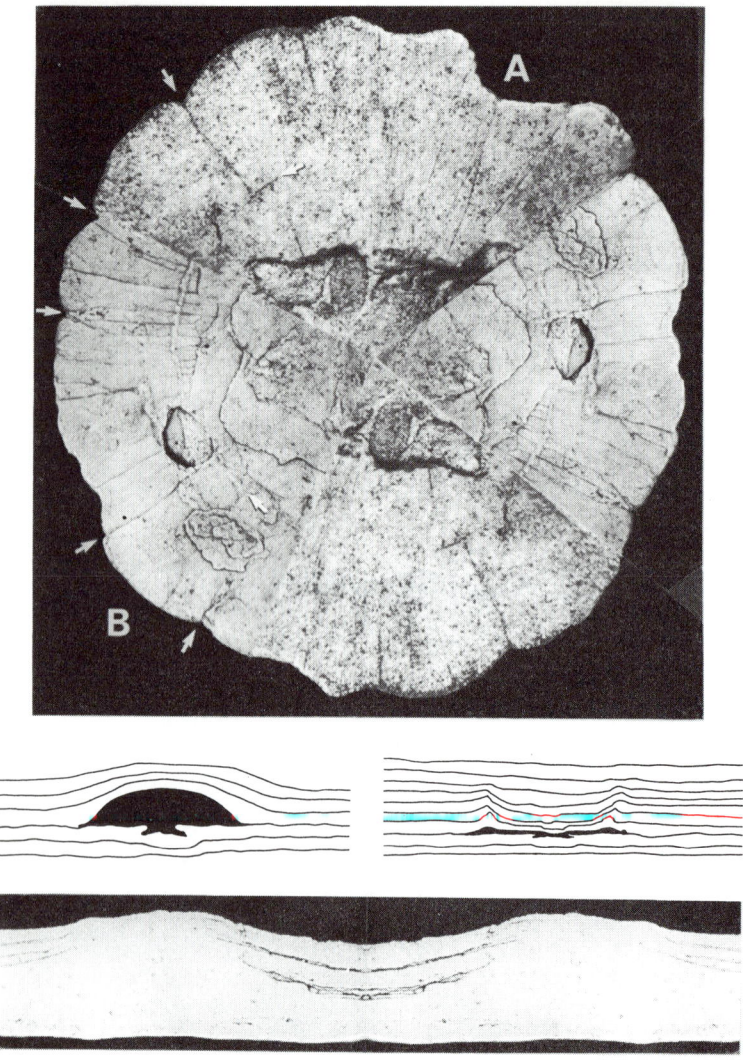

Figure 4.6 Collage made from photographs of parts of two incomplete natural casts of the late Precambrian jellyfish *Ediacaria* from South Australia.

(A is 18 cm in diameter and B 20 cm in diameter). Radial arrows mark indentations that are interpreted as the edges of lappets; the other arrows point to a concentric ridge or groove visible on these and other specimens. The drawings are from Schafer (1941; republished with permission) and they show how a crater develops above a decaying jellyfish. A similar crater (inverted by compaction) is visible in a sawn diametric section of specimen B, reflected in mirror-image for clarity. The centre of the fossil jellyfish is at the junction between the two halves of the image; dark layers are shale in a quartzose sandstone.

One of the reasons why lycopod wood is so common in nearshore sediments of Devonian age is that shipworms and other marine wood-borers had not yet evolved mechanisms to mine and digest the cellulose. Perhaps an analogous situation existed in the late Precambrian (Seilacher, 1984; Conway Morris, 1985). The large bodies of the Ediacarian animals were constructed from a new material - collagen - which may not have been easily 'biodegradable' when it first appeared. We need to know more of the evolutionary history of enzymes such as the collagenases to assess this suggestion.

It already be obvious that there may be no simple answer to Seilacher's important question about the preservation of the Ediacarian animals. Other possible factors that need to be considered are the bactericidal effects of UV light in an atmosphere with less ozone, the possibility that quartz sands in littoral environments may have contained almost anoxic pore fluids, the likely lack of sizeable scavengers and the virtual absence of sediment reworking by burrowing organisms.

Nature of the Ediacarian Animals

In a series of careful studies carried out over many years at the University of Adelaide, M.F. Glaessner and M. Wade have attempted to determine the biological affinities of the Ediacarian fossils (Glaessner, 1984). Understandably, they have tried to relate each of the Ediacarian taxa to the most similar living animals and have classified the fossils accordingly. For example, the discoidal worm *Dickinsonia* (Figure 4.2) was considered by Wade (1972b) to be a close relative of the living ectoparasitic polychaete *Spinther* and the possibility that the similarities between these two organisms could be due to convergence was rejected as 'extremely unlikely'. In contrast, Conway Morris (1979) and others have considered the similarities between *Dickinsonia* and *Spinther* to be less significant, a decision that has important repercussions on the way we view the early phylogeny of the Annelida, its higher taxa and related phyla (Runnegar, 1982a).

In Seilacher's opinion (1984) and that of Gould (1984, 1986), practically none of the Ediacarian fossils can be related to living organisms. It was a distinct episode in the history of life (the age of the 'Vendozoa') that was followed by a major extinction; the true ancestors of Cambrian and extant metazoans are not represented in the Ediacarian fossil faunas. How realistic is this view? Does it provide a better understanding of the nature of the Ediacarian animals than the approach adopted by Glaessner and Wade?

Ediacarian Cnidaria

At least two-thirds of the named taxa and more than three-quarters of the known fossils of Ediacarian animals are generally thought to be the remains of cnidarians and related extinct phyla of the coelenterate (diploblastic) grade of organisation. Broadly speaking there are two main types, disc-shaped fossils loosely called 'medusoids' and frond-like or leaf-like structures that at least superficially resemble modern sea-pens (pennatulacean anthozoans). There have been many elaborate and largely fanciful reconstructions of some of the fossils and several attempts to identify and name extinct higher taxa of the Cnidaria that are based solely on Ediacarian specimens (e.g. Fedonkin, 1985). It is not possible to examine each of these disparate proposals in this brief review; instead, I shall attempt to demonstrate that representatives of each of the three extant classes of the Cnidaria (Hydrozoa, Scyphozoa and Anthozoa) occur in the Ediacarian faunas.

Hydrozoa and Scyphozoa. Yochelson, Sturmer and Stanley (1983) have recently described a number of new specimens of an exceptionally well-preserved velellid chondrophorine, *Plectodiscus* , from the Devonian Hunsruck Sahle of Germany. The cnidarian affinities of these fossils are shown by impressions of tentacles similar to those of the tube-feet of the starfish trace fossil discussed above. The crushed floats of the Devonian chondrophorines are like those of the living 'by-the-wind sailor' *Velella* , but they are even closer to the Ediacarian genus *Ovatoscutum* (Figure 4.5) which has been referred to the Chondrophorina. This seems to be a clear case for a close relationship between an Ediacarian animal and Phanerozoic and living representatives of the cnidarian class Hydrozoa (Yochelson *et al.*, 1983).

The numerous other disc-shaped fossils are more difficult to understand. At least some are the body fossils (Figure 4.3D) or trace fossils of the holdfasts of the frond-shaped animals discussed below. Others (*Cyclomedusa, Tateana, Spriggina*) have radial and concentric structures (Jenkins, 1983) that resembles the floats of another living chondrophorine, *Porpita*, an interpretation that is supported by the impressions of tentacles on other discoidal fossils described under the generic name *Eoporpita* (Wade, 1972a; Jenkins, 1983). A third kind of structure, unreported previously, is illustrated in Figure 4.3C. The fossil is discoidal and has a well-defined edge and a series of cracks radiating in a dendritic pattern from near the centre of the disc. The nature of the cracks suggested that it was a tough, cone-shaped structure that had collapsed under the weight of overlying sediment. This interpretation was confirmed by sawing the slab to expose funnel-shaped layers above the disc and by experimentally crushing thin copper cones and domes embedded in a compactable matrix in a hydraulic press. As expected, dome-shaped models produced a series of concentric folds upon crushing whereas low, limpet-shaped cones gave a good approximation of the fossil being modelled.

The cnidarian affinity of this cone-shaped fossil is difficult to prove but there are obvious similarities in shape and ornamentation to the Ordovician fossil *Conchopeltis*. Because of the presence of tentacles, Oliver (1983) regarded *Conchopeltis* as a cnidarian and possibly a hydrozoan or scyphozoan, but he rejected previous suggestions of a relationship between *Conchopeltis* and the extinct conulariids (Conulata).

Conomedusites is another Ediacarian fossil that has been thought to be closely allied to *Conchopeltis* and hence to the Conulata (Glaessner, 1971; Glaessner, 1984, p. 116). The better specimens of *Conomedusites* are convex casts of an originally conical object that may or may not have had four-fold symmetry. Marks reminiscent of the tentacles of *Conchopeltis* are present on one specimen of *Conomedusites* (Glaessner, 1971), so it is probable that there were organisms of the general body plan of *Conchopeltis* in Ediacarian times. Such animals may have no modern counterparts but they are not easily excluded from the Cnidaria (Oliver, 1983), and they may well represent another group that persisted from Ediacarian to Cambrian and younger times.

Because so many circular or radially-symmetrical objects that were described originally as jellyfish (cnidarian medusae) have turned out to be pseudofossils, trace fossils, or the mis-identified parts of other kinds of metazoans, it is necessary to be particularly circumspect when dealing with presumed Ediacarian scyphozoans. Although many of the disc-shaped fossils resemble (and probably were) cnidarian medusae none is as unequivocal or as immediately believable as the Carboniferous sea wasp *Anthracomedusa* (Foster, 1979) or the Jurassic jellyfish *Rhizostomites* (Bayer, Boschma, Harrington, Hill, Hyman, Lecompte, Montanaro-Gallitelli, Moore, Stumm and Wells, 1956).

Many of the named taxa of Ediacarian medusoids probably represent different states of preservation of fairly simple scyphozoan medusae or their partly decayed products. For example, the much-illustrated specimens of *Mawsonites* may simply be casts of load-deformed mesogloeal cores that have separated from the cellular tissues because the mesogloea of scyphozoan medusae is much more resistant to autolysis than are the cellular layers (Chapman, 1959). Similarly, many of the smaller circular structures that lie within the margins of medusoids such as *Beltanella*, and which have been interpreted in anatomical terms, may merely be casts of bubbles of decomposition gases; these are often more distinct than the traces of organs (Schafer, 1941).

In my opinion, specimens of *Ediacaria* such as those shown in Figure 4.6 provide the best evidence for the presence of scyphozoan medusae in the late Precambrian. The animals were large (20 cm or more in diameter); the margins were clearly scalloped into structures resembling the lappets of modern scyphozoans; each 'lappet', on average, occupies about one sixteenth of the circumference of the disk; there is frequently a concentric groove or ridge about half-way to the margin as in aboral (exumbrellar) impressions of the Jurassic medusa *Rhizostomites* (Bayer *et al.*, 1956, p. 49); and there are sets of fan-shaped marks that appear to represent casts of the radial (or delta) muscles of the oral surface of the bell. Also, a cross-section of the sandstone overlying one specimen of *Ediacaria* shows that the fossil lies beneath a mud-filled crater just like collapse structures observed by Schafer (1941) in experiments with buried decaying bodies of jellyfish (Figure 4.6).

Anthozoa. The distinctive and large frond-like fossils *Charnia* (Figure 4.4), *Rangea* and *Charniodiscus* exhibit many features that relate them to living pennatulacean octocorals (Glaessner and Wade, 1966; Jenkins and Gehling, 1978; Glaessner, 1984; Jenkins, 1985). (I consider *Glaessnerina* Germs 1973 to be synonym of *Charnia* Ford 1958 - see Glaessner, 1984, p. 91 for illustrations of both forms). Given the fact that representatives of many living phyla are found in overlying Cambrian strata, the correct approach to the question of the affinities of these organisms is to disprove a relationship with the Pennatulacea before reaching the conclusion that the 'entire Ediacarian fauna...represents a unique and extinct experiment in the basic construction of living things.' (Gould, 1984, p. 16).

The similarities between the widely-distributed Ediacarian genus *Charnia* and the living pennatulacean *Virgularia* (Bayer *et al.* 1956, p. 228) make it difficult to disprove a relationship using the evidence now available. *Virgularia* has contractile polyps which disappear behind 'polyp leaves' that closely resemble the 'quilted' structure of *Charnia* fronds. I therefore agree with Jenkins (1985, p. 350) that 'there seems every reason to maintain the placement of the pennatulacean-like genera *Charniodiscus* and *Glaessnerina* within the Octocorallia and to conclude that the origins of the Anthozoa are indeed Precambrian.'

Ediacarian Segmented Animals

There are six genera of Ediacarian fossils that might represent the remains of segmented animals: *Dickinsonia* (Figures 4.2-4.3), *Marywadea* (Glaessner, 1976), *Parvancorina* (Glaessner, 1980), *Praecambridium* (Glaessner and Wade, 1971), *Spriggina* (Figure 4.3A), and *Vendia* (Glaessner, 1984). Of these, *Vendia* is suspect because it appears to be asymmetrical (Fedonkin, 1985); *Parvancorina* is uninformative because most of its features are concealed by what is regarded as a dorsal carapace; *Praecambridium* may be a juvenile *Dickinsonia* (Runnegar, 1982a); and *Marywadea* is close to if not identical with *Spriggina*. I shall therefore deal only with *Dickinsonia* and *Spriggina*.

Dickinsonia had an oval, sheet-like body that is normally preserved as concave impressions on the lower surface of sandstone beds. The animal is known from a large number of specimens that range in size from a few millimetres to over half a metre, and it is probable that at least one species of *Dickinsonia* grew to about a metre in length. A partly exposed *in situ* specimen of *Dickinsonia brachina* I discovered recently has a half-length of about 0.5 m. Most of the animals are thought to have been buried alive as they lay on the bottom because they are frequently surrounded by smooth zones with faint radial marks that were produced when the animal contracted (Figure 4.2).

The characteristic transverse markings are considered to represent the boundaries of body segments (Wade, 1972b; Runnegar, 1982a) because they increase in number and size during ontogeny and are more substantial than surface 'annulations'. The conventional view is that the larger segments are older and anterior, that the animal was strictly bilaterally symmetrical (Figure 4.3B), and that new segments were added progressively during ontogeny at or very near to the posterior end of the body. A sagittal ridge interrupts the continuity of left and right parts of the segments along the dorsal midline (Figure 4.2) so external moulds lacking this feature are thought to be casts of the ventral surface (Figure 4.3B). The prominence of this dorsal ridge on some specimens has been interpreted as evidence for the presence of a sediment-filled gut, but it may also be due to body fluids that were driven from the flanks to the midline by sediment loading.

The two Australian specimens of *Dickinsonia* shown in Figure 4.2 have about the same number of segments and are therefore thought to have been similar in size (volume). The larger individual had contracted from an original area of about 140 cm^2 as shown by the smooth zone surrounding the specimen; the smaller, contracted individual has an area of about 40 cm^2. It is clear, therefore, that *Dickinsonia* had a highly extensible and probably fluid-filled body that was capable of three-fold or four-fold changes in surface area and thickness. The grooves on the sandstone moulds that define the boundaries of the segments are thought to represent the casts of dorsoventral and radial muscles that were inserted into a collagenous body wall. For these and other reasons, *Dickinsonia* was considered by Wade (1972b) and Runnegar (1982a) to be a segmented coelomate of the annelid grade of construction, probably an early member of the phylum Annelida. This view has been rejected by Fedonkin (1981, 1985) who regards *Dickinsonia* as a primitive acoelomate flatworm and by Seilacher (1984) who considered it to be a typical member of his extinct 'Vendozoa'.

Glaessner has steadfastly rejected suggestions that *Spriggina* (Figure 4.3A) might be a primitive arthropod because he thinks that the the appendages are unjointed and bear straight acicular setae at their ends (Glaessner, 1984, pp. 61-62; personal communication). Although I have not seen all of the specimens of *Spriggina* being studied by Glaessner and Wade, I am not convinced that these interpretations are correct. It seems more likely that *Spriggina* is a stem arthropod of the tribolite-crustacean-chelicerate type.

One possible scenario for the origin of the Arthropoda is that the characteristic exoskeleton evolved independently in two different groups (Cisne, 1974). This model, which recognises two arthropod phyla (Uniramia and 'Multiramia' or Trilobita + Crustacea + Chelicerata), points to an animal like *Spriggina* as the direct common ancestor of all non-uniramian arthropod lineages. Interpreted as an arthropod, *Spriggina* has a crescent-shaped head-shield followed by about 40 segments each bearing a narrow, jointed walking legs and, presumably, gill branches. The mouth appears to have been underneath the head-shield.

In many respects, *Spriggina* may not be very different from the primitive Cambrian arthropod *Marrella* from the Burgess Shale (Whittington, 1971). *Marrella* was a blind animal that had a head-shield formed of two segments, each bearing a pair of long appendages and a pair of backwardly curving spines. Each of the other segments of the body (25) was subcircular in cross-section, did not have lateral extensions (pleurae) and bore a pair of biramous appendages. Such an animal could be derived from one like *Spriggina* by small changes in proportions following a duplication of the sprigginid head shield. The duplication may have occurred in a manner analogous to the production of four-winged flies (Diptera) by the combination of three mutants of the bithorax complex in *Drosophila* (Bender, Akam, Karch, Beachy, Peifer, Spiere, Lewis and Hogness, 1983). This may seem far-fetched, but unless we are prepared to admit the possibility of major morphological changes of this kind it will be exceedingly difficult to understand the origins and relationships of early body plans.

Incertae sedis. Some of the Ediacarian organisms are so different from all other living and fossil animals that they may need to be referred to separate extinct phyla. For example, bag-shaped objects from Namibia (*Ernietta*) and three-vaned leaf-like structures from Australia, Africa, North America and the USSR (*Pteridinium*) have been referred to the phylum Petalonamae Pflug by Jenkins (1981), and disk-shaped fossils with triradial symmetry (*Skinnera, Tribrachidium, Albumares*) are grouped in the extinct cnidarian class Trilobozoa by Fedonkin (1985). Whether these taxonomic decisions are correct or not is immaterial for the present discussion; the point is that each of these organisms had a high ratio of surface area to volume and may well have been of the diplobastic grade of organisation.

Precambrian History of the Metazoa: Indirect Evidence

Indirect evidence of the Precambrian history of the Metazoa is obtained from the traces of their activities they may have left in the sediments (trace fossils, faecal pellets, etc.), from possible effects on the non-metazoan biota and, increasingly, from the comparative biochemistry and molecular genetics of living organisms. Questions that may be addressed at present by these methods are: Have all metazoans descended from a single metazoan ancestor? In other words, is the Metazoa a monophyletic clade? Second, did the Metazoa have a short (100 Ma) or long (ca500 Ma) Precambrian history? Third, was the timing of the great metazoan radiation that produced the Cambrian fauna controlled by extrinsic factors such as the growth of the oxygen fraction of the atmosphere?

There are few, if any, sedimentary structures older than about 650 Ma that can be attributed with certainty to the activities of multicelled animals. One promising approach is the search for faecal pellets in the microfossil size range (Robbins, Porter and Haberyan, 1985), but the positive results obtained so far from rocks as old as 1900 Ma are not sufficient to establish the existence of metazoans in pre-Ediacarian time.

The oldest certain fossils are surface markings (Figure 4.7) and perhaps shallow horizontal burrows from rocks of Ediacarian age (Glaessner, 1969; Fedonkin, 1981). Older structures considered previously to be trace fossils have proved either to be body fossils of Ediacarian type (*Bunyerichnus*) or pseudofossils (*Brooksella, Canyonensis*). Vertical burrows are rare or absent from rocks of Ediacarian age and it is now certain that the primary radiation of sizeable soft-bodied burrowing organisms occurred at the

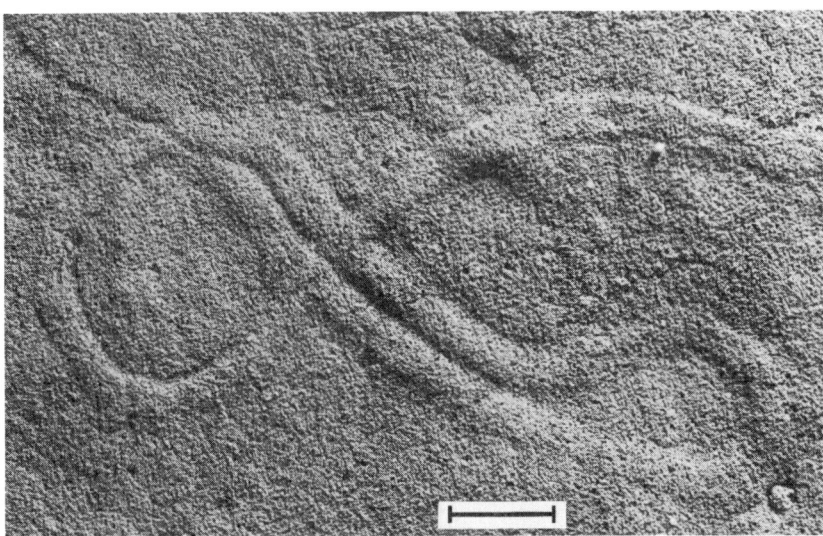

Figure 4.7 Surface trail preserved as a natural cast on the base of a sandstone bed from Ediacaria, South Australia. The animal that produced the trail moved from left to right. Scale bar equals 1 cm.

beginning of the Cambrian.

However, this does not necessarily mean that metazoans had a short Precambrian history. Comparisons of the amino acid sequences of vertebrate and invertebrate oxygen-carrying proteins point to a Precambrian history of some hundreds of millions of years (Runnegar, 1982b, 1986), and a comparable time of origin of the Metazoa is obtained from comparisons of the nucleotide sequences of collagen genes (Runnegar, 1985). Because collagen is both restricted to the Metazoa and found in all metazoan phyla, it is also possible to use collagen sequence data to show that metazoans constitute a monophyletic group (Runnegar, 1986).

The Role of Oxygen in the Early Evolution of the Metazoa

The apparent paradox of a long Precambrian history and a short Precambrian fossil record is resolved if there were some long-standing environmental constraint on maximum body size. Small, soft-bodied metazoans of the type found in modern interstitial habitats have almost no chance of fossilisation and, as a result, no fossil record. It was the rise of large, muscular and/or extensively mineralised animals that was responsible for the explosive radiation of metazoans at the beginning of the

Phanerozoic (Runnegar, 1982c). These animals were ruled for the first time by gravity rather than surface forces. They replaced ciliary motion with muscular action, used hydrostatic and mechanical skeletons to transmit muscular energy, and they acquired extra surface for respiration and digestion. These innovations allowed them to move more rapidly for longer distances, to burrow in soft substrates, to grow larger, to broaden their range of foods, and of course, to be fossilised. The history of their appearance is - as J.B.S. Haldane put it in another context - 'largely the story of the struggle to increase surface in proportion to volume', and if there was one single commodity they needed for this task it was molecular oxygen. It was used to provide energy for muscular and other metabolic functions, to manufacture the vital structural protein collage (Towe, 1970, 1981), to harden proteinaceous cuticles by sclerotisation (Brunet, 1967) and to increase body size (Runnegar, 1982a). An increased availability of oxygen also permitted animals to form continuous mineral skeletons (Rhoads and Morse, 1971), partly because less surface area was required for respiratory exchange and partly because the acidic end-products of the breakdown of glucose under aerobic conditions may inhibit the deposition of carbonate minerals (Lutz and Rhoads, 1977). It is therefore possible to reformulate the Nursall and Berkner-Marshall hypotheses in the following way:

1. The Metazoa have a long (>200 Ma) Precambrian history.
2. The proximate cause of the late Precambrian-Cambrian radiation of the Metazoa was an increased availability of dissolved O_2 in ocean waters.
3. Because dissolved oxygen in shallow ocean waters was in equilibrium with oxygen in the atmosphere at sea level, an increase in the availability of oxygen could be due to one or more of the following factors: an increase in the partial pressure of oxygen in the atmosphere; a decrease in the mean ocean temperature (and hence an increase in the solubility of oxygen); or the evolution of oxygen-carrying proteins which significantly increase capacitance of oxygen in body fluids.
4. The pre-Ediacarian metazoans were pre-adapted for the 'oxygen revolution' because they had inherited the tricarboxylic cycle and a defence against oxygen toxicity (Cu/Zn superoxide dismutase) from their remote ancestors. They had also acquired at least one gene for the structural protein collagen in order to become multicellular.
5. The first sizeable metazoans were two-dimensional animals that obtained oxygen by simple diffusion through the body wall. They became sizeable by filling their interiors with metabolically inert materials such as mesogloea or coelomic fluids. Animals with a lower ratio of surface area to volume became larger as the oxygen tension in the ocean climbed towards modern values.
6. The first active, burrowing organisms maintained a close connection with overlying sea water. Complex burrow systems would not have been constructed until the oxygen content of sea water had increased significantly.
7. Because mineral exoskeletons reduce the area available for epithelial respiration, they should appear relatively late and place more constraints on body size than would contemporaneous endoskeletons.

The best support for this model comes from the order of appearance and the geometry of different kinds of metazoans in the Ediacarian and Cambrian faunas. For example, many living 'lower invertebrates' require about 0.1 ml $O_2.g^{-1}h^{-1}$, although the amount varies with temperature and the size of the animal (Figure 4.8A). If the animal has no circulatory system oxygen must enter the body by diffusion and so it is possible to calculate the maximum distance that could be adequately supplied by diffusion for different oxygen tensions in the external medium (Raff and Raff, 1970; Alexander, 1971). Such calculations show that the maximum half-thickness or radius of an invertebrate relying solely on epithelial diffusion in a low-oxygen environment (say 0.1 PAL O_2) is about 0.2 mm. Because this value is increased by only 10 as the O_2 content of the atmosphere is raised by an order of magnitude, it is pretty clear that the maximum size of animals that rely on epithelial diffusion is severely limited even under present conditions unless the animals resort to largely anaerobic metabolism (Alexander, 1971; Tielens, van den Heuvel and van den Bergh, 1984) or adopt the strategy used by the Cnidaria. Cnidarians obtain their oxygen by epithelial diffusion and their mesogloea can be regarded as a 'metabolically inert, low density skeleton on which an actively metabolising thin sheet of tissue is spread' (Raff and Raff, 1970, p. 1004). The body wall of cnidarian is formed of only two layers of cells separated by the mesogloea and so cnidarians can inhabit extremely low-oxygen environments (Tunnicliffe, 1981).

With animals that use blood or coelomic fluids to distribute the oxygen within their bodies it is another matter (Figure 4.8B; Alexander, 1971; Runnegar, 1982a). The maximum thickness, 2s, of a sheet-like animal is given by:

$$s < K(pe - pb)/md$$

where K is the permeability constant for the diffusion of O_2 in tissues (about 8×10^{-4} $ml.g^{-1}h^{-1}$; Brown, 1984), pe is the oxygen tension in the external medium, pb the average oxygen tension in the blood, m the rate of oxygen consumption, and d the distance from the outer surface of the body to the most superficial blood vessels (about $1-3 \times 10^{-3}$ cm in molluscs and annelids, less in smaller animals).

The relationship may be used to estimate the oxygen requirements of the Ediacarian 'worm' *Dickinsonia*. The two Australian specimens of *Dickinsonia* shown in Figure 4.2 had approximately equal volumes so their relative thicknesses are given by the ratio of their areas in plan view (about $140/40$ $cm^2 = 3.5$). As many other specimens of *Dickinsonia* were folded prior to fossilisation (Figure 4.3E) it is unlikely that even large individuals were more than a few millimetres in thickness when the animals were fully expanded. Thus the volume of each of the two Australian animals (Figure 4.2) was probably about $140 \times 0.1 = 40 \times 0.35 = 14$ cm^3. This gives a body weight of about 14 g because the density of wet animal tissues is very nearly 1; it may be seen from Figure 4.8A that a 14 g specimen would have required about 0.04 ml O_2 $g^{-1}h^{-1}$ (or somewhat less if the animal were thicker).

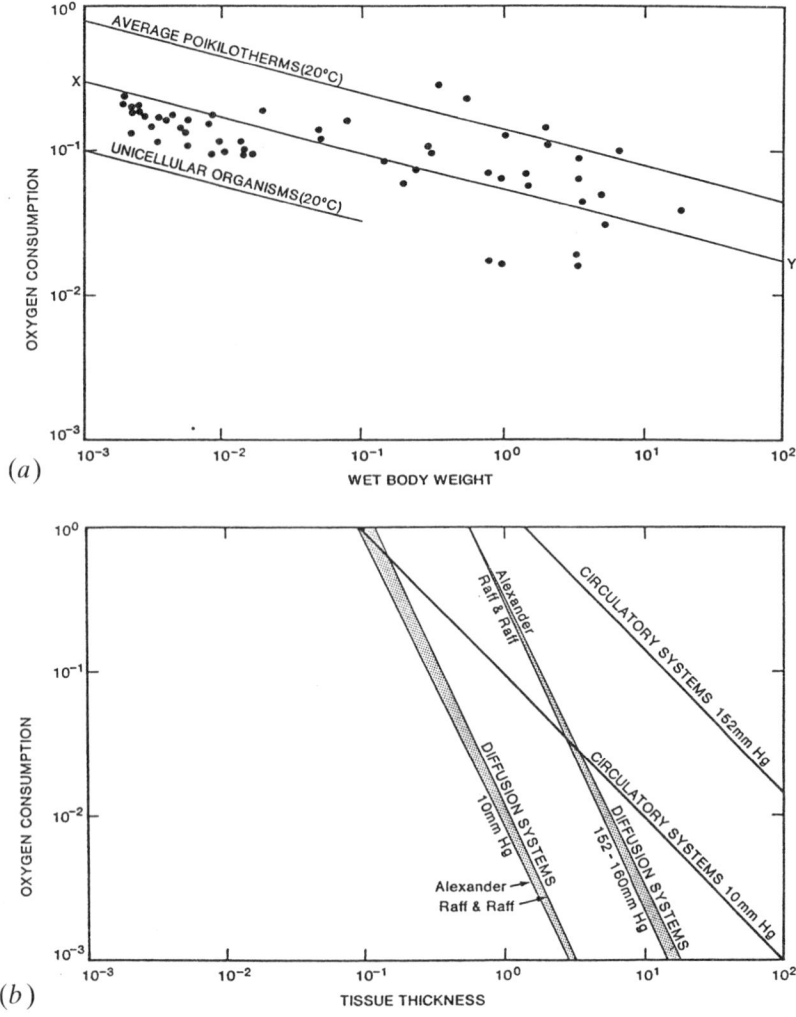

Figure 4.8 (*a*) Measured oxygen consumption (ml.g^{-1}.h^{-1}) per unit body weight (g) for a variety of living aquatic worms. The oxygen requirements of *Dickinsonia* are assumed to be shown by the line X–Y, which lies parallel to the average oxygen requirements of many living cold-blooded animals and within the range of living flatworms (left cluster of points) and living annelids (other points). (*b*) Calculated maximum possible tissue thicknesses (mm) at different oxygen tensions for animals without gills that rely on simple diffusion or have blood to distribute the oxygen. See Raff and Raff (1970) and Alexander (1971) for details. Both figures from Runnegar (1982a), republished with permission.

Using this value in equation (1) and making pe and pb equal to 0.1 PAL O2 and 0.33 pe respectively, it can be shown that if *Dickinsonia* used blood to circulate O_2 within its body, an animal about 15 cm in length and up to 2 mm in thickness when fully expanded could have survived in sea water containing only about a tenth of the present amount of dissolved oxygen (Runnegar, 1982a). On the other hand, similar calculations show that co-existing cylindrical worms would have been limited to a maximum diameter of about 3-4 mm and equidimensional animals to a size less than approximately 5 mm (Runnegar, 1982c). These estimates fit fairly well with the trace fossils found in Ediacarian strata; the trail shown in Figure 4.7 is about 4 mm in width and was obviously made by a cylindrical or equidimensional animal; some wider structures described previously from strata of Ediacarian age may not be trace fossils.

Most of the characteristic features of the Ediacarian faunas - the predominance of cnidarians and cnidarian-like animals, the geometry of animals such as *Dickinsonia* and *Spriggina*, the absence of burrows and sizeable cylindrical or equidimensional animals, and the absence of mineral skeletons - are features to be expected in low-oxygen environments (Rhoads and Morse, 1971; Tunnicliffe, 1981; Thompson, Mullins, Newton and Vercoutere, 1985). Judging from the modern analogues and from the respiratory physiology of living organisms, the evolution of the Ediacarian fauna occurred when the oxygen tension in sea water rose to a sizeable fraction of 0.1 PAL. As Glaessner (1984, p. 53) noted, the UV flux at sea level could not have been harmful if hydrozoans such as *Ovatoscutum* (Figure 4.5) formed part of the permanent surface fauna of the oceans. The calculations of Blake and Carver (1977) and Kasting and Donahue (1980) indicate that the present mid-latitude ozone column density (about 10^{19} cm^{-2}) was attained when the O_2 content of the atmosphere reached between about 10^{-2} and 0.5×10^{-1} PAL (in the range 0.05-0.25 ml l^{-1} dissolved O_2). An ozone column density near the modern value would seem to be necessary for a biologically effective UV shield because even the present UV flux excludes many bottom-dwelling reef organisms from the uppermost 5m of tropical seas (Jokiel, 1980), and because the biologically damaging dose of UV radiation is thought to increase by a factor of two as the ozone layer is depleted (Gerstl, Zardecki and Wiser, 1981). Thus a value of about 0.1 PAL O_2 during the Ediacarian period would have been sufficient to provide an adequate UV shield and also to have allowed the animals observed in the fossil record to respire. It would have been ample for the oxidation of Fe^{2+} in the well-agitated nearshore environments in which the animals lived and so the sediments formed in those settings should not differ in their oxidation states from ones deposited in comparable modern environments. However, an atmosphere containing only a tenth of the present amount of oxygen would be fatal to humans and most other advanced animals; physiologically fit humans just survive at the top of Everest where the partial pressure of O_2 is a third of the value at sea level.

By the beginning of the Cambrian the oxygen content of the atmosphere appears to have risen significantly, perhaps to about 0.2 PAL (Runnegar, 1982c). Many different kinds of organisms suddenly developed mineral skeletons at this time but it was those with large and permeable surfaces such as archaeocyaths and trilobites (Jell, 1978) that diversified first. All barriers were finally removed when the oxygen content of the atmosphere reached about a third of the present amount, because at this level even equidimensional animals such as molluscs and brachiopods could grow large.

The great diversity of respiratory proteins in living molluscs, arthropods, annelids and other invertebrate phyla suggests that at least some of these oxygen-binding compounds were not inherited from a remote common ancestor but instead evolved separately during the radiation of the Metazoa. It may be no coincidence that the oldest sizeable animal fossils postdate the late Precambrian world-wide glaciation. The lower global temperatures that must have accompanied that event may have triggered respiratory evolution - and hence larger body size - both by depressing metabolic rates and by increasing the solubility of oxygen in body fluids. The invasion of soft substrates and the formation of mineral skeletons may therefore have been delayed until this respiratory evolution had occurred and until the oxygen content of the atmosphere had increased significantly.

Summary

Although the oldest animal fossils are of late Precambrian age (Ediacarian; 600-550 million years ago), indirect evidence points to an unobserved Precambrian history of some hundreds of millions of years. One explanation of this apparent paradox is that metazoans were forced to remain too small to be fossilised for hundreds of millions of years because the low oxygen content of the atmosphere and oceans prohibited growth to larger sizes. Once the oxygen content of the atmosphere approached about a tenth of the present amount, large two-dimensional animals became possible. These were cnidarians, cnidarian-like animals or sheet-like 'worms', and they constitute what is now known as the 'Ediacarian fauna'. It was the rise of large, muscular and/or mineralised animals that marks the end of the Precambrian and the beginning of the Phanerozoic, and the best explanation for their appearance may be a significant increase in the oxygen content of the atmosphere.

References

Alexander, R.M. (1971) *Size and Shape* (Edward Arnold, London)
Anderson, M.M. and Conway Morris, S. (1982) 'A Review, with Descriptions of Four Unusual Forms, of the Soft-bodied Fauna of the Conception and St John's Groups (Late-Precambrian), Avalon Penninsula, Newfoundland', in B. Mamet and M.J. Copeland (eds.), *Third North American Paleontological Convention Proceedings, Vol. 1* (University of Montreal and Geological Survey of Canada, Ottawa), 1-8
Bayer, F.M., Boschma, H., Harrington, H.J., Hill, D., Hyman, L.H., Lecompte, M., Montanaro-Gallitelli, M., Moore, R.C., Stumm, E.C. and Wells, J.W. (1956)

Treatise on Invertebrate Paleontology, Part F, Colenterata (Geological Society of America and University of Kansas, Lawrence)

Bender, W., Akam, M., Karch, F., Beachy, P.A., Peifer, M., Spierer, P., Lewis, E.B. and Hogness, D.S. (1983) 'Molecular Genetics of the Bithorax Complex in *Drosophila melanogaster*' *Science* **221**, 23-29

Berkner, L.V. and Marshall, L.C. (1965) 'On the Origin and Rise of Oxygen Concentration in the Earth's Atmosphere', *Journal of the Atmospheric Sciences* **22**, 225-261

Blake, A.J. and Carver, J.H. (1977) 'The Evolutionary Role of Atmospheric Ozone', *Journal of the Atmospheric Sciences* **34**, 720-728

Brown, A.C. (1984) 'Oxygen Diffusion in to the Foot of the Whelk *Bullia digitalis* (Dillwyn) and its Possible Significance in Respiration', *Journal of Experimental Marine Biology and Ecology* **79**, 1-7

Brunet, P.C.J. (1967) 'Sclerotins', *Endeavour* **26**, 68-74

Carver, J.H. (1981) 'Prebiotic Atmospheric Oxygen Levels', *Nature (London)* **292**, 136-138

Chapman, G. (1959) 'The Mesogloea of *Pelagia noctiluca*', *Quarterly Journal of Microscopical Science* **100**, 599-610

Cloud, P. (1972) 'A Working Model of the Primitive Earth', *American Journal of Science* **272**, 537-548

Cloud, P. and Glaessner, M.F. (1982) 'The Ediacarian Period and System: Metazoa Inherit the Earth', *Science* **218**, 783-792

Cisne, J.L. (1974) 'Trilobites and the Origin of Arthropods', *Science* **186**, 13-18

Clemmey, H. and Badham, N. (1982) 'Oxygen in the Precambrian Atmosphere: An Evaluation of the Geological Evidence', *Geology* **10**, 141-146

Conway Morris, S. (1979) 'Middle Cambrian Polychaetes from the Burgess Shale of British Columbia', *Philosophical Transactions of the Royal Society of London B* **285**, 227-274

Conway Morris, S. (1985) 'The Ediacaran Biota and Early Metazoan Evolution', *Geological Magazine* **122**, 77-81

Dimroth, E. and Kimberley, M.M. (1976) 'Precambrian atmospheric oxygen: Evidence in the Sedimentary Distributions of Carbon, Sulphur, Uranium and Iron', *Canadian Journal of Earth Sciences* **13**, 1161-1185

Dyson, I.A. (1985) 'Frond-like Fossils from the Base of the Late Precambrian Wilpena Group, South Australia', *Nature (London)* **318**, 283-285

Fedonkin, M.A. (1981) 'White Sea Biota of the Vendian', *Trudy Akademya Nauk SSSR* **342**, 1-100 [in Russian]

Fedonkin, M.A. (1985) 'Precambrian Metazoans: The Problems of Preservation, Systematics and Evolution', *Philosophical Transactions of the Royal Society of London B* **311**, 27-45

Foster, M.W. (1979) 'Soft-bodied Coelenterates in the Pennsylvanian of Illinois', in M.H. Nitecki (ed.), *Mazon Creek Fossils* (Academic Press, New York), pp. 191-267

Gerstl, S.A.W., Zardecki, A. and Wiser, H.L. (1981) 'Biologically Damaging Radiation Amplified by Ozone Depletions', *Nature (London)* **294**, 352-354

Glaessner, M.F. (1969) 'Trace Fossils from the Precambrian and Basal Cambrian', *Lethaia* **2**, 369-393

Glaessner, M.F. (1971) 'The Genus *Conomedusites* Glaessner and Wade and the Diversification of the Cnidaria', *Palaontologische Zeitschrift* **45**, 7-17

Glaessner, M.F. (1976) 'A New Genus of Late Precambrian Polychaete Worms from South Australia', *Transactions of the Royal Society of South Australia* **100**, 169-170

Glaessner, M.F. (1980) 'Parvancorina - An Arthropod from the Late Precambrian (Ediacarian) of South Australia', *Annalen des Naturhistorischen Museums in Wien* **83**, 83-90

Glaessner, M.R. (1984) *The Dawn of Animal Life* (Cambridge University Press, Cambridge)

Glaessner, M.F. and Wade, M. (1966) 'Late Precambrian Fossils From Ediacara, South Australia', *Palaeontology* **9**, 599-628

Glaessner, M.F. and Wade, M. (1971) 'Praecambridium - A Primitive Arthropod', *Lethaia* **4**, 71-77

Gould, S.J. (1984) 'The Ediacaran Experiment', *Natural History* **2/84**, 14-23

Gould, S.J. (1986) 'A Short Way to Big Ends', *Natural History* **1/86**, 18-28

Hantzschel, W. (1975) *Treatise on Invertebrate Paleontology*, 2nd Ed. (Geological Society of America and University of Kansas, Lawrence)

Holland H.D. (1984) *The Chemical Evolution of the Atmosphere and Oceans* (Princeton University Press, Princeton)

Jell, P.A. (1978) 'Trilobite Respiration and Genal Caeca', *Alcheringa* **2**, 251-260

Jenkins, R.J.F. (1981) 'The Concept of an "Ediacaran Period" and its Stratigraphic Significance in Australia', *Transactions of the Royal Society of South Australia* **105**, 179-194

Jenkins, R.J.F. (1983) 'Interpreting the Oldest Fossil Cnidarians', *Palaeontographica Americana* **54**, 95-104

Jenkins, R.J.F. (1984) 'Ediacaran Events: Boundary Relationships and Correlation of Key Sections, Especially in "Armorica"', *Geological Magazine* **121**, 635-643

Jenkins, R.J.F. (1985) 'The Enigmatic Ediacaran (Late PreCambrian) Genus *Rangea* and related forms', *Paleobiology* **11**, 336-355

Jenkins, R.J.F. and Gehling, J.G. (1978) 'A Review of the Frond-Like Fossils of the Ediacara Assemblage', *Records of the South Australian Museum* **17**, 347-359

Jenkins, R.J.F., Ford, C.H. and Gehling, J.G. (1983) 'The Ediacara Member of the Rawnsley Quartzite: The Context of the Ediacara Assemblage (Late Precambrian, Flinders Ranges)', *Journal of the Geological Society of Australia* **30**, 101-119

Jokiel, P.L. (1980) 'Solar Ultraviolet Radiation and Coral Reef Epifauna', *Science* **207**, 1069-1071

Kasting, J.F. and Donahue, T.M. (1980) 'The Evolution of Atmospheric Ozone', *Journal of Geophysical Research* **85**, 3255-3263

Lutz, R.A. and Rhoads, D.C. (1977) 'Anaerobiosis and a Theory of Growth Line Formation', *Science* **198**, 1222-1227

Margulis, L., Walker, J.C.G. and Rambler, M. (1976) 'Reassessment of Roles of Oxygen and Ultraviolet Light in Precambrian Evolution', *Nature (London)* **264**, 620-624

Nursall, J.R. (1959) 'Oxygen as a Prerequisite to the Origin of the Metazoa', *Nature (London)* **183**, 1170-1172

Oliver, W.A. (1983) '*Conchopeltis*: Its Affinities and Significance', *Palaeontographica Americana* **54**, 141-147

Raff, R.A. and Raff, E.C. (1970) 'Respiratory Mechanisms and the Metazoan Fossil Record', *Nature (London)* **228**, 1003-1005

Rhoads, D.C. and Morse, J.W. (1971) 'Evolutionary and Ecologic Significance of Oxygen-Deficient Marine Basins', *Lethaia* **4**, 413-428

Robbins, E.A., Porter, K.G. and Haberyan, K.A. (1985) 'Pellet Microfossils: Possible Evidence for Metazoan Life in Early Proterozoic Time', *Proceedings of the National Academy of Science USA* **82**, 5809-5813

Runnegar, B. (1982a) 'Oxygen Requirements, Biology and Phylogenetic Significance of the Late Precambrian Worm *Dickinsonia*, and the Evolution of the Burrowing Habit', *Alcheringa* **6**, 223-239

Runnegar, B. (1982b) 'A Molecular-Clock Date for the Origin of the Animal Phyla', *Lethaia* 15, 199-205
Runnegar, B. (1982c) 'The Cambrian Explosion: Animals or Fossils?', *Journal of the Geological Society of Australia* 29, 395-411
Runnegar, B. (1985) 'Collagen Gene Construction and Evolution', *Journal of Molecular Evolution* 22, 141-149
Runnegar, B. (1986) 'Molecular Palaeontology', *Palaeontology* 29, 1-24
Schafer, W. (1941) 'Fossilisations - Bedingungen von Quallen und Laichem', *Senckenbergiana* 23, 189-216
Seilacher, A. (1953) 'Studien zur Palichnologie. II. Die Fossilen Ruhespuren (Cubichnia)', *Neues Jahrbuch fur Geologie und Palaeontologies, Abhandlungen* 98, 87-124
Seilacher, A. (1984) 'Late Precambrian and Early Cambrian Metazoa: Preservational or Real Extinctions?', in H.D. Holland and A.F. Trendall (eds.), *Patterns of Change in Earth Evolution*, Dahlem Konferenzen 1984 (Springer-Verlag), pp. 159-168
Thompson, J.B., Mullins, H.T., Newton, C.R. and Vercoutere, T.L. (1985) Alternative Biofacies Model for Dysaerobic Communities', *Lethaia* 18, 167-179
Tielens, A.G.M., van den Heuvel, J.M. and van den Bergh, S.G. (1984) The Energy Metabolism of *Fasciola hepatica* During its Development in the Final Host', *Molecular and Biochemical Parasitology* 13, 301-307
Towe, K.M. (1970) 'Oxygen-Collagen Priority and the Early Metazoan Fossil Record', *Proceedings of the National Academy of Sciences USA* 65, 781-788
Towe, K.M. (1978) 'Early Precambrian Oxygen: A Case Against Photosynthesis', *Nature (London)* 274, 657-661
Towe, K.M. (1981) 'Biochemical Keys to the Emergence of Complex Life', in J. Billingham (ed.), *Life in the Universe* (MIT Press, Cambridge, Massachusetts), pp. 297-306
Tunnicliffe, V. (1981) 'High Species Diversity and Abundance of the Epibenthic Community in an Oxygen-Deficient Basin', *Nature (London)* 294, 354-356
Wade, M. (1968) 'Preservation of Soft-Bodied Animals in Precambrian Sandstones at Ediacara, South Australia', *Lethaia* 1, 238-267
Wade, M. (1972a) 'Hydrozoa and Scyphozoa and Other Medusoids from the Precambrian Ediacara, South Australia', *Palaeontology* 15, 197-225
Wade, M. (1972b) '*Dickinsonia:* Polychaete Worms from the Late Precambrian Ediacara Fauna, South Australia', *Memoirs of the Queensland Museum* 16, 171-190
Walker, J.C.G., Klein, C., Schidlowski, M., Schopf, J.W., Stevenson, D.J. and Walter, M.R. (1983) 'Environmental Evolution of the Archean-Early Proterozoic Earth', in J.W.Schopf (ed.), *Earth's Earliest Biosphere* (Princeton University Press, Princeton,) pp. 260-290
Whittington, H.B. (1971) 'Redescription of *Marrella splendens* (Trilobitoidea) from the Burgess Shale, Middle Cambrian, British Columbia', *Bulletin of the Geological Survey of Canada* 209, 1-24
Yochelson, E.L., Sturmer, W. and Stanley, G.D. (1983) '*Plectodiscus discoideus* (Rauff): A Redescription of a Chondrophorine from the Early Devonian Hunsruck Slate, West Germany', *Palaeontologische Zeitschrift*, 57, 39-68

Chapter 5

Fumarate Reductase and the Evolution of Electron Transport Systems

C. A. Behm

Introduction

A fundamental and important process in the anaerobic energy metabolism of many microorganisms and groups of lower Metazoa is the reduction of fumarate to succinate by a membrane-bound electron transfer system known as the fumarate reductase complex. This reaction is common in both obligate and facultative anaerobes. In many of these organisms the reduction of fumarate appears to be accompanied by energy conservation in the form of proton motive force-driven phosphorylation of ADP. Fumarate thus serves as a terminal electron acceptor and permits the synthesis of an additional mole of ATP, an important consideration for organisms living under anoxic conditions.

This paper is concerned with the evolution, molecular nature and role of fumarate reductase. It is divided into four main parts – a brief discussion of thermodynamics and energy conservation in living organisms, followed by an examination of ideas about the evolution of the energy conservation mechanisms found in living organisms, an introduction to the molecular nature of the fumarate reductase complex as it exists in bacteria and a brief discussion of what is known of the molecular structure and function of the complex in lower metazoan organisms.

Energy conservation and bioenergetics: thermodynamic considerations

All physical and chemical processes require the application, transfer or transformation of energy. The processes of life are no exception: they can be described in terms of the laws of thermodynamics, that is in terms of the exchanges of energy, in one form or another, that occur during a specific reaction or set of reactions or activities. For the purposes of this discussion, 'energy' can be simply defined as the capacity to do work. Living organisms carry out a number of different types of work, including mechanical, osmotic, chemical and electrical work, all of which are essential to maintain life. All living organisms therefore require energy to allow them to carry out their biological functions.

The precise forms of energy available to living organisms are limited and, in general, 'usable' energy is supplied by chemical energy; other forms of energy, such as kinetic or electrical energy cannot be harnessed by living organisms. Chemical energy is converted by living organisms into the other types of energy necessary for the dynamic functions of life.

Chemical energy is stored in the reactants of chemical reactions – the substrates of enzyme-catalysed reactions – and is interconverted or transformed in living organisms without, under normal circumstances, significant loss of energy as heat. Under standard physiological conditions, the standard free energy change ($\Delta G^{o'}$) of a chemical reaction is described by the relationship

$$\Delta G^{o'} = -RT \ln K$$

where R is the gas constant, T is the absolute temperature and K the equilibrium constant of the reaction. Reactions with a negative value of $\Delta G^{o'}$ proceed spontaneously and are described as exergonic, whereas those with a positive value of $\Delta G^{o'}$ are endergonic and do not proceed spontaneously under physiological conditions without an additional driving force. Under relatively stable physiological conditions, it can be shown that the tendency for a reaction to occur spontaneously is related to the equilibrium constant. This relationship is illustrated in Table 5.1, which shows that a reaction that is freely reversible (i.e. $K = \sim 1$) has a very small free energy change, whereas large changes are associated with reactions that are substantially irreversible.

Enzymes play a central role in catalysing the chemical reactions in living organisms. They permit reactions (exergonic or endergonic) that might otherwise be

Table 5.1 The quantitative relationship between the standard free energy change ($\Delta G^{o'}$) and the equilibrium constant (K) of a reaction at 25°, pH7.0 (recalculated from Lehninger, 1971).

K	$\Delta G^{o'}$ (kJ/mol)
0.001	+17.12
0.01	+11.43
0.1	+5.69
1.0	0
10.0	−5.69
100.0	−11.43
1000.0	−17.12

extremely slow under physiological conditions to take place rapidly, by effectively lowering the activation energy of the reactants. It is important to note that enzymes do not alter the equilibrium of a chemical reaction. Thus, the thermodynamic considerations discussed above are not affected by the fact that enzymes catalyse the reaction or pathway. However, the free energy change of a specific reaction *in vivo* may be altered by the local cellular environment (e.g. pH, concentrations of reactants) and may differ from the quoted figures, which are determined under standardised conditions *in vitro*. In addition, the presence of specific cellular structures, such as membranes or protein assemblies, which may be solid or provide a non-aqueous environment, can drastically alter the reaction conditions *in vivo* such that the local free energy change is very different from that determined for the reactants in solution.

The ultimate source of energy for living organisms is sunlight, which is utilised by photosynthetic organisms to provide the energy and reducing power necessary to synthesise substrates suitable for oxidation, such as glucose. Heterotrophic organisms gain the energy they require by releasing it from organic foodstuffs. The catabolism of foods usually occurs by a process of oxidation; these pathways are exergonic: for example, the complete oxidation of glucose to CO_2 and water releases 2872 kJ/mol. The chemical energy of these fuels is not all lost as heat: some is conserved by coupling the exergonic reactions with the synthesis of small molecules, such as ATP or CoA esters, with a high free energy of hydrolysis of the functional groups. Some values for $\Delta G^{o'}$ in phosphate and CoA compounds found in cells are given in Table 5.2. These molecules have the dual function in cells of both

Table 5.2 Standard free energy of hydrolysis ($\Delta G^{o'}$) of 'high energy' phosphate compounds and CoA thioesters at 25°, pH7.0 (recalculated from Lehninger, 1971).

	$\Delta G^{o'}$ kJ/mol
Phosphoenolpyruvate	−61.96
1,3-diphosphoglycerate	−49.40
Phosphocreatine	−43.12
Acetyl phosphate	−42.29
SuccinylCoA	−33.49
AcetylCoA	−31.40
ATP	−30.56
Glucose 1-phosphate	−20.93
Fructose 6-phosphate	−15.91
Glucose 6-phosphate	−13.82
3-phosphoglycerate	−10.05
Glycerol 3-phosphate	−9.21

temporarily storing and then transferring energy trapped from dissimilation of foodstuffs. In this way, energy conserved in exergonic (usually catabolic) reactions is used to supply the driving force for endergonic (usually anabolic) reactions. Phosphagens such as phosphocreatine are used in some tissues for temporary storage of energy. Phosphocreatine is the most common phosphagen in vertebrates. In contrast, many invertebrates utilise different phosphagens, such as phosphoarginine, phospholombricine and phosphotaurocyamine; the value of $\Delta G^{o'}$ for these compounds would be of the same order of magnitude as that for phosphocreatine.

The evolution of energy transduction in heterotrophic organisms

The early evolution of life took place under anoxic conditions in a non-oxidising environment. It is postulated that organic compounds were formed in shallow seas as a result of physical and chemical processes associated with volcanic activity, intense solar irradiation, and electrical discharges in the atmosphere, acting upon free and dissolved gases such as CO_2, CH_4, NH_3, H_2. These organic molecules would have both supplied the raw materials (e.g. purine and pyrimidine bases, sugars, simple lipids, amino acids) for the development of catalytic molecules and membranous structures and, at the same time, also supplied the substrates for the catalytic molecules to act upon.

It is surmised, therefore, that the first true organisms, which appeared about 3.5-4 billion years before present, were chemotrophs, relying for energy supplies and biosynthetic substrates on the pool of organic compounds available in their environments. The development of photosynthetic processes, and hence autotrophic organisms, probably occurred later, around 3 billion years before present, from which time life was no longer dependent on exogenous organic molecules formed by non-biological processes.

The early chemotrophic organisms probably gained their energy by taking up molecules (e.g. sugars) from the environment and catabolising them to a lower energy state, during the course of which a proportion of the chemical energy available in the substrates was conserved in a form useful to the organism. This primitive bioenergetic process, *anaerobic fermentation,* is retained in all extant organisms as the pathway of glycolysis (Broda, 1975). All anaerobic fermentation processes utilise ATP for energy conservation and NADH for electron (as H^-) transfer; these small molecules perform the same roles in nearly all metabolic processes in living organisms today, which indicates that they appeared early in bioenergetic evolution.

The yield of ATP from the anaerobic fermentation of organic substrates is low. Fermentation of glucose to lactate via the glycolytic pathway, for example, yields only 2 moles of ATP/mole of glucose and much of the chemical energy is lost as heat or retained in the end-products. This yield of ATP is very low compared with the yield of 38 moles/mole glucose when glucose is fully oxidised to CO_2 and water via the tricarboxylic acid cycle and electron transport system. There are two major reasons for the low yield of ATP in glycolysis and they account for the major limitations of fermentation as an energy-yielding pathway in cells (Hall, 1971).

First, the conservation of chemical energy in the form of ATP in anaerobic fermentations requires the presence in the catabolic pathway of a phosphorylated intermediate with a sufficiently high negative value of $\Delta G^{o'}$ to permit transfer of the phosphoryl group to ADP. Compounds suitable for this purpose are listed in Table 2; they are the phosphorylated molecules located above ATP in the list. The conservation of energy by this mechanism, known as substrate-linked phosphorylation, is therefore exclusively a phosphoroclastic process. This requirement for a suitable phosphorylated intermediate means that many potential fuels are not available for fermentation processes because, for chemical and thermodynamic reasons, appropriate phosphorylated intermediates are not formed during catabolism.

Second, there is no net change of oxidation-reduction (redox) level of the substrates and products of fermentation pathways. Large quantities of energy can be liberated from organic molecules by oxidising them, but this process is not available in fermentation because, although dehydrogenation reactions may occur in the pathway, the redox differential for these is not great and the reduced electron acceptor must be reoxidised, by a later reaction within the pathway, for net carbon flux through the pathway to occur. For example, in glycolysis the reduced cofactor, NADH, produced at the reaction catalysed by glyceraldehyde 3-phosphate dehydrogenase, must be reoxidised by the lactate dehydrogenase-catalysed reaction in the classical pathway. Under the non-oxidising environmental conditions that existed during the early evolution of life, exogenous terminal electron acceptors would not have been available for this purpose.

A great advance was made during the evolution of bioenergetic processes when the reoxidation of reduced cofactors became independent of the pathway of oxidation of the substrate, i.e. when organisms developed the ability to utilise exogenous terminal electron acceptors. This permitted net oxidation of substrates and eventually bypassed the mechanistic limitations of substrate-linked phosphorylation. Initially, inorganic electron acceptors were probably used. These would have become available as the early atmosphere became more oxidising and as a result of the activities of photosynthetic bacteria; they include CO_2, reduced to CH_4, SO_4^-, reduced to S^{2-}, and NO_4^-, reduced to a variety of products including nitrite and nitrogen gas. Terminal electron acceptors such as these are utilised today by certain groups of bacteria but this process is not reported for eukaryotic organisms. At a later stage in evolution, organisms became capable of acquiring or synthesising and utilising organic terminal electron acceptors such as acetate (reduced to butyrate) or fumarate (reduced to succinate), which conferred a number of advantages and which remains common in strictly and facultatively anaerobic bacteria and eukaryotes today.

In the classical glycolytic pathway, pyruvate acts as the terminal electron acceptor. This is disadvantageous to organisms that are unable to utilise alternative electron acceptors because pyruvate is an essential substrate for biosynthetic processes in cells. Therefore, the ability to use alternative electron acceptors would make available a portion of the pyruvate pool for biosynthesis, conferring a significant selective advantage on organisms capable of doing this (Gest, 1980). This concept is summarised in Figure 5.1, where the synthesis of fumarate provides an alternative oxidant to pyruvate in the branched pathway. The incorporation of CO_2 into phosphoenolpyruvate is an essential feature of this pathway and was an important evolutionary prerequisite for the synthesis of fumarate. Fumarate is a suitable metabolite for the role of electron acceptor in this pathway, having a relatively positive redox potential in comparison with other potential acceptors (Table 5.3).

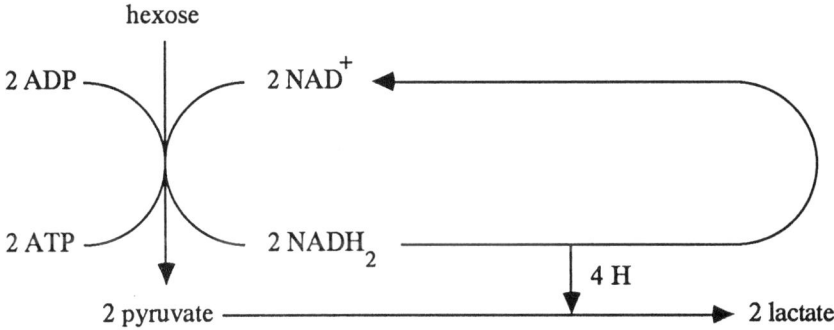

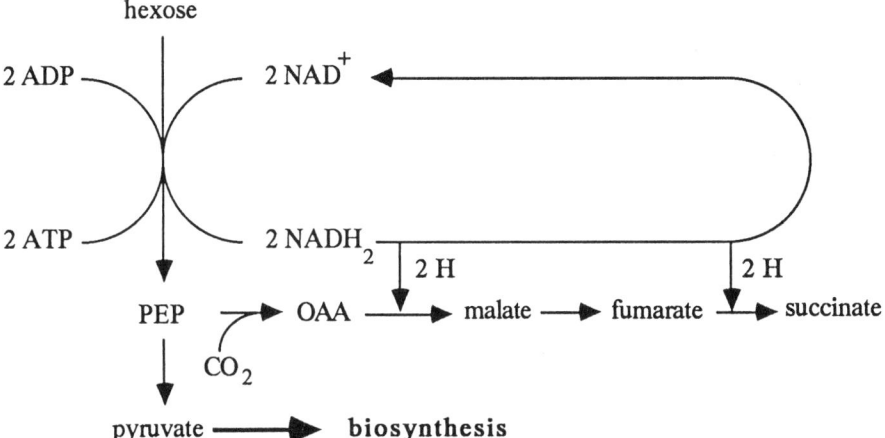

Figure 5.1 Scheme showing (*top*) redox balance in hexose fermentation and (*below*) redox balance when fumarate is synthesised (modified from Gest, 1980).

(Fumarate acts as an alternative electron acceptor to pyruvate, thus liberating a portion of the pyruvate pool for biosynthesis.)

Table 5.3 Standard redox potentials (E_o') at pH 7.0.

	E_o' (mV)	
H_2/H^+	−420	(1)
formate/HCO_3^-	−416	(1)
NADH/NAD^+	−320	(1)
lactate/pyruvate	−197	(1)
malate/oxaloacetate	−172	(1)
menaquinone ox/red	−74	(2)
rhodoquinone ox/red	−63	(3)
cytochrome b_{558} ox/red	−34	(4)
fumarate/succinate	+30	(1)
ubiquinone ox/red	+40	(2)
O_2/H_2O	+815	(1)

References: (1) Kröger, 1978; (2) Nicholls, 1982; (3) Erabi *et al.*, 1975; (4) S. Takamiya *et al.*, unpublished results quoted by Kita *et al.*, 1988a.

The redox differential between the NADH/NAD^+ couple and the fumarate/succinate couple is 350mV, which, for a 2-electron reduction, is sufficient to support the synthesis of 1 mole of ATP from ADP + P_i, which requires a differential of approximately 250mV (Thauer, Jungermann and Dekker, 1977). Hence, an additional selective advantage would have been the potential to conserve more energy from the hexose substrate in those organisms capable of controlling the electron transfer in a suitable way (see below).

The development of the pathway for the reduction of fumarate is considered to have provided the earliest evolutionary link between fermentation and anaerobic respiration (Jones 1985). The ability to synthesise succinate, the end product, was important for the development of biosynthetic capacities in the early organisms as it would have provided cells with the substrate for synthesising succinylCoA. SuccinylCoA is essential for *de novo* synthesis of certain amino acids (e.g. methionine) and is also an essential requirement for *de novo* synthesis of porphyrins, which would have been a necessary prerequisite for the development of haemoproteins, in particular the cytochromes and the chlorophylls. As the supplies of exogenous substrates and electron acceptors available to the early life forms declined, there would have been great selection pressure towards development of light-dependent electron transfer systems as sources of usable energy. After the capacity to synthesise porphyrins had developed, a change in the metal ion from Fe to Mg would have resulted in a more photochemically reactive molecule capable of trapping light energy and permitting the development of early forms of photosynthesis.

The new pathway for fumarate reduction required a new electron transport component capable of transferring electrons from NADH to the alternative acceptor. It is thought that soluble flavoproteins probably fulfilled this role initially, and that there was no energy conservation associated with these reductions. The reduction of fumarate, described by the reaction

$$\text{fumarate}^{2-} + \text{NADH} + \text{H}^+ \rightarrow \text{succinate}^{2-} + \text{NAD}^+ \quad (\Delta G^{o\prime} = -67.70 \text{ kJ/mol})$$

is theoretically sufficiently exergonic to support the synthesis of a mole of ATP (requiring 50.24 kJ/mol under cellular conditions) (Gnaiger, 1977; Kröger, 1978). The development of energy conservation associated with this reaction, i.e. redox-linked conservation or oxidative phosphorylation, by the mechanism of chemiosmotic coupling as it occurs in many modern organisms (Mitchell, 1970), required three separate processes to come together (Wilson and Lin, 1980). These were (i) association of the fumarate reductase enzyme with the cell membrane and a sequential transfer of electrons between carriers of suitable intermediate redox potential within the membrane so that the process could be stabilised and better controlled, (ii) the development of molecules in the cell membrane capable of catalysing redox-driven vectorial transfer of protons to the exterior, resulting in the formation of a transmembrane electrochemical gradient for protons, and (iii) the development of a reversible proton-translocating ATPase complex in the cell membrane.

The accumulation of intracellular protons during fermentation is a problem that has been solved by cells by developing mechanisms for export of protons from the cell (Hochachka and Mommsen, 1983). The major mechanisms include the export of protonated acid end-products, membrane permeases catalysing antiport exchange of protons with other molecules, intracellular buffering, and membrane H^+-ATPases which export protons at the expense of ATP hydrolysis. The development of proton pumping associated with electron transport, i.e. redox-driven proton translocation, would have spared ATP hydrolysis and hence provided a selective advantage to cells possessing this ability. Similarly, cells with a reversible membrane H^+-ATPase capable of responding to a transmembrane pH gradient by acting in reverse (i.e. importing protons and synthesising ATP once a critical potential difference of 200-240mV had been established) would have had a significant selective advantage. Additional electron transport and proton-translocating molecules that became included in the prokaryotic membrane included quinones, iron-sulphur proteins and cytochromes, which have redox potentials intermediate between the electron donor(s) and the acceptor, fumarate. This scheme for the proposed evolution of anaerobic respiration is outlined in Figure 5.2. A number of authors (Gest, 1980; Wilson and

96   The Evolution of Electron Transport Systems

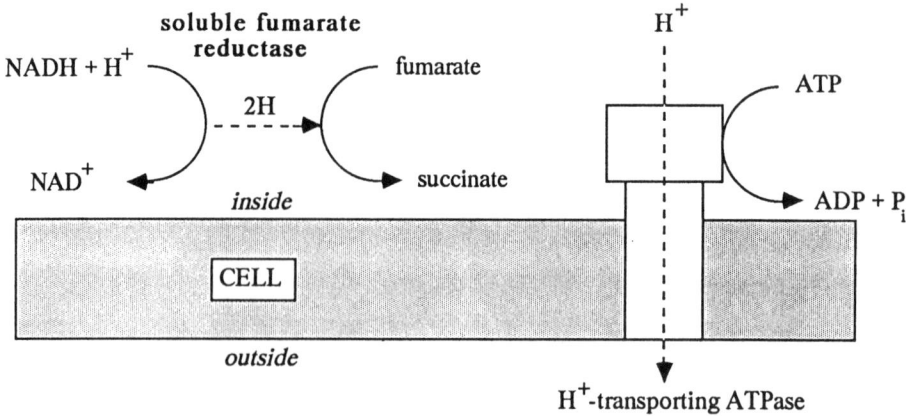

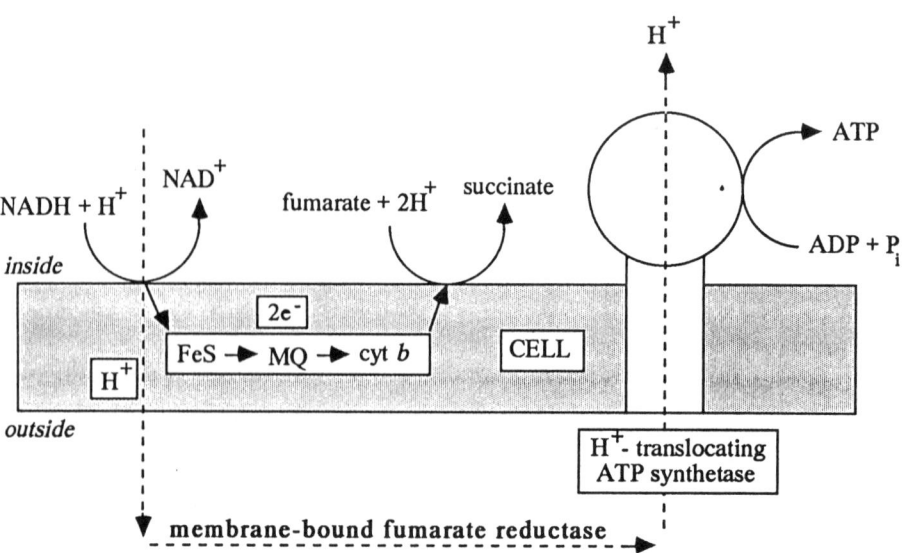

Figure 5.2 A scheme for the evolution of anaerobic respiration. Above, soluble fumarate reductase and proton-exporting ATPase in early cells. Below, fumarate reductase associated with the cell membrane, together with proton export; proton import by a proton-transporting ATP synthetase.

Lin, 1980) have pointed out that electron transfer sequences similar to that outlined for fumarate reductase are found in anaerobic photosynthetic bacteria where, in association with chlorophylls and a primary electron acceptor, they catalyse the process of photophosphorylation. It has been argued (Gest, 1980) that the development of Mg-porphyrins by modification of Fe-porphyrins and the association of these in the prokaryotic membrane with an electron transfer-phosphorylation system like that of fumarate reductase could have led to the development of the first photosynthetic processes in living organisms.

It appears likely that aerobic, cytochrome-dependent respiration, utilising O_2 as the terminal electron acceptor when it became abundant, evolved from the fermentation and 'pre-respiration' processes described above, via a modification of photosynthetic electron transport in the ancestors of the purple non-sulphur bacteria, the putative ancestors of mitochondria (Broda, 1975; Broda and Peschek, 1979; Gest, 1980; Dickerson, 1980). As O_2 concentrations increased, there would have been selection pressure for the combination of existing pathways to form the oxidative TCA cycle as a means of supplying electrons to the electron transport system to reduce (and hence detoxify) oxygen. Succinate dehydrogenase probably evolved from fumarate reductase under this selection pressure (Jones, 1985; Weitzman, 1985).

The molecular nature of fumarate reductase

The reduction of fumarate, catalysed by a distinct fumarate reductase, was first recognised in the anaerobic bacterium *Veillonella alcalescens* (formerly *Micrococcus lactilyticus*) (Peck, Smith and Gest, 1957). In eukaryotes it catalyses the reversible reduction of fumarate to succinate. In eukaryotes the electron donor is normally NADH; the enzyme is inactive with NADPH. The reductase activity appears to be favoured by the more favourable kinetic properties of the enzyme acting in this direction. The activity and kinetic properties of the enzyme have been examined in a wide variety of organisms but, to date, the enzyme has been purified from only a small number. A representative list of K_m values for fumarate reduction and succinate oxidation is given in Table 5.4. It should be noted that most of these determinations have been made on impure preparations and that different assay methods for the enzyme result in quite different activities. For example, the spectrophotometric assay significantly underestimates activity in comparison with the radioisotope assay (Fry and Brazeley, 1984), and the presence of O_2 to act as alternative electron acceptor may interfere in some cases. It has been customary to calculate SDH/FR ratios for the reversible reaction, as a guide to the normal direction of catalysis *in vivo,* but the reliability of such a ratio is clearly compromised by the limitations noted above in the enzyme assays and must therefore be interpreted with extreme caution. Competitive inhibition of fumarate reductase activity by succinate has been widely reported. The molecular properties of enzymes that have been purified are summarised in Table 5.5.

Table 5.4 Apparent K_m values for fumarate and succinate and SDH/FR activity ratios of fumarate reductase from different organisms.

		K_m: fumarate	succinate	SDH/FR
Bacteria	E. coli (anaerobic)	0.420 [1]	0.017 [1]	0.06 [2]
	Veillonella alcalescens	<0.3 [3]	5.3 [3]	0.03 [4]
				0.09 [5]
	Wolinella succinogenes	0.1 [6]	8.9 [6]	0.11 [7]
		0.35 [7]	20 [7]	
	Propionibacterium pentosaceum	0.7 [3]	2.2 [3]	3 [8]
	Desulfovibrio multispirans	2.5 [8]	N.A.	0
Yeasts	Saccharomyces cereviseae	0.21 [9]	N.A.	0
Protozoa	Trypanosoma brucei	0.016 [10]	–	–
Nematodes	Ascaris – muscle	3.09 [11]	0.067 [12]	0.05 [14]
	1.71 [12] 1.89 [13]	–		
	Ascaris – eggs	–	–	1.05 [11]
	Trichinella spiralis	2.1 [15]	1.82 [15]	1.95 [15]
	Ascaridia galli	0.31 [16]	1.35 [16]	–
	Nippostrongylus brasiliensis	0.27 [16]	0.50 1.47 [16]	–
Trematodes	Fasciola hepatica	–	–	2.3 [17]
	Paragonimus westermani	–	0.021 [13]	–
Annelids	Tubifex sp.	0.027 [18]	–	–
	Arenicola marina	0.025 [19]	–	–
Molluscs	Mytilus edulis	0.063 [20]	0.10 [20]	–
Mammals	Beef heart mitochondria 32 [11]	–	0.044 [13]	62 [3]
	Rat liver mitochondria	–	0.54 [16]	20 [11]

All K_m values are mM; N.A. = no activity; – = not reported
References: 1. Dickie and Weiner (1979); 2. Lara, 1959; 3. Singer, 1971; 4. Warringa et al., 1958; 5. Warringa and Giuditta, 1958; 6. Kenney and Kröger, 1977; 7. Unden et al., 1980; 8. He et al., 1986; 9. Muratsubaki and Katsume, 1982; 10. Turrens, 1987; 11. Kita et al., 1988a; 12. Comley and Wright, 1981; 13. Ma et al., 1987; 14. Oya and Kita, 1989; 15. Rodriguez-Caabeiro et al., 1985; 16. Fry and Brazeley, 1984; 17. Barrett, 1978; 18. Schöttler, 1977; 19. Schroff and Schöttler, 1977; 20. Holwerda and De Zwaan, 1980

Table 5.5 Molecular properties of purified fumarate reductase from different organisms.

Organism	Apparent Molecular Weight of Subunits (kDa)	Cytochrome	Flavin
E. coli	69, 27, 15, 13 (1)	b_{556} (2)	Covalent
W. succinogenes	79, 31, 25 (3)	b_{560} (-15mV) (4)	Covalent
B. subtilis	65, 28, 19 (5)	b_{560}	Covalent
A. suum	68, 26, 15, 13.5 (6)	b_{558} (-34mV)	Covalent
P. westermani	69, 27, 14.5, 12 (7)	b_{556}	Covalent
D. multispirans	45, 32, 30, 27 (8)	not reported	Non-covalent
S. cereviseae	60 (9)	not reported	Non-covalent

References: (1) Cole et al., 1985; (2) Kita et al., 1988a; (3) Unden et al., 1980; (4) Unden and Kröger, 1981; (5) Hederstedt et al., 1979; (6) Oya and Kita, 1989; (7) Ma et al., 1987; (8) He et al., 1986; (9) Muratsubaki and Katsume, 1982

(a) *The soluble enzyme*

Soluble fumarate reductase is found as a flavoprotein in a number of bacteria and yeasts. In general, the reaction catalysed by this enzyme is irreversible. The cytoplamic enzyme from anaerobically-grown cells of the yeast *Saccharomyces cereviseae* contains non-covalent FAD (Muratsubaki and Katsume, 1982). It consists of a single polypeptide of molecular weight 60kD, with no non-haem iron. The enzyme from the anaerobic bacterium *V. alcalescens* also contains non-covalent FAD and has no cytochrome *b* (Warringa, Smith, Giuditta and Singer, 1958; Wilson and Lin, 1980).

(b) *The membrane-bound enzyme*

The particulate fumarate reductase complex in all organisms examined is composed of a primary dehydrogenase (e.g. NADH or α-glycerophosphate dehydrogenase), a *b*-type cytochrome, a membrane-soluble quinone and the flavoenzyme fumarate reductase. The presence of the cytochrome appears to be essential for oxidative phosphorylation to take place, but fumarate reduction can occur in its absence (Haddock and Jones, 1977). In bacteria, menaquinone or desmethylmenaquinone are the functional quinone species, whereas in eukaryotes rhodoquinone is most commonly found.

Because of its low redox potential (see Table 5.3), ubiquinone does not support the reduction of fumarate.

(i) The bacterial enzyme

The enzyme has been purified from several bacteria, in particular *Escherichia coli*, which has received the most attention and which will be emphasised in this discussion.

Structure: In *E. coli* the reductase is a tetramer composed of four equimolar subunits, designated FRD A, B, C, D. Subunits A and B are hydrophilic, extrinsic to the membrane and are catalytic, whereas subunits C and D are very hydrophobic and intrinsic to the membrane and appear both to interact with quinones and to have an anchoring role (Cole, Condon, Lemire and Weiner, 1985; Weiner, Cammack, Cole, Condon, Honoré, Lemire and Shaw, 1986). The catalytic site faces the cytoplasm and is located on the A (flavoprotein) subunit. It contains an essential cysteine residue (Cys-247) which is analogous to a thiol residue in the active site of succinate dehydrogenase from *E. coli*. The enzyme from *E. coli* can be visualised by transmission electron microscopy as 4-5nm knobs protruding into the cytoplasm from the bacterial inner membrane (Cole *et al.*, 1985). There is evidence to show that the B subunit is enveloped by the outer A subunit. The C and D subunits provide binding sites for the A and B subunits to form the stable tetramer.

Electron transport: In *E. coli*, all four subunits are required for electron transport, though the reduction of fumarate to succinate occurs in the AB dimer. Three Fe/S centres, designated Centres 1-3, are present in the B subunit from *E. coli* and several other bacteria; this is also the case for succinate dehydrogenase from *E. coli*, other bacteria and bovine heart (Cole *et al.*, 1985). The proposed path of electrons in fumarate reductase from *E. coli* is illustrated in Figure 5.3, where Fe/S Centres 1 and 3 are ordered according to their mid-point potentials (-20 and -70mV respectively).

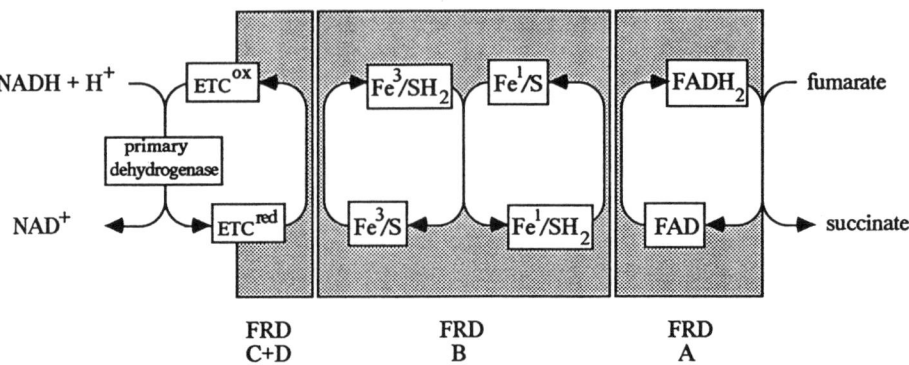

Figure 5.3 Electron transport in fumarate reductase of *E. coli* (redrawn from Cole, *et al.* 1985). $ETC^{red/ox}$, electron transport components, reduced/oxidised; FAD, flavin adenine dinucleotide.

The role of Centre 2 (omitted from the Figure) is unclear as it has a very low midpoint potential (-330mV). Subunit A contains 1 mole of covalently bound flavin as FAD; this is also found in subunit A of succinate dehydrogenase from *E. coli*. The FAD attachment site in both enzyme subunits is conserved. Subunit C has been identified as cytochrome b_{556} in *E. coli* (Murakami, Kita, Oya and Anraku, 1985). The enzyme from two other bacterial species, *Wolinella succinogenes* and *Bacillus subtilis* also contains *b*-type cytochromes. Fumarate reductase from these two species has a different subunit structure from that of *E. coli* and does not contain subunits homologous to the C and D subunits, although a cytochrome *b* subunit is present in both enzymes and may function in membrane attachment of the membrane-extrinsic subunits (Holmgren, Hederstedt and Rutberg, 1974; Unden, Albracht and Kröger, 1984).

Molecular genetics: In *E. coli* the genes coding for fumarate reductase, located in the *frd* operon, are distinct from those encoding succinate dehydrogenase (found in the *sdh* operon). Expression of the *frd* operon is induced by the absence of oxygen. The *frd* genes are present, highly conserved, in all members of the Enterobacteriaceae. Four genes are present in the operon, *frdA, frdB, frdC* and *frdD*, corresponding, respectively, to the subunits of the enzyme. In *E. coli* there is a single promoter for the operon and expression under aerobic conditions is repressed at the transcription level by a specific repressor molecule (Cole *et al.*, 1985). A possible positively regulating protein, *fnr,* which has the structure of a DNA-binding protein, has also been identified in *E. coli*. It promotes the expression of a range of anaerobically-expressed enzymes, including fumarate reductase. Interestingly, the *fnr* protein has a number of cysteine residues clustered at the amino terminus that may function in detecting changes in redox potential associated with the onset of anoxic conditions. A structural change induced by reducing conditions is hypothesised to permit binding of cAMP to the *fnr* protein, followed by binding to DNA and transcription of anaerobically-expressed genes (Cole *et al.*, 1985).

Relationship to succinate dehydrogenase: Although succinate dehydrogenase and fumarate reductase are similar in structure and function in both *E. coli* and certain other bacterial species (see Cole *et al.*, 1985), and although the two enzymes show high homology at the amino acid level in the catalytic subunits, they are nonetheless immunologically distinct proteins with separate genetic opera(ons?) controlled by different regulatory elements. It has been speculated (Cole *et al.*, 1985) that the catalytic subunits of the two enzymes are derived from common ancestral genes and that the hydrophobic membrane-anchoring subunits, although not genetically related, have the same function in each enzyme and, indeed, are responsible for the respective and different catalytic properties of the two enzymes.

(ii) *Fumarate reductase in eukaryotes*
Fumarate reductase activity is found in a variety of eukaryotic organisms that are anaerobic or facultatively anaerobic in their lifestyles. Electron transfer from NADH to fumarate has been recorded in Protozoa (e.g. *Trypanosoma cruzi*, Boveris, Hertig and Turrens, 1986; *T. brucei*, Turrens, 1987), nematodes (e.g. *Ascaris suum*, Kmetec and Bueding, 1961; *Haemonchus contortus*, Prichard, 1973; Bryant and Bennet, 1983; *Ascaridia galli, Nippostrongylus brasiliensis*, Fry and Brazeley, 1984), platyhelminths (e.g. *Fasciola hepatica*, Barrett, 1978; *Paragonimus westermani*, Takamiya, Furushima and Oya, 1986), oligochaete and polychaete annelids (e.g. *Tubifex* sp., Schöttler, 1977; *Arenicola marina*, Schroff and Schöttler, 1977), intertidal bivalves (e.g. *Mytilus edulis*, Holwerda and De Zwaan, 1979) and terrestrial gastropods (e.g. *Achatina achatina*, Umezurike and Chilaka, 1982). These organisms all accumulate succinate and/or propionate or their derivatives as end-products of carbohydrate metabolism. The coupled synthesis of ATP by a presumed site 1 phosphorylation has been demonstrated in some species. Fumarate reductase activity, coupled to the synthesis of ATP, has also been observed in beef heart mitochondria, where it appears to be a reversal of the reduction of NAD^+ by succinate (Haas, 1964). Utilisation of fumarate as a terminal electron acceptor is thus widespread amongst eukaryotes, but the process has received remarkably little attention except in parasitic organisms and molluscs.

Relatively little is known of the structure and function of fumarate reductase in eukaryotes as the enzyme has been purified from only two organisms – both parasites – the intestinal nematode *Ascaris suum* and the lung fluke *Paragonimus westermani* (Takamiya *et al.*, 1986; Oya and Kita, 1989; Ma, Kita, Hamajima and Oya, 1987). The enzyme from most Metazoa that have been examined contains covalently-bound FAD and is associated with a *b*-type cytochrome.

The most probing work to date has been that of Oya's laboratory, carried out on the enzyme from mitochondria of muscle tissue of the adult stage of the parasitic nematode *Ascaris suum*. These workers have purified respiratory complex II (succinate-ubiquinone reductase) from this tissue and shown that it possesses high fumarate reductase activity. Its subunit composition and kinetic parameters are included in Tables 5.4 and 5.5. The activity of the complex is not inhibited by malonate, the classical inhibitor of succinate dehydrogenase (Kita, Takamiya, Furushima, Ma, Suzuki, Ozawa and Oya, 1988a). Complex II was shown to be a major component of the respiratory chain of *Ascaris* mitochondria, in comparison with the content of complex III which was relatively minor (Kita, Takamiya, Furushima, Ma and Oya, 1988b).

Cytochrome b_{558} is an essential component of NADH-fumarate electron transfer in *Ascaris* (Oya and Kita, 1989). Indeed, it is the major cytochrome component in mitochondria from this species (Takamiya, Furushima and Oya, 1984). It can be reduced by exogenous NADH, α-glycerophosphate or succinate and is rapidly reoxidised by fumarate. It is not sensitive to inhibition by CO and does not appear to be an *o*-type cytochrome (Takamiya *et al.*, 1986). Similar *b*-type cytochromes are known from a range of organisms that utilise the phosphoenolpyruvate-succinate pathway in energy metabolism (Oya and Kita, 1989). In *Ascaris* this cytochrome is localised in complex II, tightly bound to the inner mitochondrial membrane. It is comprised of two subunits of 15 and 13.5kD. A low-potential *b* cytochrome is also a component of complex II in mammalian mitochondria but its precise function is not clear (Hatefi and Galante, 1980).

Antibodies (polyclonal and monoclonal) prepared against the largest (FAD) subunit of *Ascaris* fumarate reductase cross-react with the corresponding subunit from complex II from the lung fluke *P. westermani* and also from bovine heart mitochondria, preparations of human mitochondria, and the cytoplasmic membrane of bacteria grown either aerobically or anaerobically (Oya and Kita, 1989). This means that these subunits are all related, at least at the topographic (i.e. epitope) level. Antibodies prepared against the other three subunits of *Ascaris* complex II did not react, however, indicating that, as in bacteria, the smaller subunits are more species-specific (Kita *et al.*, 1988a). Nonetheless, there was considerable homology at the amino acid level between the 26kD subunit from *Ascaris* and the Fe/S subunit from beef heart.

The 26kD subunit from *Ascaris* contains 3 Fe/S clusters (Hata-Tanaka, Kita, Furushima, Oya and Itoh, 1988) of which the third, designated S3, differs in properties from S3 of bovine heart complex II in that it is not fully reduced by excess succinate and is not inhibited by TTFA. Isolated Complex II from *P. westermani*, on the other hand, is sensitive to TTFA, but less sensitive than that from bovine heart (Ma *et al.*, 1987).

The proposed pathways of electron transport in *Ascaris* muscle mitochondria are summarised in Figure 5.4. Rhodoquinone has been shown to replace ubiquinone in *Ascaris* and in other parasitic helminths. Cytochrome b_{558} has been shown spectroscopically to be directly involved in electron transport by complex II. It is not clear at present for this or any other eukaryotic species whether the fumarate reductase activity is catalysed by a separate enzyme from the (relatively low) succinate dehydrogenase activity, as in *E. coli*, or whether both activities are catalysed by the

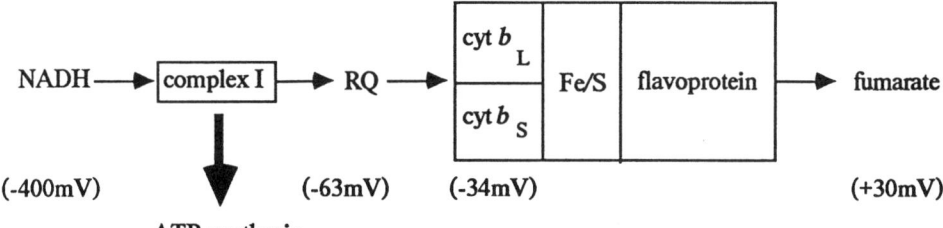

Figure 5.4 Scheme for electron transfer in the fumarate reductase system of *Ascaris* muscle mitochondria (modified from Kita *et al.*, 1988a). Mid-point potentials are in parentheses. RQ, rhodoquinone-9; cyt $b_{L,S}$, large and small subunits of cytochrome b_{558}, respectively; Fe/S, iron-sulphur subunit.

same protein. The precise role(s) of the small subunit of the complex is not known at present, but it appears that this subunit is species-specific and therefore probably has an important role in assembly or regulation of the catalytic activity of the complex.

The relationship between fumarate reductase and succinate dehydrogenase: The question of the relationship between the two enzyme activities in the mitochondria of eukaryotic organisms is an open one at present. It appears in *Ascaris* that the activities are catalysed by the same protein in the eggs, which have aerobic metabolism and in the anaerobic adults, as the molecular weight and antigenic properties of all the subunits are similar (Kita *et al.*, 1988a). In *A. galli* the properties of succinate dehydrogenase and fumarate reductase are also very similar (Fry and Brazeley, 1984). On the other hand, in *N. brasiliensis* two types of succinate dehydrogenase have been observed, one apparently associated with the classical aerobic electron transport pathway and the other with the 'alternative' pathway, which has both succinate dehydrogenase and fumarate reductase activity, depending on the ambient oxygen concentration. In either case, the direction of catalysis may depend on the availability of oxygen to act as terminal electron acceptor, via the classical or alternative oxidase pathways, summarised in Figure 5.5 (Köhler and Bachmann, 1980; Köhler, 1985). This question will ultimately be resolved when molecular and genetic studies of fumarate reductase and its genes are undertaken.

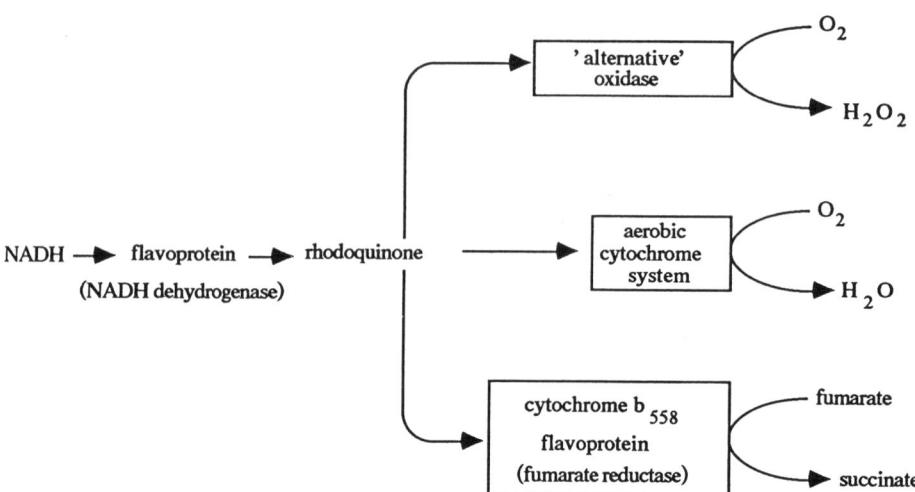

Figure 5.5 Branched electron transport in parasitic helminths (modified from Bryant and Behm, 1989).

Conclusions

There are many questions to be answered about fumarate reductase in the lower Metazoa. For example, what is the molecular and evolutionary relationship of this enzyme to the TCA cycle enzyme, succinate dehydrogenase; is there more than one type of fumarate reductase expressed in an organism's life cycle; can it be induced by anoxic conditions; if so, what are the environmental stimuli; how is differential expression of the gene(s) induced and controlled during a life cycle; how have the gene(s) evolved among the lower Metazoa; how is assembly of the mitochondrial protein controlled, given that nuclear genes are involved? Fumarate reductase plays such a vital role in the metabolism of anaerobic and facultatively anaerobic invertebrates that it is worthwhile seeking answers to these questions. The answers will give clues to the evolutionary development of these bioenergetic processes in invertebrates, they will allow us to understand better the biology of these organisms and they may identify points of attack for invertebrate pests such as internal parasites.

Abbreviations. CoA, coenzyme A; FAD, flavinadenine dinucleotide; FR, fumarate reductase; SDH, succinate dehydrogenase; TCA, tricarboxylic acid; TTFA, thenoyltrifluoroacetone.

References

Barrett, J. (1978) 'Activation of succinate dehydrogenase from adult *Fasciola hepatica* (Trematoda)' *Parasitology* 76, 269-275

Boveris, A., Hertig, C.M. and Turrens, J.F. (1986) 'Fumarate reductase and other mitochondrial activities in *Trypanosoma cruzi*' *Molecular and Biochemical Parasitology* 19, 163-169

Broda, E. (1975) *The Evolution of the Bioenergetic Process.* Pergamon, Oxford

Broda, E. and Peschek, G.A. (1979) 'Did respiration or photosynthesis come first?' *Journal of Theoretical Biology* 81, 201-212

Bryant, C. and Behm, C.A. (1989) *Biochemical Adaptation in Parasites.* Chapman and Hall, London

Bryant, C. and Bennet, E-M. (1983) 'Observations on the fumarate reductase system in *Haemonchus contortus* and their relevance to anthelmintic resistance and to strain variations of energy metabolism' *Molecular and Biochemical Parasitology* 7, 281-192

Cole, S.T., Condon, C., Lemire, B.D. and Weiner, J.H. (1985) 'Molecular biology, biochemistry and bioenergetics of fumarate reductase, a complex membrane-bound iron-sulphur flavoenzyme of *Escherichia coli*' *Biochimica et Biophysica Acta* 811, 381-403

Comley, J.C.W. and Wright, D.J. (1981) 'Succinate dehydrogenase and fumarate reductase activity in *Aspiculuris tetraptera* and *Ascaris suum* and the effect of the anthelmintics cambendazole, thiabendazole, and levamisole' *International Journal for Parasitology* 11, 79-84

Dickerson, R.E. (1980) 'Evolution and gene transfer in purple photosynthetic bacteria' *Nature* 283, 210-212

Dickie, P. and Weiner, J.H. (1979) "Purification and characterization of membrane-bound fumarate reductase from anaerobically-grown *Escherichia coli*' *Canadian Journal of Biochemistry* 57, 813-821

Erabi, T., Higuti, T., Kakuno, T., Yamashita, J., Tanaka, M. and Horio, T. (1975) "olarographic studies on ubiquinone-10 and rhodoquinone bound with chromatophores from *Rhodospirillum rubrum*' *Journal of Biochemistry* 78, 795-801

Fry, M. and Brazeley, E.P. (1984) 'NADH-fumarate reductase and succinate dehydrogenase activities in mitochondria of *Ascaridia galli* and *Nippostrongylus brasiliensis*' *Comparative Biochemistry and Physiology* 77B, 143-150

Gest, H. (1980) 'The evolution of biological energy-transducing systems' *FEMS Microbiology Letters* 7, 73-77

Gnaiger, E. (1977) 'Thermodynamic considerations of invertebrate anoxibiosis' In: *Application of Calorimetry in Life Sciences* (Eds I. Lamprecht and B. Schaarschmidt) pp. 281-303. Walter de Gruyter, Berlin

Haas, D.W. (1964) 'Phosphorylation coupled to the oxidation of NADH of fumarate in digitonin fragments of beef-heart mitochondria' *Biochimica et Biophysica Acta* 92, 433-439

Haddock, B.A. and Jones, C.W. (1977) 'Bacterial respiration' *Bacteriological Reviews* 41, 47-99

Hall, J.B. (1971) 'Evolution of the prokaryotes' *Journal of Theoretical Biology* 30, 429-454

Hata-Tanaka, A., Kita, K., Furushima, R., Oya, H. and Itoh, S. (1988) 'ESR studies on iron-sulphur clusters of complex II in *Ascaris suum* mitochondria which exhibits strong fumarate reductase activity' *FEBS Letters* 242, 183-186

Hatefi, Y. and Galante, Y.M. (1980) 'Isolation of cytochrome-b_{560} from Complex II (succinate ubiquinone oxidoreductase) and its reconstitution with succinate dehydrogenase' *Journal of Biological Chemistry* 255, 5530-5537

He, S.H., Der Vartanian, D.V. and LeGall, J. (1986) 'Isolation of fumarate reductase from *Desulfovibrio multispirans*, a sulphur-reducing bacterium. *Biochemical and Biophysical Research Communications* 135, 1000-1007

Hederstedt, L., Holmgren, E. and Rutberg, L. (1979) 'Characterization of a succinate dehydrogenase complex solubilized from the cytoplasmic membrane of *Bacillus subtilis* with the nonionic detergent Triton X-100' *Journal of Bacteriology* 138, 370-376

Hochachka, P.W. and Mommsen, T.P. (1983) 'Protons and anaerobiosis' *Science* 219, 1391-1397

Holmgren, E., Hederstedt, L. and Rutberg, L. (1974) 'Role of heme in synthesis and membrane binding of succinic dehydrogenase in *Bacillus subtilis*' *Journal of Bacteriology* 138, 377-382

Holwerda, D.A. and De Zwaan, A. (1979) 'Fumarate reductase of *Mytilus edulis* L.' *Marine Biology Letters* 1, 33-40

Holwerda, D.A. and De Zwaan, A. (1980) 'On the role of fumarate reductase in anaerobic carbohydrate catabolism of *Mytilus edulis*' *Comparative Biochemistry and Physiology* 67B, 447-453

Jones, C.W. (1985) 'The evolution of bacterial respiration' In: *Evolution of Prokaryotes* (Eds K.H. Schleifer and E. Stackebrandt), *FEMS Symposia* No. 29, pp.175-204. Academic Press, London

Kenney, W.C. and Kröger, A. (1977) 'The covalently bound flavin of *Vibrio succinogenes*

succinate dehydrogenase' *FEBS Letters* **73**, 239-243
Kita, K., Takamiya, S., Furushima, R., Ma, Y-C., Suzuki, H., Ozawa, T. and Oya, H. (1988a) 'Electron-transfer complexes of *Ascaris suum* muscle mitochondria. III. Composition and fumarate reductase activity of Complex II' *Biochimica et Biophysica Acta* **935**, 130-140
Kita, K., Takamiya, S., Furushima, R., Ma, Y. and Oya, H. (1988b) 'Complex II is a major component of the respiratory chain in the muscle mitochondria of *Ascaris suum* with high fumarate reductase activity' *Comparative Biochemistry and Physiology* **89B**, 31-34
Kmetec, E. and Bueding, E. (1961) 'Succinic and reduced diphosphopyridine nucleotide oxidase systems of *Ascaris* muscle' *Journal of Biological Chemistry* **236**, 584-591
Köhler, P. (1985) 'The strategies of energy conservation in helminths' *Molecular and Biochemical Parasitology* **17**, 1-18
Köhler, P. and Bachmann, R. (1980) 'Mechanisms of respiration and phosphorylation in *Ascaris*' *Molecular and Biochemical Parasitology* **1**, 75-90
Kröger, A. (1978) 'Fumarate as terminal acceptor of phosphorylative electron transport' *Biochimica et Biophysica Acta* **505**, 129-145
Lara, F.J.G. (1959) 'The succinate dehydrogenase of *Propionibacterium pentosacem*' *Biochimica et Biophysica Acta* **33**, 565-567
Lehninger, A.L. (1971) *Bioenergetics. The Molecular Basis of Biological Energy Transformation.* 2nd Ed. W.A. Benjamin, Menlo Park, Ca
Ma, Y-C, Kita, K., Hamajima, F. and Oya, H. (1987) 'Isolation and properties of Complex II (succinate-ubiquinone reductase) in the mitochondria of *Paragonimus westermani*' *Japanese Journal of Parasitology* **36**, 107-117
Mitchell, P. (1970) 'Membranes of cells and organelles: morphology, transport and metabolism' *Symposia of the Society for General Microbiology* **20**, 121-166
Murakami, H., Kita, K., Oya, H. and Anraku, Y. (1985) 'The *Escherichia coli* cytochrome b_{556}-gene, CYBA is assignable as 5DHc in the succinate dehydrogenase cluster' *FEMS Microbiology Letters* **30**, 307-311
Muratsubaki, H. and Katsume, T. (1982) 'Purification and properties of fumarate reductase from Baker's yeast' *Agricultural Biology and Chemistry* **46**, 2909-2917
Nicholls, D.G. (1982) *Bioenergetics. An Introduction to the Chemiosmotic Theory.* Academic Press, London
Oya, H. and Kita, K. (1989) 'The physiological significance of complex II (succinate-ubiquinone reductase) in respiratory adaptation' In: *Comparative Biochemistry of Parasitic Helminths* (Eds E-M. Bennet, C.A. Behm and C. Bryant) pp.35-52. Chapman and Hall, London
Peck, H.D., Smith, O.H. and Gest, H. (1957) 'Comparative biochemistry of the biological reduction of fumarate' *Biochimica et Biophysica Acta* **25**, 142-147
Prichard, R.K. (1973) 'The fumarate reductase reaction of *Haemonchus contortus* and the mode of action of some anthelmintics' *International Journal for Parasitology* **3**, 409-417
Rodriguez-Caabeiro, F., Criado-Fornelio, A. and Jimenez-Gonzalez, A. (1985) 'A comparative study of the succinate dehydrogenase-fumarate reductase complex in the genus *Trichinella*' *Parasitology* **91**, 577-583
Schöttler, U. (1977) 'The energy-yielding oxidation of NADH by fumarate in anaerobic mitochondria of *Tubifex* sp.' *Comparative Biochemistry and Physiology* **58B**, 151-156
Schroff, G. and Schöttler, U. (1977) 'Anaerobic reduction of fumarate in the body wall musculature of *Arenicola marina* (Polychaeta)' *Journal of Comparative*

Physiology 116, 325-336
Singer, T.P. (1971) 'Evolution of the respiratory chain and its flavoproteins' In: *Biochemical Evolution and the Origin of Life* (Ed. E. Schoffeniels) pp.203-223. North-Holland, Amsterdam
Takamiya, S., Furushima, R. and Oya, H. (1984) 'Electron transfer complexes of *Ascaris suum* muscle mitochondria: I. Characterization of NADH-cytochrome c reductase (Complex I-III) with special reference to cytochrome localization' *Molecular and Biochemical Parasitology* 13, 121-134
Takamiya, S., Furushima, R. and Oya, H. (1986) 'Electron-transfer complexes of *Ascaris suum* muscle mitochondria. II. Succinate-coenzyme Q reductase (complex II) associated with substrate-reducible cytochrome b-558' *Biochimica et Biophysica Acta* 848, 99-107
Thauer, R.K., Jungermann, K. and Dekker, K. (1977) 'Energy conservation in chemotrophic anaerobic bacteria' *Bacteriological Reviews* 41, 100-180
Turrens, J.F. (1987) 'Possible role of the NADH-fumarate reductase in superoxide anion and hydrogen peroxide production in *Trypanosoma brucei*' *Molecular and Biochemical Parasitology* 25, 55-60
Umezurike, G.M. and Chilaka, F.C. (1982) 'Succinate-DCPIP and NADH-fumarate oxidoreductases in submitochondrial particles of the giant African snail (*Achatina achatina*) foot muscle' *Comparative Biochemistry and Physiology* 71B, 181-185
Unden, G. and Kröger, A. (1981) 'The function of the subunits of the fumarate reductase complex of *Vibrio succinogenes*' *European Journal of Biochemistry* 120, 577-584
Unden, G., Albracht, S.P.J. and Kröger, A. (1984) 'Redox potentials and kinetic properties of fumarate reductase complex from *Vibrio succinogenes*' *Biochimica et Biophysica Acta* 767, 460-469
Unden, G., Hackenberg, H. and Kröger, A. (1980) 'Isolation and functional aspects of the fumarate reductase involved in the phosphorylative electron transport of *Vibrio succinogenes*' *Biochimica et Biophysica Acta* 591, 275-288
Warringa, M.G.P.J. and Giuditta, A. (1958) 'Studies on succinic dehydrogenase - IX. Characterization of the enzyme from *Micrococcus lactilyticus*' *Journal of Biological Chemistry* 230, 111-123
Warringa, M.G.P.J., Smith, O.H., Giuditta, A. and Singer, T.P. (1958) 'Studies on succinic dehydrogenase. VIII. Isolation of a succinic dehydrogenase-fumaric reductase from an obligate anaerobe' *Journal of Biological Chemistry* 230, 97-109
Weiner, J.H., Cammack, R., Cole, S.T., Condon, C., Honoré, N., Lemire, B.D. and Shaw, G. (1986) 'A mutant of *Escherichia coli* fumarate reductase decoupled from electron transport' *Proceedings of the National Academy of Sciences, U.S.A.* 83, 2056-2060
Weitzmann, P.D.J. (1985) 'Evolution in the citric acid cycle' In: *Evolution of Prokaryotes* (Eds K.H. Schleifer and E. Stackebrandt), *FEMS Symposia* No. 29, pp.253-275. Academic Press, London
Wilson, T.H. and Lin, E.C.C. (1980) 'Evolution of membrane bioenergetics' *Journal of Supramembrane Structure* 13, 421-446

Chapter 6

Metazoan Adaptations to Hydrogen Sulphide

Russell D. Vetter, Mark A. Powell and George N. Somero

Introduction

Twenty years ago Fenchel and Riedl (1970) focused attention on the unique biota of the anaerobic sulphide ecosystem, the thiobios, which exists below the surface of most marine sediments. Their results suggested that this ecosystem was separate from the overlying oxygenated sediments and that the metazoans living there contained specific adaptations to survive in the complete absence of oxygen. Following this study there has been a burst of research and controversy concerning the extent to which the animals of the thiobios are truly independent of aerobic processes (Boaden and Platt, 1971; Maguire and Boaden, 1975; Reise and Ax, 1979, 1980). While this controversy and a general renewed interest in the process of anaerobic metabolism have led to the elucidation of a number of unusual biochemical pathways of anaerobic metabolism, which are the subjects of other chapters in this volume, the study of the 'thio' in thiobios has been virtually ignored. Few studies have investigated how animals living in hypoxic environments can adapt to potentially lethal concentrations of hydrogen sulphide. The biochemical steps in sulphide oxidation are virtually unknown, and the possibility that metazoans may exploit the tremendous energy potential of reduced sulphur compounds has only recently been considered.

A major stimulus to the investigation of the physiological and biochemical characteristics of animals living in high-sulphide environments was provided by the discovery in 1977 (Corliss, Dymond, Gordon, Edmond, von Herzon, Ballard, Green, Williams, Bainbridge, Crane and van Andel, 1979) of rich communities of large tube worms and bivalve molluscs at the deep-sea hydrothermal vents. At the depths of these vents, approximately 2500 m, too little photosynthetically generated food enters to sustain the dense assemblages of life clustered around the vent effluents. Thus, attention was quickly focused on the possible role played by hydrogen sulphide, which is present at concentrations of up to several hundred micromolar in the vent waters (Edmond, von Damm, McDuff and Measures, 1982), in providing an energy source for driving net carbon fixation. The large sizes of the vent animals and the development of new techniques for studying the metabolism of sulphide and other reduced sulphur compounds have enabled marine biologists to study in detail the manners in which these animals detoxify, transport and exploit the energy of sulphide, (reviewed by Somero, Childress and Anderson, 1989). In this chapter we draw together recent data on these diverse aspects of sulphide relationships in marine animals and develop, to the extent currently possible, some general models to explain how animals from high-sulphide habitats tolerate and exploit the toxic sulphide molecule. We hope to infect the reader with our enthusiasm for this new and exciting aspect of life without oxygen, and to point out important questions about the metabolism of sulphide that remain scientific challenges.

Sources of Hydrogen Sulphide

Hydrogen sulphide can be produced by geothermal activity, as occurs at hydrothermal vents (Edmond *et al.*, 1982); through the microbial degradation of organic matter containing reduced thiols (putrefication) (see National Research Council, 1979); or through microbial sulphate reduction, whereby sulphate reducing bacteria in hypoxic environments utilise sulphate as a terminal electron acceptor during the heterotrophic oxidation of organic matter (see Peck and LeGall, 1982).

By far the most common source of hydrogen sulphide is sulphate reduction in organic rich hypoxic marine environments where abundant seawater sulphate (28 mM) allows high rates of sulphate reduction. Intertidal habitats including mudflats and saltmarshes, which are classic areas for the collection of organisms with high capacities for anaerobic metabolism, have some of the highest rates of sulphide production. In a typical saltmarsh up to twelve times as much of the primary productivity is degraded via anaerobic sulphate reducers as by aerobic processes (Howarth and Teal, 1980). The resulting hydrogen sulphide concentrations in these sediments vary with water-flow, organic content and season but, in general, sulphide is near zero at the sediment surface or redox discontinuity and increases with depth to the high μM or low μM range. Luther and co-workers reported maximal sulphide concentrations of 3.36 mM at 30-33 cm in a Delaware saltmarsh and 2.61 mM at 23-28 cm in a Massachusetts saltmarsh (Luther, Church, Scudlark and Cosman, 1986). These levels are actually much higher than those surrounding animals at the Galapagos hydrothermal vents where hydrogen sulphide concentrations vary from 0 to 110μM (Johnson, Beehler, Sakamoto-Arnold and Childress, 1986).

Hydrogen sulphide is also produced in the mammalian gut via the microbial degradation of protein and in ruminants by the additional process of sulphate reduction. High concentrations have occasionally been detected in the guts of human and ruminants but the accuracy of the measurements and whether this is a pathological or normal condition have been disputed (National Research Council, 1979).

In the context of anaerobiosis, it is important to emphasise that, in the presence of oxygen, sulphide oxidises rapidly without a need for biological catalysts (Cline and

Richards, 1969; Chen and Morris, 1972). Because of this rapid oxidation, hydrogen sulphide can only accumulate in those environments that are low in oxygen (Jorgensen, 1982). Thus the ability of an animal to exploit successfully an anaerobic habitat may be linked to its ability to adapt to hydrogen sulphide as well as to its capacity to generate ATP via pathways independent of oxygen.

Forms of Hydrogen Sulphide

In the following discussion the term 'sulphide' refers to all three forms of hydrogen sulphide: hydrogen sulphide gas, H_2S; the bisulphide anion HS^- (pK_1 7.02); and the sulphide anion S^{2-} (pK_2 <12) (see Millero, Plese and Fernandez, 1988). At typical environmental pH's the predominant form is the bisulphide anion. However, the existence of a significant proportion of the total sulphide pool as the gas should not be overlooked since it is probably this form that diffuses across membranes and is poisonous to aerobic respiration. At pH 7.8, 10% of total sulphide is in the gaseous phase while at pH 6.8, 50% is gaseous. Because the blood of marine animals typically has a lower pH than seawater, and because the intracellular water generally is about 0.5 pH unit acidic to the blood, the fraction of total sulphide present as H_2S will increase from seawater to blood to intracellular water.

Effects of Sulphide on Organisms

At the physiological level hydrogen sulphide is recognised as an extremely potent neurotoxin which results in the cessation of pulmonary function. At the biochemical level there are two well documented effects of hydrogen sulphide: inhibition of cytochrome c oxidase and binding to methemoglobin (National Research Council, 1979). Binding to cytochrome c oxidase occurs at the haem site of cytochrome aa_3 and results in the immediate, but reversible, inhibition of aerobic respiration. As stated above, environmental pH and the pH of blood and intracellular fluids can have important effects on toxicity since even small decreases in pH will have large effects on the proportion of sulphide in the permeable undissociated form. In addition, it is the undissociated form of sulphide which actually binds to the cytochrome haem and causes inhibition (Nicholls, 1975). An increase in toxicity with decreasing environmental pH has been demonstrated for a number of aquatic fish and invertebrates (Broderius and Smith, 1976; Smith, Oseid, Adelman and Broderius, 1976; National Research Council, 1979).

Although the toxic effects of sulphide discussed above would appear to be of relevance only in the context of aerobic metabolism, it is not necessarily true that anaerobic organisms face no threat from sulphide. Sulphide is a very strong nucleophile which can potentially reduce protein disulphide bridges (cf. Arp, Childress and Vetter, 1987). To our knowledge this facet of sulphide toxicity has not been investigated, either in aerobic or anaerobic organisms, but the potentially disruptive effects of sulphide on proteins suggests that, even under conditions of anoxia, defensive strategies against sulphide accumulation within cells may be adaptive.

Most studies of the effects of hydrogen sulphide have been carried out with animals from highly aerobic environments and few have attempted to determine what is unique about the animals dwelling in anaerobic habitats. Patel and Spencer (1963) were probably the first to examine the effects of hydrogen sulphide exposure on invertebrates that inhabit high sulphide habitats. When they exposed the burrow dwelling polychaete *Arenicola marina* to seawater stripped of oxygen and bubbled with hydrogen sulphide in closed jars for 24 hours the animals were unaffected. Powell, Crenshaw and Reiger (1979, 1980) subsequently demonstrated that the meiofauna of reducing sediments differed from their counterparts living in well oxygenated sediments in their resistance to and metabolism of sulphide.

Mechanisms for Protecting Against Sulphide

Potential adaptations to high sulphide concentrations include the following:

1 Evolution of sulphide insensitive cytochrome c oxidase systems.
2 Exclusion of sulphide at the body wall because of impermeability or a layer of sulphide oxidising bacteria.
3 Reliance on anaerobic metabolism without attempting to defend the cytochrome oxidase system.
4. Sulphide binding proteins with subsequent unloading at sites of detoxification or upon return to low external sulphide conditions.
5. Symbiotic association with chemoautotrophic bacteria capable of sulphide oxidation.
6. Eukaryotic sulphide detoxification (oxidation) with the potential utilisation of energy released in sulphide oxidation to drive the generation of ATP.

Some of these potential strategies have been observed, some remain hypothetical, and more than one adaptation can occur in a single organism.

Passive Defence Against Sulphide

Passive defence strategies include the first three alternatives mentioned above. They are passive in the sense that these solutions to the problem of sulphide toxicity require little if any expenditure of metabolic energy to transport or process sulphide. Of course, a total shift to anaerobic ATP generation (strategy 3) is costly in terms of the decrease in metabolic efficiency it entails.

While the evolution of sulphide insensitive cytochrome oxidase systems would appear to be a useful strategy we have found no evidence that metazoans have taken this course. The cytochrome oxidase of the hydrothermal vent tube worm *Riftia pachyptila* is strongly inhibited by sulphide concentrations of <10 µM and is protected by its remarkable sulphide binding haemoglobin (Powell and Somero, 1986b; Arp and Childress, 1983; Arp *et al*, 1987). Cytochrome c oxidase preparations from the hydrothermal vent crab *Bythograea thermydron,* the shallow water crab *Menippe mercenaria* and the bivalve mollusc *Mercenaria mercenaria* were all inhibited by sulphide at nanomolar to low micromolar concentrations (Hand and Somero, 1983). Cytochrome c oxidase from six different sulphide tolerant and intolerant estuarine fish species all had K_i's for sulphide in the range 330-210 nM (Bagarinao and Vetter, 1989). The levels of sulphide required to inhibit the cytochrome c oxidase of marine species from oxygen depleted, sulphide rich habits are in the same range as those reported for mammalian cytochrome c oxidases (National Research Council, 1979). Presumably the requirement of the haem iron centre in cytochrome aa_3 for the binding of oxygen from the surrounding solution makes this cytochrome uniquely accessible to sulphide. Thus, the basic steric requirements of this protein may preclude evolutionary modifications that remove or reduce its sensitivity to sulphide.

Surface impermeability to sulphide is another potential strategy that may not, in fact, be tenable. Powell *et al.* (1979) exposed meiofauna from oxic and anoxic environments to ^{35}S-sulphide and determined tissue distribution of ^{35}S compounds using autoradiography. There was no evidence that any of the 19 species representing four phyla (the nematodes, gnathostomulids, turbellarians and gastrotrichs) were able to exclude sulphide completely at the body wall. Meiofauna from the sulphide zone did appear to have less internal sulphide and were more resistant to sulphide toxicity than their surface dwelling counterparts, however. Animals with a hard exoskeleton such as decapod crustaceans are also incapable of excluding sulphide at the body wall (Vetter, Wells, Kurtsman and Somero, 1987). When the hydrothermal vent crab *Bythograea thermydron* and three other shallow dwelling benthic crab species were exposed to environmentally realistic levels of sulphide they all showed evidence of sulphide or its metabolites at high concentrations in the blood. Presumably the undissociated form of sulphide freely diffuses across areas where the exoskeleton is thin, e.g. the gills.

Based on the above studies, it appears clear that most animals that have adapted to high sulphide environments rely upon active schemes of detoxification rather than passive adaptations such as sulphide resistant enzymes or permeability barriers. The various ways that sulphide detoxification occurs have only begun to be investigated, but it is already clear that a wide variety of mechanisms are employed, including specific sulphide binding proteins and enzymatic oxidation mechanisms which may be found in the animals' cells or, in many cases, within endosymbiotic sulphur bacteria.

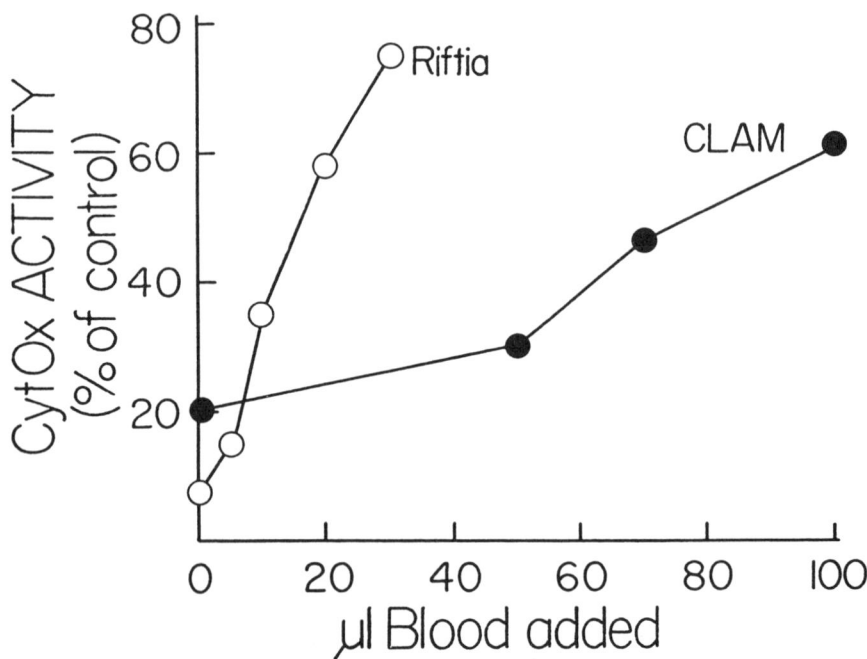

Figure 6.1 The abilities of bloods of the hydrothermal vent tube worm (*Riftia pachyptila*) and clam (*Calyptogena magnifica*) to reverse sulphide inhibition of cytochrome c oxidase (CytOx) activity. Each curve shows how additions of small aliquots of blood to the *in vitro* enzymatic assay system titrate the inhibitory effects of sulphide on CytOx activity. Control activity refers to the CytOx activity measured in the absence of added sulphide. The curve labelled *Riftia* illustrates the ability of *Riftia* whole blood to reverse the inhibitory effects of hydrogen sulphide (5mM in assay medium) on the activity of CytOx isolated from plume tissue of the worm. The curve labelled clam represents the effects of *Calyptogena* blood on the sulphide inhibited CytOx of clam heart; the sulphide concentration in the CytOx assay system was 500μM. See Powell and Somero (1986b) for details of the experimental protocols. The figure is from Powell and Somero (1986b) with the permission of the publisher.

Sulphide Binding Proteins

Blood-borne sulphide-binding proteins are now known for at least three invertebrate phyla, the mollusca (Arp, Childress and Fisher, 1984), the pogonophora (Arp and Childress, 1981, 1983; Arp, Childress and Vetter, 1987) and the echiura (Arp, personal communication). Figure 6.1 shows how aliquots of blood from the pogonophoran tube worm *R. pachyptila* and the hydrothermal vent clam *Calyptogena magnifica* counteract the inhibition by sulphide of cytochrome c oxidase of these two species. The sulphide binding proteins in the bloods of these two vent species do not achieve their protection of cytochrome c oxidase activity by catalysing the oxidation of sulphide. Rather, these proteins tightly bind sulphide and prevent, on the one hand, poisoning of respiration and, on the other hand, the uncatalysed oxidation of sulphide, which would occur rapidly in the blood. Oxidation of sulphide in the blood could lead to the generation of elemental sulphur granules, which might have serious effects on circulatory flow in smaller diameter vessels. The binding proteins thus achieve three important functions: the protection of aerobic respiration in the presence of high ambient sulphide concentrations, the prevention of oxidation of sulphide in the blood, and the transport of sulphide to internal centres where enzymatic systems are present to allow exploitation of its energy (see below).

These blood components have probably evolved separately in each group. The sulphide-binding protein in *R. pachyptila* is the high molecular weight extracellular haemoglobin present in high concentration in both vascular and coelomic blood (Arp *et al.*, 1987). In contrast, in *C. magnifica*, the haemoglobin occurs in erythrocytes and the sulphide-binding protein occurs in the serum (Arp *et al.*, 1984). The nature of the sulphide-binding protein of *C. magnifica* is currently under investigation. In the case of *R. pachyptila*, we have shown that the sulphide binding mechanism is not associated with the haem moiety of the haemoglobin and is not a methemoglobin phenomenon (Arp *et al.*, 1987). The most likely mechanism for the reversible binding of sulphide to protein is a thiol/disulphide exchange reaction whereby a reduced thiol group carries out a nucleophilic attack on a protein disulphide which results in a reduced cysteine moiety and a cysteine-thiol mixed disulphide. This mechanism is well known for glutathione-protein mixed disulphides (Gilbert, 1984). We have shown that unlike most haemoglobins the haemoglobin of *R. pachyptila* contains disulphide bridges and that these disulphides are easily reducible by free low molecular weight thiols (Arp *et al.*, 1987).

It is not known whether sulphide binding proteins occur widely among invertebrates from high sulphide environments, e.g. meiofaunal species, or whether intracellular as well as extracellular sulphide binding proteins are important in the transport and detoxification of sulphide. The biochemistry of sulphide binding proteins thus is an area ripe for further investigation in both comparative and mechanistic directions.

Sulphide-binding proteins are, by their very nature, only a partial solution to the problem of sulphide toxicity. Once saturated with sulphide, the sulphide-binding proteins cannot protect against additional sulphide entering the organism. Under these conditions, the sulphide-binding proteins must either deliver the sulphide to another protein system which has a sulphide oxidising capacity, or the animal must return to a low-sulphide environment where sulphide can be released back to the environment down a concentration gradient. Because sulphide poisoning of mitochondrial respiration is reversible, the use of sulphide binding proteins as a short term expedient for protecting respiration when both sulphide and oxygen are present seems feasible. For some organisms, then, the sulphide binding protein may have as its sole function the temporary removal of sulphide from the body fluids, with subsequent release of sulphide to the ambient seawater occurring when sulphide is low or absent in the environment. However, sulphide-binding proteins appear to play an important additional role, that of transport of sulphide to internal sites where the energy of sulphide can be exploited.

Sulphide Metabolism with the Aid of Symbiotic Bacteria

Specific sites of sulphide oxidation have been found in almost all invertebrates from high sulphide environments that have been so far studied. Some of the most potent sites of sulphide oxidation are specific tissues or regions of the body that have been modified to facilitate the growth of symbiotic chemoautotrophic bacteria capable of oxidising reduced sulphur compounds such as sulphide. The number of phyla in which this type of association has now been recognised and the different locations and specialised structures within these animals for sustaining these symbioses are truly remarkable (Fisher, 1989).

Among the pseudocoelomates, the nematode, *Eubostrichys dianeie*, houses sulphide oxidising bacteria in a mucus web over the exterior of its body. This web of bacteria is thought to decrease the influx of sulphide (Powell *et al.*, 1979). Within the annelids, associations with sulphur oxidising bacteria have been described for both oligochaetes and polychaetes. The polychaete worm *Alvinella pompeijana*, which inhabits the mineral precipitate chimneys of sulphide-laden hydrothermal vents, maintains an epicuticular layer of sulphur containing filamentous bacteria (Desbruyeres and Laubier, 1980; Laubier, Desbruyeres and Chassard-Bouchard, 1983). There is evidence that these bacteria may provide a nutritional as well as a detoxification function for the animals (Desbruyeres, Gaill, Laubier, Prieur and Rau, 1983). In the marine oligochaetes *Phallodrilus leukodermatus* and *P. planus* there is a rich layer of chemoautotrophic sulphur metabolising bacteria located between the cuticle and the epidermis (Giere, 1981; Felbeck, Liebezeit, Dawson and Giere, 1983). Here the bacteria are internal but not intracellular. They are located below the relatively inert cuticle and in a position to shield the internal organs from the influx of sulphide. Dubulier (1986) has recently described filamentous epibacteria which are deeply embedded at one end in the cuticle of the posterior region of the annelid *Tubificoides benedii*. These worms occur in sulphide rich sediments and the bacteria resemble known filamentous sulphur bacteria. However, no direct evidence of sulphide oxidation was presented for this species.

In the Phylum Mollusca, associations are rare among the gastropods but common among the bivalves. The undescribed archaegastropod limpet from the hydrothermal vents has been shown by de Burgh and Singla (1984) to contain abundant filamentous sulphur bacteria which are intercalated between the gill lamellae. This is the only animal that has been shown to endocytose directly external sulphur bacteria and digest them intracellularly. While this mechanism clearly has a nutritional potential we suggest that it may also serve to keep the bacterial colony in a growth state which is more able to oxidise sulphur than senescent colonies which are filled with elemental sulphur. This is the case for laboratory cultures (Keunen and Beudeker, 1982). Recently a gastropod from hydrothermal vents in the western North Pacific has been found which contains a highly developed sulphur symbiosis in the gill. The bacteria are intracellular (Stein, Cary, Hessler, Ohta, Vetter, Childress and Felbeck, 1988).

In the bivalves, there are specific intracellular (endosymbiotic) strains of sulphur bacteria housed in modified cells called bacteriocytes (Cavanaugh, 1983; Felbeck, 1983; Vetter, 1985). This endosymbiotic association is now known for at least 15 species in the superfamily Lucinacea, four species of Solemyidae, four species of Vesicomyidae and at least three Mytilidae. In the bivalves, the primary role of the bacteria is not the detoxification of sulphide to protect the animal's respiration. Rather, evidence is accumulating that, in at least some of the symbiont-containing bivalves, the initial step(s) in sulphide oxidation occur in the animal compartment of the symbiosis, and that the animal delivers a more oxidised and less toxic form of sulphur to the bacteria where it is further metabolised. Powell and Somero (1985a) have shown that, in the gutless protobranch clam *Solemya reidi*, sulphide oxidising activity occurs in the animal portion of the gill and in the foot tissue. In the palial fluid of *S. reidi* and in the lucinid clam *Lucinoma annulata* there are millimolar concentrations of circulating thiosulphate but virtually no free or bound sulphide (Vetter, Matrai, Javor and O'Brien, 1989; Cary, Vetter and Felbeck, 1989). These findings suggest that some molluscs living in high sulphide environments have evolved the ability to oxidise sulphide and that the incorporation of chemoautotrophic bacteria which utilise thiosulphate is a secondary adaptation to exploit what would have been a waste product. In this scheme the degeneration of the feeding appendages and gut with subsequent reliance on CO_2 fixation by bacteria as a source of nutrition (Felbeck, Childress and Somero, 1981) would be a tertiary consequence.

Solemya reidi is the most extensively studied bivalve sulphide symbiosis. Observations that sulphide oxidation occurs in pigment granules and mitochondria in the animal cells (Powell and Somero, 1985; Powell and Somero, 1986b) and the discovery of high concentrations of thiosulfate in the blood (Powell, Vetter and Somero, 1987; Vetter, Matrai, Javor and O'Brien, 1989) have been followed by detailed studies showing how mitochondria oxidise sulphide to thiosulfate and the

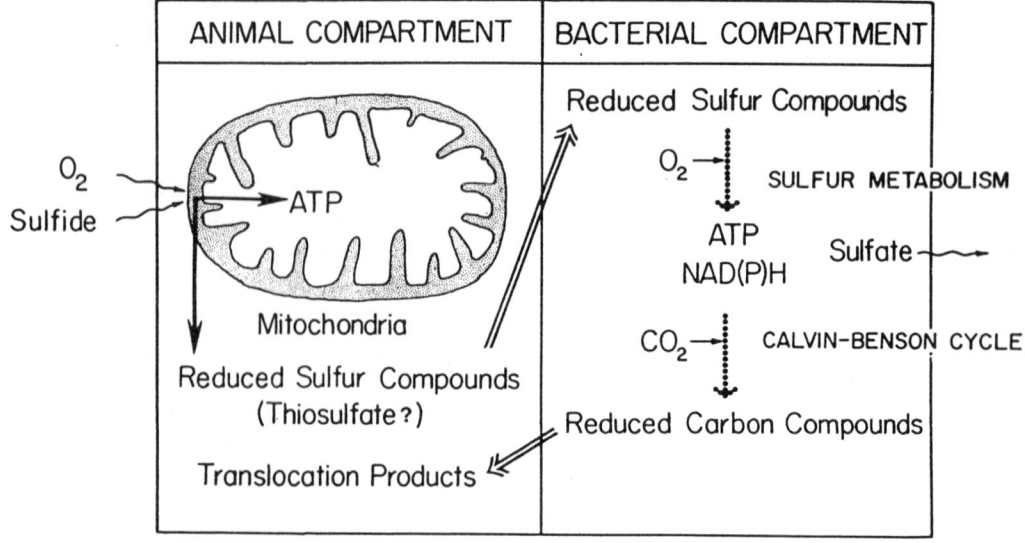

Figure 6.2 A diagrammatic illustration of the symbiosis of the gutless protobranch clam *Solemya reidi* and its gill-housed sulphur bacteria symbionts. This model is proposed to apply as well to at least some of the other bivalve-sulphur bacteria symbiosis discussed in this article. The initial oxidation of sulphide is thought to occur primarily in the animal compartment of this type of symbiosis (Powell and Somero, 1985). The oxidation of sulphide commences in the mitochondria, leading to the production of ATP via oxidation phosphorylation (Powell and Somero, 1986a), and the production of one oxidised sulphur product (thiosulphate), which is transported to the gill bacteria for further oxidation. The bacteria may oxidise thiosulphate and other reduced sulphur compounds (including sulphide) to produce sulphate, which is transported to the environment. Some of the energy released during sulphur metabolism is trapped as ATP and NAD(P)H, which are used in part to drive net CO_2 fixation *via* the Calvin-Benson cycle. Some fraction of this fixed carbon is translocated to the host, where it may contribute importantly to the animal's nutritional needs. The figure is from Somero (1987) with permission of the publisher.

whole animal studies of the metabolism of sulphide and thiosulfate (Anderson, Childress and Favuzzi, 1987). We now have a basic understanding of how sulphide oxidation occurs in *Solemya reidi* (figure 6.2).

Using different concentrations of hydrogen sulphide and oxygen in their incubation media, Anderson and colleagues demonstrated that sulphide concentrations up to 100 µM led to net CO_2 fixation. Higher concentrations of sulphide led to decreased rates of oxygen consumtion and cessation of net CO_2 fixation, due to inhibition by sulphide of aerobic respiration. Anaerobic end-products characteristic of molluscan metabolism were noted only when high concentrations of sulphide were present in the medium (Anderson, 1989). When oxygen was abundant, free sulphide in the blood was negligible, whereas thiosulphate rose to substantial concentrations, and was maintained at steady state level in clams with viable symbiont populations. When oxygen was limiting, free sulphide and thiosulphate were found in the blood. Thus, even in metazoans dependent on sulphide as a source of nutrition, the level of sulphide which is toxic (>100 µM) is very low. Isolated mitochondria begin to show toxic effects of sulphide at 50 µM but continue rapidly to oxidise sulphide to thiosulphate at that concentration (Vetter, Matrai, Javor and O'Brien, 1989). these findings indicate tht the animal is able to oxidise sulphide to thiosulfate without the help of bacteria and that the bacteria utilise a waste-product of animal metabolism. This is not surprising since bacterial oxidation would be ineffective at protecting the foot and mantle which re in direct contact with sulphide unless there was a sulphide binding and transport system. However, *Solemya* blood does not bind sulphide. The bacteria may help to protect the animal at higher sulphide concentrations. In addition to mitochondrial sulphide oxidation the bacteria are able to oxidise sulphide and could strip sulphide from the blood when the animal's detoxification systems are overloaded. At higher sulphide (200 µM), the bacteria can shunt sulphide into elemental sulphur, a temporary, non-toxic repository for excess sulphide (D. Wilmot, unpublished data).

The type of sulphide metabolism strategy adopted by an invertebrate may be partly a consequence of the animal's fundamental body plan. The body plan of bivalves with large areas of exposed mantle, gill and foot tissue would make a peripheral defence, such as a continuous coating of sulphide oxidising bacteria, much more difficult than for animals with a tubular body plan and a hard exoskeleton. Consequently, it appears that most molluscs living in high sulphide habitats have adopted sulphide detoxification schemes that rely on enzymes localised within the animal cells to catalyse at least the initial oxidation (detoxification) of sulphide penetrating the exposed body surfaces (Powell and Somero, 1985, 1986b).

In animals with the pogonophoran body plan, e.g. *R. pachyptila,* a very different strategy for sulphide metabolism has been found (Figure 6.3). Even though the outer cell layers of the body wall musculature have the ability to oxidise sulphide (Powell and Somero, 1986b) and serve as a peripheral defence against sulphide poisoning, by far the greatest share of sulphide oxidation occurs in the trophosome, an internal organ which contains extremely high numbers of sulphide oxidising

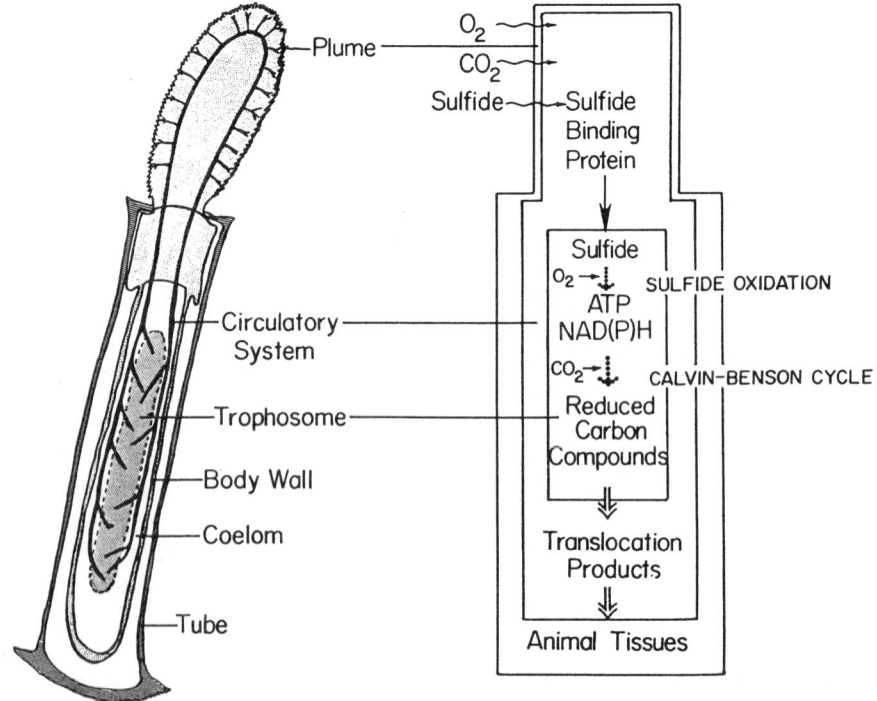

Figure 6.3 A diagrammatic illustration of the symbiosis between the vestimentiferan tube worm *Riftia pachyptila* and its trophosome-housed sulphur bacteria symbionts. The left hand drawing illustrates the key components of the worm's anatomy. The animal lacks a mouth and gut, and the plume serves as the primary site of exchange of gases and solutes with the ambient seawater. The bacteria-containing trophosome fills much of the coelomic cavity. The symbionts receive sulphide, oxygen, carbon dioxide, and other nutrients *via* the well-developed circulatory system of the worm. The right hand drawing illustrates the biochemical and physiological organisation of this symbiosis. Sulphide, oxygen, and CO_2 are carried to the trophosome bacteria by the blood. The abundant haemoglobin in the vascular and coelomic spaces carried both oxygen and sulphide (Arp *et al.*, 1987). The symbionts have high capacities for oxidising sulphide (Powell and Somero, 1986b) and for net fixation of Co_2 *via* the Calvin-Benson cycle (Felbeck, 1981; Felbeck *et al.*, 1981). Some of the reduced carbon compounds synthesised in the bacteria are released to the animal ('translocation products'), and these may provide a substantial fraction of the worm's requirements for reduced carbon compounds. The figure is from Somero (1987) with permission of the publisher.

bacteria (Cavanaugh, Gardner, Jones, Jannasch and Waterbury, 1981; Felbeck, 1981; Felbeck et al., 1981; Powell and Somero, 1986b). In the context of sulphide detoxification, the trophosome bacteria protect the worm's respiratory functions by oxidising the sulphide carried to the trophosome by the haemoglobin (sulphide binding protein). As long as the sulphide-binding protein is not saturated by sulphide, the animal tissues will be protected.

The trophosome bacteria function not only as a sulphide detoxifiers, of course. The energy released in sulphide oxidation generates ATP and reducing power, and some of these energy currencies are spent to drive the net fixation of CO_2 via the Calvin-Benson cycle (Felbeck, 1981). Some of the reduced carbon compounds synthesised within the symbionts are translocated to the host to supply part, and possibly the major part, of its reduced carbon requirements.

The bacterial symbionts in the trophosome of *Riftia pachyptila* have several adaptations to protect the animal from sulphide toxicity. They oxidize only sulphide; they have a V_{max} for sulphide that is at least an order of magnitude greater than free-living sulphur oxidising bacteria; and they can shunt excess sulphide into non-toxic elemental sulphur deposits. Elemental sulphur which can constitute up to 20% dry wt. of the trophosome is further oxidised when blood sulphide levels decrease (Wilmot and Vetter, 1989).

In summary, the chemoautotrophic sulphur bacteria symbionts of marine invertebrates living in diverse high sulphide environments appear to serve their hosts in at least two important ways. In many cases, the bacteria are instrumental in detoxifying sulphide and protecting the animal's respiratory capacities. In addition, the energy released during this 'detoxification' process may be trapped in biologically useful forms and may be used to generate reduced carbon compounds that are translocated to the animal host.

Where the bacteria form a peripheral coating around the animal the protective function may be the only role played by the bacteria, and the bacteria may be solely responsible for oxidising sulphide. In other cases, however, the animal portion of the symbiosis is actively involved in oxidising sulphide, e.g. in *S. reidi* (Figure 6.2), and the bacteria enter into the chain of sulphur metabolising reactions subsequent to the initial step or steps.

Metazoan Sulphide Metabolism

The startling discovery of symbioses between metazoans and sulphur oxidising bacteria has perhaps overshadowed the finding that non-symbiont containing metazoans from high sulphide environments and the animal portion of these same symbioses clearly possess specific sulphide detoxification schemes. While early studies attempted to explain the oxidation of sulphide by animals as a fortuitous result of metal ion or haem catalysis (Baxter and Van Reen, 1958; Sorbo, 1958), evidence is accumulating that specific sulphide oxidising enzyme systems, are found in many animal cells.

Using the sulphide specific reduction of the dye benzyl viologen as an indicator of sulphide oxidising activity, Powell and Somero (1985), have been able to examine the relative sulphide oxidising potentials and tissue localisation of detoxification processes in metazoans that either contain or lack sulphur bacteria symbionts. In symbiont containing species, sulphide oxidising enzymes are present in the superficial cell layers of symbiont-free tissues (the foot and mantle tissues of the bivalves, and the body wall of *R. pachyptila*) and, in the case of *R. pachyptila*, in the symbiont containing organ, the trophosome (Powell and Somero, 1986b). Our studies of marine animals lacking sulphur bacteria symbionts have focused on the decapod crustaceans (Vetter *et al.*, 1987) and fish (Bagarinao and Vetter, 1989). Many decapods occur in high sulphide environments, and the comparatively large sizes of the animals have allowed repetitive blood sampling for enzymatic activity determinations on isolated tissues. In a systematic survey of five tissues of the hydrothermal vent crab *Bythograea thermydron* and a comparison of muscle and hepatopancreas in seven other species, the sulphide oxidising potential was always concentrated in the hepatopancreas. The hepatopancreas was typically an order of magnitude greater in sulphide oxidising ability than all other tissues. Muscle and gill tissues had sulphide oxidising capacities, expressed on the basis of protein concentration, that were the same as the non-specific oxidation rates of solutions of proteins such as albumin (Vetter *et al.*, 1987). When fish from different habitats ranging from saltmarsh tidal creeks to bays to the open ocean were tested for their ability to withstand sulphide, there were clear species differences, with those organisms inhabiting sulphidic environments showing the greatest resistance to sulphide poisoning (Bagarinao and Vetter, 1989). Sulphide-oxidising activity correlated with hemoglobin content. Sulphide never accumulated in the blood but thiosulfate reached levels of above 2 mM.

The above data notwithstanding, the demonstration of a specific, in-vivo sulphide detoxification system is extremely difficult. Such a system would have to maintain sulphide levels well below the K_i of cytochrome oxidase (pM to very low nM sulphide). We would expect that this system would be sensitive to denaturation and produce a specific water-soluble sulphur metabolite. The benzyl viologen method typically uses mM sulphide concentrations and anaerobic conditions and cannot be used at the lower sulphide concentrations where toxic effects are first observed at the whole animal level (Vetter *et al.*, 1987; Anderson *et al.*, 1988; Bagarinao and Vetter, 1989). Since much of the activity remains after boiling, the assay probably measures a mixture of specific and non-specific sulphide catalysts including heme iron. The only way to study sulphide oxidation at physiologically realistic sulphide levels is with radiolabelled sulphide and HPLC detection (Vetter, Matrai, Javor and O'Brien, 1989). To date we have found only one sulphide oxidising system which has sufficient activity and affinity to keep sulphide from inhibiting cytochrome oxidase. It is located in the the intermembrane space of the mitochondrion of sulphide resistant metazoans and depends upon the very same electron transport system which is poisoned by sulphide (Powell and Somero, 1986a). The mitochondrial oxidation system was probably overlooked because it requires intact mitochondria and is inhibited above about 50 μM sulphide. In mitochondria isolated from *Solemya reidi* and the sulphide tolerant fish *Fundulus parvipinnis*, the mitochondrial system is completely inhibited by boiling and cyanide and produces only thiosulfate as a soluble metabolite (Vetter, Matrai, Javor and O'Brien, 1989; T. Bagarinao, unpublished results). The physiological function of non-specific, heme-catalysed sulphide oxidation (Powell and Somero, 1985; Powell and Arp, 1989) is not completely known but it may be important in oxidising high concentrations of sulphide which would overwhelm the mitochondrial sulphide oxidising system.

Perhaps the most interesting aspect of metazoan sulphide oxidation is the finding that electrons liberated in the oxidation of sulphide can be funnelled into the electron transport chain resulting in ATP production (Powell and Somero, 1986a; Figure 6.2). While many of the details concerning this pathway are not known, we do know that the number of ATPs produced per oxygen consumed is near 1:1, suggesting that electrons enter at the level of cytochrome c. Since cytochrome c is the only portion of the electron transport chain that projects into the intermembranous space of the mitochondrion, this also suggests that the sulphide oxidase enzyme is located on the outer surface of the inner membrane. A similar phenomenon has been described for the soil amoeba *Acanthamoeba castellanii* (Lloyd, Kristensen and Degn, 1981). While they did not demonstrate ATP production directly, they showed that at low concentrations sulphide clearly stimulated oxygen consumption and that inhibition of electron transport with NaN_3 prevented sulphide consumption.

The ability of a metazoan to generate ATP from an inorganic energy source such as sulphide is not without precedent. The enzyme sulphite oxidase, which has been purified from a variety of organisms including humans, is also located in the intermembranous space of mitochondria and also generates ATP via cytochrome c (Rajagopalan, 1980).

One of the major objectives of our current research is to determine how widespread sulphide dependent ATP production is among the different Metazoa. Since we have clearly demonstrated that Metazoa from sulphide rich habitats have sulphide oxidising capacities that are far in excess of those of animals from normal aerobic habitats, we believe there is a high likelihood that other sulphide biome metazoans may link sulphide oxidation to electron transport, that is, to ATP production.

The second objective of our current research is to understand the complete *in vivo* pathway of sulphide oxidation. We know that the primary product released into the blood in a variety of sulphide tolerant animals is thiosulphate; sulphite appears at only much lower concentrations (Vetter *et al.*, 1987, 1989). This suggests that the overall detoxification requires several steps (enzymes?). While sulphite once produced can non-enzymatically carry out a nucleophilic attack on a putative polysulphide intermediate, this reaction could also be carried out by rhodanese (Sorbo, 1960) an enzyme whose true function is not fully understood (Cerletti, 1986). Currently, we have no plausible enzymatic mechanism for the production of sulphite.

Summary

The recent discoveries that metazoans from hypoxic, high sulphide environments possess specific biochemical adaptations to overcome the potentially toxic effects of hydrogen sulphide, and that certain of these animals have found ways to exploit the tremendous energy potential of sulphide suggest that the sulphur biochemistry of these organisms will turn out to be as exciting and diverse as the different anaerobic carbon pathways discussed by the other authors in this volume. Aside from understanding the specific sulphur biochemistry of these organisms, an appreciation of the potential costs and benefits of sulphide adds a new perspective to the study of other aspects of anaerobic metabolism. For example, in environments where some oxygen is present, is aerobic metabolism 'defensible' in terms of potential rates of sulphide influx and what are the additional costs of this defence in terms of energy or additional oxygen required for detoxification? Are there toxic effects of hydrogen sulphide, as a strong nucleophile, that require its metabolism even when organisms rely entirely on anaerobic metabolism? Conversely, if ATP production from sulphide oxidation turns out to be a widespread phenomenon among animals from high sulphide environments, we will have to factor this component of energy metabolism into the overall energy budgets of these organisms.

In summary, then, the study of the detoxification and exploitation of sulphide by animals from sulphide rich marine environments must be appreciated as a necessary and important complement to the well developed studies of the pathways of carbon and nitrogen metabolism in many of these same animals. Mechanisms for detoxifying sulphide may be instrumental in allowing these animals to invade and thrive in low-oxygen, high-sulphide waters, and the possible contribution made by the ATP generated during sulphide oxidation may require a careful reanalysis of the energy budgets of these organisms.

Acknowledgement

The studies of the authors were supported by National Science Foundation Grants OCE83-11259 and PCM83-00983.

References

Anderson, A.E. (1989) *Physiological Studies of* Solemya reidi, *a Highly Integrated Marine Invertebrate-Bacteria Symbiosis.* Doctoral Thesis (University of California, Santa Barbara)

References

Anderson, A.E., Childress, S.J. and Favuzzi, J.A. (1987) 'Net Uptake of CO_2 Driven by Sulphide and Thiosulphate Oxidation in the Bacterial Symbiont Containing Clam, *Solemya reidi* ' *Journal of Experimental Biology* **133**, 1-

Arp, A.J. and Childress, J.J. (1981) 'Blood Function in the Hydrothermal Vent Ventimentiferan Tube Worm'. *Science* **219**, 342-344

Arp, A.J. and Childress, J.J. (1983) 'Sulfide Binding by the Blood of the Hydrothermal Vent Tube Worm *Riftia pachyptila*'. *Science* **219**, 295-297

Arp, A.J., Childress J.J., and Fisher, C.R. Jr. (1984) 'Metabolic and Blood Gas Transport Characteristics of the Hydrothermal Vent Bivalve *Calyptogena magnifica*'. *Physiological Zoology* **57**, 648-662

Arp, A.J., Childress, J.J., and Vetter, R.D. (1987) 'Sulfide Binding Protein in the Blood of *Riftia pachyptila* is the Extracellular Haemoglobin'. *Journal of Experimental Zoology* **128**, 139-158

Bagarinao, E. and Vetter, R.D. (1989) 'Sulfide Tolerance and Detoxification in Shallow-water Marine Fishes'. *Marine Biology* (in press).

Baxter, C.F. and Van Reen R. (1958) 'The Oxidation of Sulfide to Thiosulfate by Metallo-Protein Complexes and by Ferritin'. *Biochimica et Biophysica Acta* **28**, 573-578

Boaden, P.J.S. and Platt, M. (1971) 'Daily Migration Patterns in an Intertidal Meiobenthic Community'. *Thalassia Jugoslavica* **7**, 1-12

Broderius, S.J. and Smith L.L. Jr. (1976) 'Effect of Hydrogen Sulfide on Fish and Invertebrates Part II. Hydrogen Sulfide Determination and Relationship Between pH and Sulfide Toxicity'. EPA-600/3-76-062a

Cary, S.C., Vetter, R.D., and Felbeck, H. (1989) 'Habitat Characterization and Nutritional Strategies of the Endosymbiont-Bearing Bivalve *Lucinoma aequizonata*' *Marine Ecology Progress Series* (in press)

Cavanaugh, C.M. (1983) 'Symbiotic Chemoautotrophic Bacteria in Marine Invertebrates from Sulphide-Rich Habitats'. *Nature (London)* **302**, 58-61

Cavanaugh, C.M., Gardner, S.L., Jones, M.L., Jannasch, H.W., and Waterbury, J.B. (1981) 'Procaryotic Cells in the Hydrothermal Vent Tube Worm *Riftia pachyptila* Jones: Possible Chemoautotrophic Symbionts'. *Science* **213**, 340-342

Cerletti, P. (1986) 'Seeking a Better Job for an Under-Employed Enzyme: Rhodanese'. *Trends in Biochemical Science* **11**, 369-372

Chen, K.Y. and Morris, J.C. (1972) 'Kinetics of Oxidation of Aqueous Sulfide by O_2'. *Environmental Science and Technology* **6**, 529-537

Cline, J.D. and Richards, B.R. (1969) 'Oxygenation of Hydrogen Sulfide in Sea Water at Constant Salinity, Temperature and pH'. *Environmental Science and Technology* **3**, 838-843

Corliss, C.M., Dymond, J., Gordon, L.I., Edmond, J.M., von Herzon, R.P., Ballard, R.D., Green, K., Williams, D., Bainbridge, A., Crane, K., and van Andel T.H. (1979) 'Submarine Hydrothermal Springs on the Galapagos Rift'. *Science* **203**, 1073-1083

de Burgh, M.E. and Singla, C.L. (1984) 'Bacterial Colonization and Endocytosis on the Gill of a New Limpet Species from a Hydrothermal Vent'. *Marine Biology* **84**, 1-6

Desbruyeres, D.P. and Laubier, L. (1980) '*Alvinella pompejana* gen. Ampharetidae Aberrant des Sources Hydrothermales de la Ride Es-Pacifique'. *Oceanologica Acta* **3**, 267-274

Desbruyeres, D., Gaill, F., Laubier, L., Prieur, D., and Rau, G.H. (1983) 'Unusual Nutrition of the "Pompeii worm" *Alvinella pompejana* (Polychaetous Annelid) from a Hydrothermal Vent Environment: SEM, TEM, ^{13}C and ^{15}N evidence'. *Marine Biology* **75**, 201-205

Dubulier, N. (1986) 'Association of Filamentous Epibacteria with *Tubificoides benedii* (Oligochaeta:Annelida)'. *Marine Biology* **92**, 285-288

Edmond, J.M., von Damm, K.L., McDuff, R.E., and Measures, C.I. (1982) 'Chemistry of Hot Springs on the East Pacific Rise and Their Effluent Dispersal'. *Nature (London)* **297**, 187-191

Felbeck, H. (1981) 'Chemoautotrophic Potential of the Hydrothermal Vent Tube Worm, *Riftia pachyptila* Jones (Ventimentifera)'. *Science* **213**, 336-338

Felbeck, H. (1983) 'Sulfide Oxidation and Carbon Fixation by the Gutless Clam *Solemya reidi:* An Animal-Bacterial Symbiosis'. *Journal of Comparative Physiology B* **152**, 3-11

Felbeck, H., Childress, J.J., and Somero, G.N. (1981) 'Calvin-Benson Cycle and Sulphide Oxidation Enzymes in Animals from Sulphide-Rich Habitats'. *Nature (London)* **293**, 291-293

Felbeck, H., Liebezeit G., Dawson, R., and Giere, O. (1983) 'CO_2 Fixation in Tissues of Marine Oligochaetes (*Phallodrilus leukodermatus* and *P. planus*) Containing Symbiotic, Chemoautotrophic Bacteria'. *Marine Biology* **75**, 187-191

Fenchel, T.M. and Riedl, R.J. (1970) 'The Sulfide System: A New Biotic Community Underneath the Oxidized Layer of Marine Sand Bottoms'. *Marine Biology* **7**, 255-268

Fisher, C.R., Jr (1989). 'Marine Invertebrates and their Chemolithoautotropic Symbionts' in *Critical Reviews in Aquatic Sciences*, (CRC Press, Boca Raton, in press)

Giere, O. (1981) 'The Gutless Marine Oligochaete *Phallodrilus leukodermatus.* Structural Studies on an Aberrant Tubificid Associated with Bacteria'. *Marine Ecology Progress Series* **5**, 353-357

Gilbert, H. (1984) 'Redox Control of Enzyme Activities by Thiol/disulfide Exchange'. *Methods in Enzymology* **107**, 330-351

Hand, S.C. and Somero, G.N. (1983) 'Energy Metabolism Pathways of Hydrothermal Vent Animals: Adaptations to a Food-Rich and Sulfide-Rich Deep Sea Environment'. *Biological Bulletin* **165**, 167-181

Howarth, R.W. and Teal, J.M. (1980) 'Energy Flow in a Salt Marsh Ecosystem: The Role of Reduced Inorganic Sulphur Compounds'. *American Naturalist* **116**, 862-872

Johnson, K.S., Beehler, C.L., Sakamoto-Arnold, C.M., and Childress, J.J. (1986) '*In situ* Measurements of Chemical Distributions in a Deep-Sea Hydrothermal Vent Field'. *Science* **231**, 1139-1141

Jorgensen, B.B. (1982) 'Ecology of the Bacteria of the Sulphur Cycle With Special Reference to Anoxic-Oxic Interface Environments'. *Philosophical Transactions of the Royal Society of London, Series B* **298**, 543-560

Keunen, J.G. and Beudeker R.F. (1982) 'Microbiology of Thiobacilli and Other Sulphur-Oxidizing Autotrophs, Mixotrophs and Heterotrophs'. *Philosophical Transactions of the Royal Society of London, Series B* **298**, 113-116

Laubier, L., Desbruyeres, D. and Chassard-Bouchaud, C. (1983) 'Microanalytical Evidence of Sulphur Accumulation in a Polychaete from Deep Sea Hydrothermal Vents', *Marine Biology Letters* **4**, 113-116

Lloyd, D., Kristensen, D., and Degn, H. (1981) 'Oxidative Detoxification of Hydrogen Sulphide Detected by Mass Spectrometry in the Soil Amoeba *Acanthamoeba castellanii*'. *Journal of General Microbiology* **126**, 167-170

Luther, G.W. III, Church, T.M., Scudlark, J.R., and Cosman, M. (1986) 'Inorganic and Organic Sulphur Cycling in Salt-Marsh Pore Waters'. *Science* **232**, 746-749

Maguire, C. and Boaden, P.J.S. (1975) 'Energy and Evolution in the Thiobios: An Extrapolation from the Marine Gastrotrich *Thiodasys sterreri'*. *Cahiers de Biologie Marine* **16**, 635-646

Millero, F.J., Plese, T., and Fernandez, M. (1988) 'The Dissociation of Hydrogen Sulfide in Seawater'. *Limnology and Oceanography* **33**, 269-274

National Research Council, Division of Medical Science, Subcommittee on Hydrogen Sulfide. (1979) *Hydrogen Sulfide*. (University Park Press, Baltimore)

Nicholls, P. (1975) 'The Effect of Sulphide on Cytochrome aa$_3$ Inosteric and Allosteric Shifts of the Reduced Peak'. *Biochimica et Biophysica Acta* **396**, 24-35.

Patel, S. and Spencer, C.P. (1963) 'The Oxidation of Sulphide by the Haem Compounds from the Blood of *Arenicola marina*'. *Journal of the Marine Biological Association of the U.K.* **43**, 167-175

Peck, H.D. and LeGall J. (1982) 'Biochemistry of Dissimilatory Sulphate Reduction'. *Philosophical Transactions of the Royal Society of London, Series B* **298**, 443-465

Powell, M.A., and Arp, A.J. (1989) 'Hydrogen Sulfide Oxidation by Abundant Nonhemoglobin Heme Compounds in Marine Invertebrates from Sulfide-Rich Habitats' *Journal of Experimental Zoology* **249**, 121-132

Powell, E.N., Crenshaw, M.A., and Reiger, R.M. (1979) 'Adaptations to Sulfide in the Meiofauna of the Sulfide System. 1. ^{35}S Accumulation and the Presence of a Sulfide Detoxification System'. *Journal of Experimental Marine Biology and Ecology* **37**, 57-76

Powell, E.N., Crenshaw, M.A., and Reiger, R.M. (1980) 'Adaptations to Sulfide in Sulfide-System Meiofauna. Endproducts of Sulfide Detoxification in Three Turbellarians and a Gastrotrich'. *Marine Ecology Progress Series* **2**, 169-177

Powell, M.A. and Somero, G.N. (1985) 'Sulfide Oxidation Occurs in the Animal Tissue of the Gutless Clam, *Solemya reidi*'. *Biological Bulletin* **169**, 164-181

Powell, M.A. and Somero, G.N. (1986a) 'Hydrogen Sulfide Oxidation is Coupled to Oxidative Phosphorylation in Mitochondria of *Solemya reidi*'. *Science* **233**, 563-566

Powell, M.A. and Somero, G.N. (1986b) 'Adaptations to Sulfide by Hydrothermal Vent Animals: Sites and Mechanisms of Detoxification and Metabolism'. *Biological Bulletin* **171**, 274-290

Powell, M.A., Vetter, R.D., and Somero, G.N. (1987) 'Sulfide Detoxification and Energy Exploitation by Marine Animals' in P. Dejours, L. Bolis, C.R. Taylor, E.R. Weibel (eds) Fidia Research Series, IX (Liviana Press, Padova)

Rajagopalan, K.V. (1980) 'Sulfite oxidase' in Michael P. Coughlan (ed), *Molybdenum and Molybdenum-containing Enzymes* (Pergamon Press, Oxford, England)

Reise, K. and Ax, P. (1979) 'A Meiofauna "thiobios" Limited to the Anaerobic Sulfide System of Marine Sand Does Not Exist'. *Marine Biology* **54**, 225-237

Reise, K. and Ax, P. (1980) 'Statement on the Thiobios-Hypothesis'. *Marine Biology* **58**, 31-32

Smith, L.L. Jr., Oseid, D.M., Adelman, I.R., and Broderius, S.J. (1976) 'Effect of Hydrogen Sulfide on Fish and Invertebrates. Part I. Acute and Chronic Toxicity Studies'. EPA 600/3-76-062a.

Somero, G.N. (1987) 'Exploitation of Hydrogen Sulfide by Animal-Bacteria Symbioses'. *News of Physiological Sciences* **2**, 3-6

Somero, G.N., Childress, S.S., and Anderson, A.E. (1989) Transport, Metabolism, and Detoxification of Hydrogen Sulfide in Animals from High Sulfide Environments. Critical Reviews in Aquatic Sciences. CRC Press. (Boca Raton, in press)

Sorbo, B. (1958) 'On the Formation of Thiosulfate from Inorganic Sulfide by Liver Tissue and Heme Compounds'. *Biochimica et Biophysica Acta* **27**, 324-329

Sorbo, B. (1960) 'On the Mechanism of Sulfide Oxidation in Biological Systems'. *Biochimica et Biophysica Acta* **38**, 349-351

Stein, J.L., Cary, S.C., Hessler, R.R., Ohta, S., Vetter, R.D., Childress, J.J., and Felbeck, H. (1988) 'Chemoautotrophic symbiosis in a Hydrothermal Vent Gastropod'. *Biological Bulletin* **174**, 373-378

Vetter, R.D. (1985) 'Elemental Sulphur in the Gills of Three Species of Clams Containing Chemoautotrophic Symbiotic Bacteria: A Possible Inorganic Energy Storage Compound'. *Marine Biology* **88**, 33-42

Vetter, R.D., Matrai, P.A., Javor, B., and O'Brien, J. (1989) 'Reduced Sulphur Compounds in the Marine Environment: Analysis by High-Performance Liquid Chromatography' in E Saltsman and W. Cooper (eds), *Biogenic Sulphur in the Environment, American Chemical Society Symposia* **Series 393**, 243-261

Vetter, R.D., Wells, M.E., Kurtsman, A.L., and Somero, G.N. (1987) 'Sulfide Detoxification by the Hydrothermal Vent Crab *Bythograea thermydron* and Other Decapod Crustaceans'. *Physiological Zoology* **60**, 121-137

Wilmot, D.B. and Vetter, R.D. (1989) 'The Bacteria symbiont from the Hydrothermal Vents Tubeworm, *Riftia pachyptila*, is a Unique Sulfide Specialist' Abstracts of the 89th Annual Meeting of the American Society of Microbiologists p 224

Chapter 7

Interstitial Meiofauna

Warwick L. Nicholas

Introduction

The interstitial fauna are animals that live within the labyrinth of spaces between the particles which constitute marine and limnetic sediments. The availability of oxygen is often a limiting factor in their biology. They are smaller than the animals that burrow into sediments, displacing the sediments in the process, and can be distinguished by their ability to pass through fine mesh sieves. The meiofauna (or meiobenthos) are conventionally separated from the macrofauna by passing through a 2 mm mesh and from the microfauna by retention on a 30-50 µm mesh. Some Protozoa are larger than the smallest Metazoa, and lie within these limits, but will not be discussed. Many metazoan phyla are represented in the meiofauna, but because the Nematoda are the most numerous and the most studied they will receive most attention here.

The terms meiofauna and interstitial fauna are commonly applied to aquatic animals, but many essentially aquatic Metazoa, from the same taxonomic groups, inhabit the pore spaces in terrestrial soils. They are dependent for activity on a film of moisture on the soil particles, so that it is not useful to draw a sharp distinction between aquatic and terrestrial meiofauna. When soils are flooded with water, oxygen, which diffuses very much more slowly in water than air, becomes depleted through bacterial action, and the interstitial fauna may be exposed to anoxia. Unlike the meiofauna of permanent bodies of water, the soil meiofauna can often tolerate drying in an inactive 'cryptobiotic' state.

The interstitial fauna may respond to oxygen deficiency in a number of ways. They may avoid anoxic conditions by migration or they may limit its effect by physiological adaptations. Anaerobic respiration is one possible adaptation to anoxia and is more likely to be found where anoxic conditions are extreme and prolonged. In fresh water and terrestrial environments anoxic conditions are often temporary, as in flooded soils, or seasonal, as in lakes which become thermally stratified. When anoxic conditions are not indefinitely prolonged, the fauna may survive anoxia in an inactive dormant state. In sub-littoral marine sediments anoxic conditions may be permanent and invariant just a millimetres below the surface of the mud.

Size and Aerobic Respiration

Because meiofauna are so small, their aerobic respiration in well oxygenated water is not limited by the diffusion of oxygen into their tissues. When the partial pressure of oxygen in the water falls below some critical value this will no longer be true. The question of what this critical value may be has been examined theoretically for nematodes by Atkinson (1976, 1980), using equations derived by Hill (1929). The appropriate equation determines the radius of a cylinder that will permit oxygen to diffuse to the centre at an adequate rate, assuming a constant metabolic rate, a constant partial pressure in the surrounding medium and a cylindrical body. The equation is

$$r_o = (4 ky_o/a)$$

where r_o is the radius (cm), k is the diffusion coefficient of oxygen in tissue (taken by Atkinson to be 8.4×10^{-4} cm^2/atm/h); y_o the partial pressure of oxygen (atm) in the surrounding medium; and a the rate of metabolism in cm^3 O$_2$/cm^3 tissue/h.

The rate of metabolism is related to the body size in nematodes, as in all poikilotherms, by an equation:

$$Y/X = ax(b-1)$$

where Y is oxygen consumption, X is weight and a and b are constants. However, there have been differing estimates of b (Nicholas, 1984). Atkinson used published data on oxygen consumption in free-living nematodes, and $b = 0.79$, to derive a, the rate of oxygen use for nematodes of differing size.

With these values of a, Atkinson graphed the relationship between maximum oxygen uptake and body radius for varying partial pressures of oxygen using Hill's equation. Atkinson does not give the actual figures assumed for consumption and size. A fairly typical marine nematode might use 1 nl/h at 20° or 0.6 nl/mg wet weight/h with a radius of 25 μm (Nicholas, 1984). The result of these calculations was that oxygen diffusion was not likely to limit aerobic respiration, given a body radius of 50 μm or less, with a partial pressure of oxygen equal to or greater than 15 mm Hg (about 10% of atmospheric air). The argument necessarily makes a number of simplifying assumptions. Among these are to take the value of oxygen permeability from the tissues of higher animals, to assume that the critical value was determined by oxygen used at the core of the animal (the gut) rather than the body wall, and to ignore the effects of body-movement on the circulation of body fluids. Haemoglobin, present in tissues of some nematodes, may facilitate oxygen diffusion. Both diffusion and oxygen consumption vary with temperature.

There is little relevant experimental data on free-living nematodes, or other meiofauna, and what there is comes from typically aerobic soil nematodes. In *Caenorhabditis briggsae,* growth is impaired with partial pressures of oxygen (pO_2) less than 30 mm Hg and ceases at 1.2 mm Hg (Nicholas and Jantunen, 1964). With *Pelodera punctata,* from sewage sludge, activity required a pO_2 greater than 5 mm Hg (Abrams and Mitchell, 1978), equivalent to about 0.7% atmospheric oxygen. *C. briggsae* is a regulator (as distinct from a conformer) in that oxygen uptake is independent of pO_2 until it falls to a critical low value, below which oxygen uptake decreases very rapidly with decreasing pO_2. This is typical of respiration in which oxygen consumption is dependent on cytochromes (Bryant, Nicholas and Jantunen, 1967). For *C. briggsae,* the critical pO_2 is about 5% atmospheric oxygen. Cooper and Van Gundy (1970) also found 5% oxygen a critical value for *Caenorhabditis* sp. and *Aphelenchus avenae.* For *Caenorhabditis elegans* it is about 3.6% O_2 (or 27 mm Hg) (Anderson and Dusenbery, 1977).

The Marine Meiofauna

Marine sediments contain very dense meiofaunal populations which, though the individuals are small, add up to significant biomass. The Nematoda usually accounting for about 90% of the numbers of individuals and a large fraction of the biomass. Many estimates have been made of nematode population densities and biomass from a wide variety of habitats (Platt and Warwick, 1980; Nicholas, 1984) and some examples are given in Table 7.1. The meiofauna, predominantly the nematodes, can also account for a large fraction of the total ATP in marine sediments, exceeding by a large factor bacterial ATP. At an intertidal and a subtidal location off the coast of South Carolina, Sikora, Sikora, Erkenbrecher and Coull (1977) found nematodes accounted for between 68% and over 90% of the total ATP in the sediment.

The marine meiofauna is also taxonomically very diverse with many different species living in close association. Some examples of the contributions made by major taxa to meiofaunal populations are given in Table 7.2. The Nematoda are the most diverse, though some of the other taxa, although present in smaller numbers, may still be very rich in species. Thus in the littoral sediments in the North Sea studied by Reise and Ax (1979) there were 35 species of Turbellaria, and six species of Gnathostomulida. In a sheltered bay on the Irish coast, Boaden and Platt (1971) found 21 species of Nematoda, four species of Crustacea, 27 species of Turbellaria, five species of Annelida and seven species of Gastrotricha. The Gnathostomulida are apparently all members of the meiofauna. Ostracoda, Archiannelida, Tardigrada, and Rotifera are also frequently present.

Table 7.1 A few examples selected to show the densities of marine meiofauna from a variety of habitats.

habitat	numbers $10cm^2$	biomass $10cm^2$ (mg dry wt)
1. Littoral, exposed beach, Scotland	500 - 6700	0.31 - 1.1
2. Littoral, sheltered beach, Ireland	740	?
3. Estuary, mangrove, South Africa	840 - 5300	0.233 - 0.37
4. Shallow, sub-littoral, Mediterranean	4480	3.76
5. Deep-sea, W. Atlantic	40 - 849	?
6. Deep-sea, W. Pacific	37 - 114	0.012 - 0.478

1. McIntyre and Murison, 1973; 2. Boaden and Platt, 1971; 3. Dye, 1983; 4. Boucher, 1972; 5. Tietjen, 1971; 6. Shirayama, 1984a

Table 7.2 A few examples showing the percentage of the population made up of the predominate taxa in a variety of habitats.

Taxon	Littoral Ireland[1]	Mangrove S. Africa[2]	Sub-Littoral Scotland[3]	N. Sea[4]	Mediterranean[5]	Deep Sea E. Atlantic[6]
Nematoda	77.2	80.4	76	95	84.3	88.3
Copepoda	2.5	1.9	7.6	2.2	11.2	3.9
Nauplii	-	-	2.7	1.2	0.8	5.4
Other Crustacea	-	-	-	-	0.6	0.2
Turbellaria	8.9	2.8	6.3	-	-	-
Gastrotricha	3.2	0.4	6.6	-	<0.1	-
Kinorhyncha	-	2.7	-	0.7	2.9	0.3
Annelida	-	5.1	0.8	0.5	-	1.1
Others	7.2	6.3	0.57	0.5	-	0.2

1. Boaden and Platt 1971; 2. Dye 1983; 3. McIntyre and Murison 1973; 4. Juario 1975; 5, Boucher 1972; 6. Dinet and Vivier 1977

The Sulphide Biome and Thiobios.

The availability of oxygen and the physical dimensions of the sedimentary particles dominate the ecology of the interstitial fauna. In the ocean these are, in turn, both dependent on wave action. On high-energy beaches between 20 and 200 tons of clean sea water may pass through each linear metre every day (Fenchel and Riedl, 1970) oxygenating the sand to a considerable depth. The sand contains diverse fauna, reviewed by Swedmark (1964) and McIntyre (1969) which includes the smallest representatives of most of the metazoan phyla. However, in the absence of strong waves the circulation of oxygenated water is greatly reduced and smaller sedimentary particles are deposited, further restricting the penetration of oxygen into the sediment. For these reasons an anoxic zone develops below the surface of the sediment of vast extent throughout the world's oceans, as well as in sheltered coastal waters and estuaries. Its fauna differs from that of exposed sandy beaches.

Fenchel and Riedl (1970) showed that this anoxic zone, which they termed the sulphide biome, was inhabited by a rich microflora and meiofauna. Its fauna consists of many protozoa as well as many of the simplest Metazoa, the latter constituting its meiofauna. Boaden and Platt (1971) suggested calling this association the thiobios, a term which has met general acceptance. Some biologists have speculated that its fauna may have had a long separate evolutionary history dating back to times when oxygen was less abundant in the atmosphere than it is today. However, Powell, Bright, Woods and Gittings (1983) have redefined the thiobios in terms of the presence of sulphide rather than the lack of oxygen, for reasons discussed below, though the two are usually associated.

The sulphide biome usually has a sharp upper boundary where the redox electrode potential, E_h, falls rapidly with depth in the sediment from positive values to low negative values (Fenchel, 1969). The boundary is called the Redox Discontinuity Zone, RDZ. The sediment below is usually black, due to iron sulphide, smells of H_2S, and is characterized by the presence of reduced compounds, CH_4, NH_3, H_2S as well as CO_2. The pH is usually about 6.8. Anaerobic bacteria produce reduced gases in decomposing organic matter. Some, such as *Desulphovibrio* and *Desulphomaculum*, liberate H_2S. Because the reduced gases react with oxygen, creating a chemical oxygen demand, and because oxygen diffuses slowly into the sediments, in the absence of turbulent wave action, conditions become anoxic below the RDZ (Fenchel, 1969; Fenchel and Riedl, 1970).

The microbiological and physical chemistry of oxygen, sulphur and hydrogen sulphide in shallow sub-tidal sediments have been investigated by Danish workers (Jorgensen, 1977; Troelsen and Jorgensen, 1982). In these temperate latitudes there is a seasonal cycle, the top 3-4 cm of the silt/clay sediments becoming oxic in winter, but with anoxic conditions reaching the surface in summer. Oxic sediments have redox potentials, E_h, of +300 to +200 mV. Sulphate is reduced in the range +200 to -100 mV and H_2S accumulates below -100mV. H_2S may reduce sediments to -350 mV (Fenchel, 1969). Marine and estuarine sediments contain a lot of iron, which becomes reduced to Fe_2S or FeS by microbial action and is re-oxidised by oxygen diffusing into the sediments, creating a significant additional chemical demand for oxygen (Board, 1976).

Oxygen is generally considered to be absent where the E_h has a negative value. However, most electrodes used to measure the potential are large relative to the crumb structure of the sediments, and have a significant oxygen demand. Within the crumbs, often faecal pellets, oxygen may be absent and H_2S present, while in the surrounding interstitial water some oxygen may be present (Jorgensen, 1977). Mangrove mud-flats in southern Africa have such a high chemical oxygen demand that oxygen is not present more than a few millimetres from the surface of the mud (Dye, 1983).

The Existence of an Anaerobic Marine Meiofauna

The first person to claim that the marine meiofauna could tolerate anaerobic conditions was probably Moore (1931), who experimented with nematodes from the sub-littoral sediments of Clyde Loch, off the coast of Scotland. Nematodes, abundant in the top 2-9 cm of mud, survived for many days in closed containers with de-oxygenated sea water, and at least two weeks after flushing with hydrogen. Survival does not of course necessarily mean anaerobic metabolism.

It seems universally true that meiofaunal densities are greatest in the top few centimetres of sediments, with densities declining with depth. There are examples of such distributions in littoral sediments (Boaden and Platt, 1971; Platt, 1977), sub-littoral sediments (Boucher, 1972; Juario, 1975; Tietjen,1969) and deep ocean sediments (Tietjen, 1971; Shirayama, 1984b), from many parts of the world and many more could be quoted. Although the fact that densities decline with depth in the sediments suggests dependence on oxygen, it must be remembered that available food also declines with depth, whether this comes from primary production in the photic zone or the slow sedimentation of organic matter in deeper water.

Some species, however, reach their highest densities in the oxygen deficient sulphide-rich regions. Bouman (1983) found that some nematodes extended to 30 cm in the highly reduced muds in the Ems estuary. Long thin forms predominated. Blome (1983) found 17% of the nematode fauna lived below the RDZ in sandy beaches in the North Sea. Ott and Schiemer (1973) found that the nematodes restricted to the upper 1 cm of oxic intertidal sands in North Carolina were quite distinct from those rarely found at the surface. Of 165 species, 50%, representing most families except Enoploidea, were found below the 100 mV isopleth. On a Bermuda beach Schiemer (1973) found two species of Gnathostomulida, one that lived at the surface, the other below the RDZ. Mangrove mud-flats in southern Africa, though anoxic almost to the surface, have a rich meiofauna, mostly nematodes (Dye, 1982; and see Tables 7.1 and 7.2). Some nematodes, turbellarians and gastrotrichs show daily vertical migrations correlated with day night or tidal cycles (Boaden and Platt, 1971; Rieger and Ott, 1971).

Studies of the meiofauna inhabiting a submarine canyon draining a brine-seep adjacent to The East Flower Garden Bank in The Gulf of Mexico have proved extraordinarily interesting (Powell, Bright, Woods and Gittings, 1983). The dense brine which collects in Gollum's Lake is anoxic, sulphide-rich and contains no living animals. It flows out of the lake down a canyon entraining as it does so oxygenated sea water from above the canyon so that, unlike the lake, the sulphide-rich water also contains some oxygen. A gradient of decreasing sulphide extends along a 96 m canyon. A meiofauna of Gnathostomulida, Turbellaria, Gastrotricha and Nematoda, taxonomically different from that inhabiting the oxybiotic regions alongside the canyon, is found in the canyon. At some places gnathostomulids and gastrotrichs exceed nematodes in density. For the thiobios inhabiting this unique habitat there is a requirement for sulphide and not the absence of oxygen.

The phylum Gnathostomulida is largely if not entirely thiobiotic, as would appear to be some families of Turbellaria - that is, Solenofilomorphidae and Retronectidae (Powell *et al*., 1983). Crustacea, such as Copepoda and Ostracoda, appear to be intolerant of sulphide. It has been shown that meiobenthic invertebrates do not exclude sulphide from their tissues (Powell, Crenshaw and Rieger, 1979) so that thiobiotic species must be able to detoxify sulphide entering their tissues. Jensen (1986) studied the nematode fauna within the East Flower Garden canyon and of the adjacent oxybiotic region. The nematodes found in the canyon belong to many taxa represented by other species in the adjacent region. This argues against the hypothesis that the thiobios is a relict fauna with a long separate evolutionary history, at least so far as the nematodes are concerned. The thiobiotic species tend to be longer and thinner than their oxybiotic relatives and Jensen suggests that they may feed on dissolved organic matter.

Physiological Responses to Oxygen Deprivation in Marine Nematodes

Observations by Moore (1931) on the tolerance of nematodes to anoxic conditions have been repeated by other researchers. *Enoplus communis,* a large intertidal nematode inhabiting seaweed becomes paralysed when deprived of oxygen, but recovers when oxygen is readmitted (Wieser and Kanwisher, 1959). Though adaptive in such a habitat, anoxic paralysis in marine muds is likely to entrap meiofauna. Wieser and Kanwisher (1961) showed that a large number of nematodes from a salt marsh tolerated oxygen deprivation for many days in mud flushed with nitrogen *in vitro,* some for up to 50 days, and many were observed to remain active. Ott and Schiemer (1973) carried out similar experiments with salt marsh nematodes. Several species from the surface layers only tolerated anoxia for several hours, but 50% of four other species from the anoxic layers survived for 3-12 days.

Experiments on aerobic respiration, by Cartesian Diver micro-respirometry, show that nematodes from well oxygenated habitats have higher rates of respiration than those from poorly oxygenated ones, which may possibly indicate a partial reliance on anaerobic respiration, even when oxygen is not limiting (Ott and Schiemer, 1973; Warwick and Price, 1979). The aerobic respiration rate of two species of Gnathostomulida from poorly oxygenated sub-tidal sands in Bermuda is similarly low (Schiemer, 1973). A difficulty with these experiments is that the organism is exposed to unnaturally high oxygen during the experiment. What is required is a method of measuring anaerobic metabolism in the laboratory of comparable sensitivity to the Cartesian Diver for aerobic respiration, but no such method has been developed.

An interesting comparison was made by Atkinson (1975, 1977, 1980) between two intertidal species of *Enoplus* with respect to their feeding activity and haemoglobin content. *E. communis* lives amongst mussel beds (*Mytilus*) on a rocky shore, and is therefore less likely to be deprived of oxygen than *E. brevis*, an inhabitant of estuarine muds. *E. brevis* concentrates in the upper 2 cm of the mud, but may occur deeper. In both species, reducing the partial pressure of oxygen in sea water reduces oxygen consumption and feeding rate (the latter measured by the rate of assimilation of a dye, amaranth), but more so in *E. communis* than in *E. brevis*.

Oxygen consumption by *E. communis* fell to 51% when the partial pressure was reduced from 140 to 20 mm Hg, and to 12% at 12 mm Hg. In *E. brevis* the corresponding figures were 78% and 18%. Feeding rates were maintained by *E. brevis* at pressures down to 20 mm Hg, but fell progressively and more drastically *in E. communis*. It seems clear that *E. brevis* can continue to feed and respire aerobically in less oxygen than *E. communis*. Both species contain haemoglobin in their tissues, but only *E. brevis* contains haemoglobin in the oesophagus (syn. pharynx), and this oesophageal haemoglobin is fully loaded with oxygen at 20 mm Hg and unloaded at 5 mm Hg. The presence of this oesophageal haemoglobin facilitates the diffusion of oxygen into the oesophagus, which must pump continually when feeding, at low partial pressures of oxygen.

Anaerobic Fresh water Meiofauna

It has been suggested that the meiofauna of many lakes must be exposed to anoxia because of thermal stratification. However, thermal stratification is usually seasonal, so that meiofauna could respire aerobically when oxygen was present and survive anoxia in quiescent or cryptobiotic conditions, in a comparable way to that which enables many small soil Metazoa to tolerate seasonal extremes of cold or aridity.

The hypolimnion of Lake Tiberias, Israel, is oxygen-free for 8 months of the year. A nematode, *Eudorylaimus andrassy* and a tubificid Oligochaete, *Euilyodrilus heuscheri* were found at high densities throughout the period (Por and Masry, 1968). Some of the time a rhabdocoel turbellarian and an insect larva (Chironomidae) (Por and Masry, 1968) were present. The nematode and the tubificid survived for 6 months sealed in the mud, after flushing with oxygen-free nitrogen. When opened, the tubificid had apparently made burrows in the mud, which smelt of H_2S. A related nematode, *Dorylaimus* sp., from African papyrus swamps survived 86 days in a quiescent state in anoxic conditions (Banage, 1966).

In Lake Neusiedlersee in Austria, the nematode, *Tobrilus gracilis* is largely restricted to the deeper layers of fine mud, from 2 to 8 cm, where conditions are reducing. In contrast, a variety of tardigrades, naiadids, gastrotrichs, rotifers and other nematodes occurs in oxic conditions. Another nematode, *Paraplectonema*, extends into the anoxic zone. However, *Tobrilus gracilis* does take up oxygen when present in a Cartesian Diver microrespirometer, though perhaps at a lower rate than some other nematodes of comparable size. Two very interesting papers on the cytology of this nematode are discussed in the section on Cytology and Ultrastructure.

Sewage sludge becomes anoxic except at the surface. *Pelodera punctata* can survive for 14 days without oxygen (Abrams and Mitchell, 1978), and requires very low levels of oxygen for growth. *Mesodiplogaster lheritieri* in culture swallows air at the surface, presumably obtaining oxygen in this way (Klinger and Kunz, 1974).

Anaerobic Terrestrial Nematodes

Soils may become anaerobic when flooded, and nematodes, amongst the commonest interstitial fauna, survive anoxia in a quiescent state. *Caenorhabditis briggsae,* a soil nematode, survived between one and two days in a quiescent state in water equilibrated with oxygen-free nitrogen or hydrogen (Nicholas and Jantunen, 1964). *Acrobeloides beutschlii,* a very common soil species, recovered after 8 days in oxygen-free nitrogen (Nicholas, 1962). *Aphelenchus avenae,* another very widespread soil nematode, is more tolerant than *Caenorhabditis* to anoxia (Cooper and van Gundy, 1970). Under anaerobic conditions, it enters a cryptobiotic state from which it can recover after at least 90 days. Some isolates also possess limited anaerobic metabolism.

Anaerobic Metabolism

Direct evidence of anaerobic respiratory metabolism in free-living meiofaunal inhabitants of anoxic sediments is lacking. Their small size makes biochemical work very difficult because sufficient tissue, free from microbial contaminants, is so difficult to get. Many nematodes can be cultured in large numbers in the laboratory; some of them axenically, but these are not the species characteristic of anoxic zones. Those species on which a lot of biochemical work has been done seem to have strictly aerobic metabolism (Nicholas, 1984), with the exception of some very interesting work on *Aphelenchus avenae*. This species resembles a number of parasitic helminths with anaerobic respiratory metabolism, as its electron transport system, possesses cytochrome o and in its response to inhibitors (Mendis and Evans, 1984a,b,c). However, *A. avenae* is a fungal feeding terrestrial nematode, and is not a representative of the thiobios.

In the absence of any biochemical evidence of anaerobic metabolism in meiofauna from anoxic sediments, it is interesting to consider the indirect evidence from cytology and ultrastructure.

Cytology and Ultrastructure

One might expect an animal respiring anaerobically to have fewer mitochondria, or mitochondria with fewer cristae than a comparable aerobe. Generally animal tissues with relatively low aerobic capacity show such features. The mitochondria of two acoel Turbellaria were compared by Duffy and Tyler (1984) electron microscopically. Both came from the Atlantic Coast of the USA. *Solenofilomopha funilis* is a thiobiotic species collected at and below the RDZ of intertidal muds. *Otocelis* sp. occurs in well-oxygenated wave-beaten rocky shores. Contrary to expectation, the epidermis and parenchyma, but perhaps not the muscles, of *S. funilis* had more mitochondria with greater surface to volume ratio than *Otocelis*. The density of cristae within the mitochondria was greater in *S. funilis* in all three tissues and the matrix was more electron dense. The authors suggest that *S. funilis* might be able to make more efficient use of traces of oxygen, but also hypothesise that the mitochondrial enzymes of *S. funilis*, including those located in the cristae, may be involved in anaerobic metabolism, rather than the aerobic metabolism as in aerobic animals.

Electron microscope studies by Jensen, so far published only as an abstract of a paper read at the First International Congress of Nematology (1984, Guelph) are of considerable interest. He found symbiotic external prokaryotes associated with some nematodes from thiobiotic marine sands, and endozoic bacteria-like symbionts in the hypodermis of others. Such symbionts apparently detoxify sulphide ions as sulphur and may as well contribute to the nutrition of the nematodes. He noted, as have other authors, that thiobiotic meiofauna are characteristically longer and thinner than aerobic species, presumably facilitating diffusion into their tissues.

Hydrogen sulphide inhibits many enzymes, particularly those dependent on iron. The sulphide ion is bound by cytochrome aa_3 blocking the terminal step in aerobic respiration (Nicholls, 1975). Hydrogen sulphide penetrates cells slowly and is oxidised by oxygenated haemoglobin to sulphur providing a possible mechanism for protecting cytochrome oxidase and for the deposition of intracellular sulphur (Evans, 1967). However, this would require oxygenated haemoglobin and is difficult to reconcile with the anoxic conditions in which the nematodes live and the presence of inclusions containing iron and other metals which presumably enter the tissues under reducing conditions.

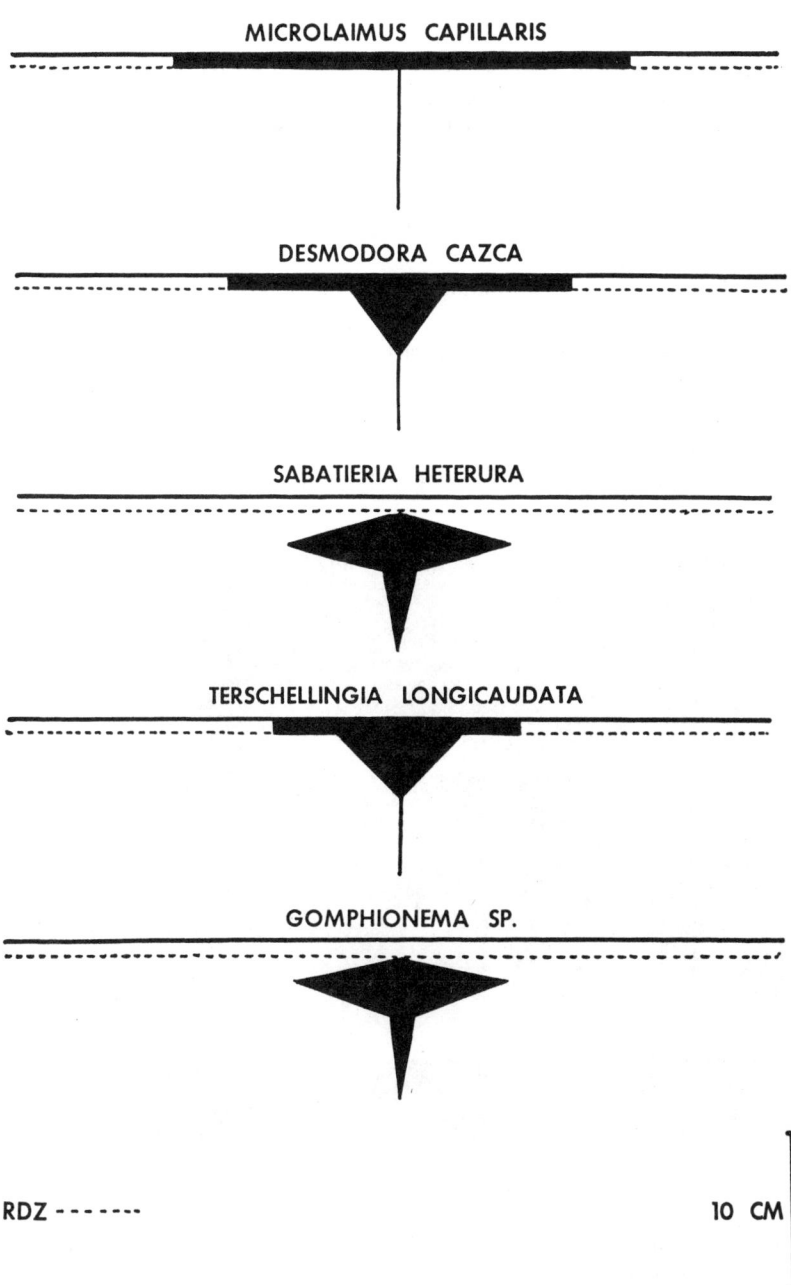

Figure 7.1 The percentage distribution with depth of four species of nematodes in the mud at the low tide mark in the Clyde Estuary, NSW. The surface of the mud and the approximate level of the redox discontinuity zone is shown.

140 *Interstitial Meiofauna*

Nuss and Trimkowski (1984) identified sulphur as a major constituent of crystalloid inclusions in the muscles of *Tobrilus gracilis,* a fresh water species from anoxic muds. They used auger electron spectroscopy and energy-dispersive X-ray spectrometry to identify sulphur in inclusions by electron microscopy. These inclusions, which are visible under the light microscope, accumulate in the muscles in summer, but are scarce in winter (Nuss, 1984). Nuss suggests that this was a means of detoxifying sulphide.

Nuss's observations have stimulated an investigation by electron microscopy of the inclusions that accumulate in thiobiotic nematodes from the Clyde estuary in NSW (Nicholas, Goodchild and Stewart, 1987). Many inclusions occur in the muscles and intestinal cells of *Sabatieria heterula, Terschellingia longicaudata* and *Sphaerolaimus* sp., but are not found in such nematodes as *Desmodora cazca* which inhabit the more superficial muds (Figures 7.1 and 7.2B). The inclusions are visible in the light microscope. X-ray dispersive analysis of inclusions which appear electron

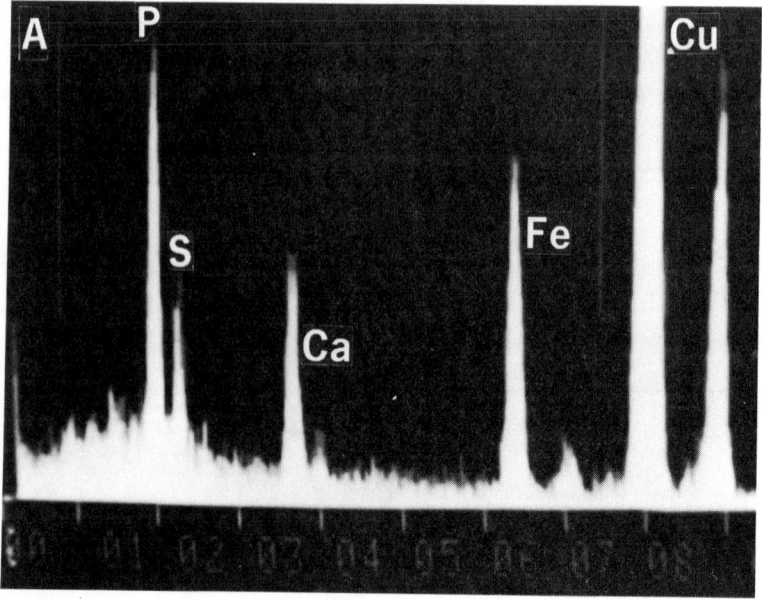

Figure 7.2A Energy dispersive analysis from an intestinal inclusion in *Terschellingia longicaudata* (Nematoda : Linhomoeidae). The spectrum comes from a thin unstained epoxy resin transmission electron microscope section (STEM). The secondary X-ray energy spectrum shows the Ka peaks of phosphorus, sulphur, calcum iron and copper. The Kb peaks of calcium, iron and copper are also evident. The copper peaks come from the instrument (Nicholas, Goodchild and Stewart, 1987).

dense with transmission microscopy shows that sulphur is a major constituent. Their composition varies within the same tissue, and in different individuals of the same species, and may include metals such as iron and manganese (Figure 7.2A). The chemical composition of thick sections of the same three species has also been studied by X-ray dispersive analysis (SEM) (Table 7.3).

Conclusion

Dense populations of meiofauna inhabit marine and fresh water environments. They are abundant throughout the world's oceans. Though small their great numbers add up to significant biomass. Most metazoan phyla are represented in the meiofauna though the Nematoda are quantitatively dominant. Except where waves disturb the sediment, microbial decomposition renders the sediments anoxic and highly reducing close to the surface of the sediment. Reduced gases, notably H_2S, accumulate.

The marine meiofauna is most abundant in the top oxic centimetre of sediment, declining with depth, and some taxa, for example most Crustacea, are restricted to the superficial zone. Other taxa are more widely distributed, with species characteristically inhabiting the anoxic, highly reduced zones, the sulphide-biome. Nematoda, Turbellaria, Gastrotricha and Gnathostomulida are the most important

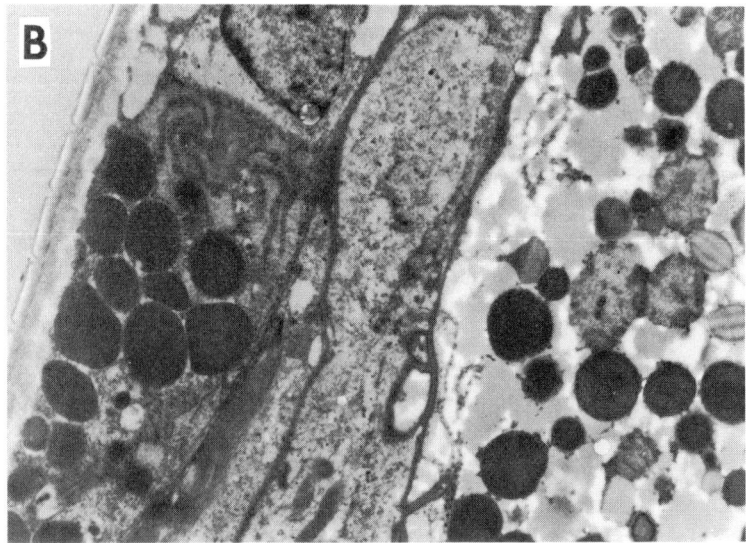

Figure 7.2B Electron micrograph (TEM) of transverse section of *Terschellingia longicaudata*, showing electron dense inclusions in intestinal and hypodermal cells. Fixed in osmium and stained with lead and uranium.

Table 7.3 Localised concentrations of elements in the tissues of two thiobiotic marine nematodes. Energy dispersive X-ray analysis with the scanning electron microscope.

Element	Si	S	K	Ca	Fe
Ka energy, KeV	1.75	2.31	3.34	3.68	6.44
	Peak to Background Ratios*				
Terschelingia longicaudata					
specimen A					
spot 1	-	22	9	-	-
spot 2	20	30	33	-	-
spot 3	6	21	59	15	11
specimen B					
spot 1	11	53	-	13	68
spot 2	18	49	-	19	50
spot 3	14	10	-	8	-
Sabatieri wieseri					
specimen A					
spot 1	6	10	4	11	6
spot 2	8	14	-	11	-
spot 3	14	10	5	14	-
Embedding resin	10	-	11	-	-

* only those significantly above background are listed - that is, peak energy greater than three times the square root of Bremsstrahlung energy. Nematodes were fixed in unbuffered glutaraldehyde, embedded in epoxy resin; 5mm sections, Magnification 2000, 30KV beam, 2nm spot, take off angle 30°

members of this thiobios. The presence of sulphide may be more significant for some meiofauna than the associated lack of oxygen, particularly for the Gnathostomulida and some Turbellaria. The thermal stratification of lakes similarly produces anoxic sediments rich in H_2S inhabited by some meiofauna.

A sharp biological distribution between aquatic and terrestrial interstitial fauna cannot be made. Many of the taxa characteristic of aquatic meiofauna occur widely in soils, dependent on a surface film of water for activity. When soils are flooded they are exposed to anoxia. Soil nematodes generally become quiescent when oxygen is deficient.

Direct biochemical evidence of anaerobic respiratory metabolism is lacking. New methods of culturing these thiobiotic organisms in the laboratory are required. Many species tolerate prolonged exposure to anaerobic conditions sometimes remaining active, suggesting a degree of anaerobic metabolism. The inhibitory effects of the H_2S on electron transport systems require biochemical adaptations in these organisms. Sulphide may be required by some Gnathostomulida, Turbellaria, Gastrotricha and Nematoda. Some thiobiotic nematodes accumulate sulphur inclusions and heavy metals within their cells. Symbiotic prokaryotes occur in some thiobiotic nematodes.

References

Abrams B.I. and Mitchell M.J. (1978) 'Role of Oxygen in Affecting Survival and Activity of *Pelodera punctata* (Rhabditidae) from Sewage Sludge'. *Nematologica* 24, 456-462

Anderson G.L. and Dusenbery D.B. (1977) 'Critical Oxygen Tension of *Caenorhabditis elegans* '. *Journal of Nematology* 9, 253-254

Atkinson H.J. (1975) 'The Functional Significance of the Haemoglobin in a Marine Nematode, *Enoplus brevis* (Bastian)'. *Journal of Experimental Biology* 62, 1-9

Atkinson H.J. (1976) 'The Respiratory Physiology of Nematodes'. In: *The Organisation of Nematodes* Ed. N.A. Croll, pp. 243-272, (Academic Press, NY)

Atkinson H.J. (1977) 'The role of pharyngeal haemoglobin in the feeding of the marine nematode, *Enoplus brevis*'. *Journal of Zoology, London* 18, 465-471

Atkinson H.J. (1980) 'Respiration in nematodes'. In: *Nematodes as Biological Models* Vol. 2. Ed. B.N. Zuckerman, pp. 101-142 (Academic Press, NY)

Banage W.B. (1966) 'Survival of a swamp nematode (*Dorylaimus* sp.) under anaerobic conditions'. *Oikos* 17, 113-120

Blome D. (1983) 'Okologie der Nematoda eines Sandstrades der Nordseeinsel Sylt'. *Mikrofauna des Meeresbodens* 88, 517-590

Boaden P.J.S. and Platt H.M. (1971) 'Daily migration patterns in an intertidal meiobenthic community'. *Thalassia Jugoslavica* 7, 1-12

Board P.A. (1976) 'Anaerobic regulation of atmospheric oxygen'. *Atmosphere and Environment* 10, 339-342

Boucher G. (1972) 'Distribution quantitative et qualitative des nematodes d'une station de vase terrigene cotiere de Banguls-sur-Mer'. *Cahiers de Biologie Marine* 13, 457-474

Bouman L.A. (1983) 'A survey of nematodes from the Ems estuary. Part II. Species assemblages and associations'. *Zoologigische Jahrbucker, Systematik, Okologie und Geographie der Tiere* 110, 345-396

Bryant C., Nicholas W.L. and Jantunen R. (1967) 'Some aspects of the respiratory metabolism of *Caenorhabditis briggsae* (Rhabditidae)'. *Nematologica* 13, 197-209

Cooper Jr. A.F. and van Gundy S.D. (1970) 'Metabolism of glycogen and neutral lipids by *Aphelenchus avenae* and *Caenorhabditis* sp. in aerobic, microaerobic and anaerobic environments'. *Journal of Nematology* 2, 305-315

Dinet A. and Vivier M.H. (1977) 'La meiobenthos abyssal du Golfe de Gascogne'. *Cahiers de Biologie Marine* 18, 85-97

Duffy J.E and Tyler S. (1984) 'Quantitative differences in mitochondrial ultrastructure of a thiobiotic and an oxybiotic turbellarian'. *Marine Biology* 83, 95-102

Dye A.H. (1982) 'Oxygen consumption by sediments in a Southern African mangrove swamp'. *Estuarine, Coastal and Shelf Science* 17, 473-478

Dye A.H. (1983) 'Composition and seasonal fluctuations of meiofauna in a Southern African mangrove estuary'. *Marine Biology* 73, 165-170

Evans C.L. (1967) 'The toxicity of hydrogen sulphide and other sulphides'. *Quarterly Journal of Experimental Physiology* 52, 231-248

Fenchel T. (1969) 'The ecology of marine microbenthos IV. Structure and function of the benthic ecosystem, its chemical and physical factors and the microfauna communities with special reference to ciliated protozoa'. *Ophelia* 6, 1-182

Fenchel T.M. and Riedl R.J. (1970) 'The sulphide system: a new biotic community underneath the oxidised layer of marine sand bottoms'. *Marine Biology* 7, 255-268

Hill A.V. (1929) 'The diffusion of oxygen and lactic acid through tissues'. *Proceedings of the Royal Society, Series B* 104, 39-96.

Jensen P (1986) 'Nematode fauna in the sulphide-rich brine seep and adjacent bottoms of the East Flower Garden, NW Gulf of Mexico IV, Ecological aspects'. *Marine Biology* 92, 489-503

Jorgensen B.B. (1977) 'Bacterial sulfate reduction within microniches of oxidised marine sediments'. *Marine Biology* 41, 7-17

Juario J.V. (1975) 'Nematode species composition and seasonal fluctuations of a sublittoral meiofauna community in the German Bight'. *Veroffentlichungen des Instituts fur Meeresforschung in Bremerhaven* 15, 283-357

Klinger J. and Kunz P. (1974) 'Investigations with a saprozoic nematode, *Mesodiplogaster lheritieri,* on a possible respiratory function of air swallowing'. *Nematologica* 20, 52-60

McIntyre A.D. (1969) 'Ecology of marine meiobenthos'. *Biological Reviews* 44, 245-290

McIntyre A.D. and Murison D.J. (1973) 'The meiofauna of a flatfish nursery ground'. *Journal of the Marine Biological Association UK* 53, 93-118

Mendis A.H.W. and Evans A.A.F. (1984a) 'Substrates respired by mitochondrial fractions of two isolates of the nematode *Aphelenchus avenae* and the effects of electron transport inhibitors'. *Comparative Biochemistry and Physiology* 78B, 373-378

Mendis A.H.W. and Evans A.A.F. (1984b) 'First evidence for the occurrence of cytochrome-o in a free-living nematode'. *Comparative Biochemistry and Physiology* 78B, 729-735

Mendis A.H.W. and Evans A.A.F. (1984c) 'Major volatile metabolites produced by two isolates of *Aphelenchus avenae* under aerobic and anaerobic conditions'. *Comparative Biochemistry and Physiology* 78B, 737-739

Moore H.B. (1931) 'The muds of the Clyde Sea area III'. *Journal of the Marine Biological Association, UK* 17, 325-358

Nicholas W.L. (1962) 'A study of *Acrobeloides* (Cephalobidae) in laboratory culture'. *Nematologica* 8, 99-109

Nicholas W.L. (1984) *The Biology of Free-living Nematodes* 2nd ed. (Oxford Univ. Press)

Nicholas W.L. and Jantunen R. (1964) '*Caenorhabditis briggsae* (Rhabditiae) under anaerobic conditions'. *Nematologica* 10, 409-418

Nicholas W. L., Goodchild D. J. and Stewart A. (1987) 'The mineral composition of intracellular inclusions in nematodes from thiobiotic mangrove mud-flats'. *Nematologica* 33, 167-179

Nicholls P. (1975) 'The effect of sulphide on cytochrome aa$_3$. Isosteric and allosteric shifts of the reduced a-peak'. *Biochimica et Biophysica Acta* 396, 24-35

Nuss B. (1984) 'Ultrastrukturelle und Okophysiologische Untersuchungen an kristalloiden Einschlussen der Muskeln eines sulfidtoleranten limnischen

Nuss B. (1984) 'Ultrastrukturelle und Okophysiologische Untersuchungen an kristalloiden Einschlussen der Muskeln eines sulfidtoleranten limnischen Nematoden (*Tobrilus gracilis*)'. *Veroffenlichungen des Meeresforschung in Bremerhaven* 20, 3-15

Nuss B. and Trimkowski V. (1984) 'Physikalische Mikroanolysen an kristalloiden Einschlusser bei *Tobrilis gracilis* (Nematoda, Enoplida)'. *Veroffenlichungen des Meeresforschung in Bremerhaven* 20, 17-27

Ott J and Schiemer F. (1973) 'Respiration and anaerobiosis of free-living nematodes from marine and limnic sediments'. *Netherlands Journal of Sea Research* 1, 233-243

Platt H.M. (1977) 'Vertical and horizontal distribution of free-living nematodes from Strangford Lough, Northern Ireland'. *Cahiers de Biologie Marine* 18, 261-273

Platt H.M. and Warwick R.M. (1980) 'The significance of free-living nematodes to the littoral ecosystem'. In*The Shore Environment Vol. 2: Ecosystems* Eds. H.J. Price, D.E.G. Irvine and W.F. Farnham, pp. 729-759 (Academic Press, London)

Por F.D. and Masry D. (1968) 'Survival of a nematode and an oligochaete species in the anaerobic benthal of Lake Tiberias'. *Oikos* 19, 388-391

Powell E. N., Crenshaw M.A. and Reiger R. M. (1979) 'Adaptations to sulfide in the meiofauna of sulfide systems I. ^{35}S-sulfide accumulations and the presence of a sulfide detoxification system'. *Journal of Experimental Marine Biology and Ecology* 37, 57-76

Powell E. N., Bright T. J., Woods A. and Gittings S. (1983) 'Meiofauna and the thiobios in the East Flower Garden Brine Seep' *Marine Biology* 73, 269-283

Reise K. and Ax P. (1979) 'Meiofaunal "thiobios" limited to the anaerobic sulphide system of marine sand does not exist'. *Marine Biology* 54, 225-237

Rieger R. and Ott J (1971) 'Gezeittendedingte Wanderungen von Turbellarien und Nematoden eines nordadriatischen Sandstrandes' *Vie Milieu, Supplement* 22, 425-447

Schiemer F. (1973) 'Respiration rates of two species of Gnathostomulids'. *Oecologia* 13, 403-406

Shirayama Y. (1984a) 'The abundance of deep sea meiobenthos in the Western Pacific in relation to environmental factors'. *Oceanologica Acta* 7, 113-121

Shirayama Y. (1984b) 'Vertical distribution of meiobenthos in the sediment profile in bathyal, abyssal and hadal deep seas systems of the Western Pacific'. *Oceanologica Acta* 7, 123-129

Sikora J.P., Sikora W.B., Erkenbrecher C.W. and Coull B.C. (1977) 'Significance of ATP, carbon and caloric content of meiobenthic nematodes in partitioning benthic biomass'. *Marine Biology* 44, 7-14

Swedmark B. (1964) 'The interstitial fauna of marine sand'. *Biological Reviews* 39, 1-41

Tietjen J.H. (1969) 'The ecology of shallow water meiofauna in two New England estuaries'. *Oecologia* 2, 251-291

Tietjen J.H. (1971) 'Ecology and distribution of deep-sea meiobenthos off North Carolina'. *Deep-sea Research* 18, 941-957

Troelsen H. and Jorgensen B.B. (1982) 'Seasonal dynamics of elemental sulphur in two coastal sediments'. *Estuarine Coastal and Shelf Science* 15, 255-266

Warwick R.M. and Price R. (1979) 'Ecological and metabolic studies on free-living nematodes from an estuarine mud-flat'. *Estuarine and Coastal Marine Science* 9, 257-271

Wieser W. and Kanwisher J. (1961) 'Ecological and physiological studies on marine nematodes from a small salt marsh near Woods Hole, Massachusetts'. *Limnology and Oceanography* 6, 262-270

Wieser W. and Kanwisher J. (1959) 'Respiration and anaerobic survival in some weed-inhabiting invertebrates'. *Biological Bulletin Marine Biological Laboratory, Woods Hole* 117, 594-600

Chapter 8

Parasitic Helminths

J. Barrett

Introduction

Parasitic helminths (cestodes, digeneans, nematodes and acanthocephalans) can inhabit virtually any tissue within their hosts. Some parasitic sites such as the lumen of the intestine and the bile duct may have very low oxygen tensions (microaerobic), other sites such as the blood stream and lungs are unambiguously aerobic. Characteristically parasitic helminths break down carbohydrate to reduced organic end-products, usually acids, but occasionally alcohols (Barrett, 1981). The pathways of carbohydrate breakdown are essentially anaerobic and the parasites have an absolute dependency on carbohydrate either as glycogen or exogenous glucose as their sole energy source. In adult parasitic helminths there is no active β-oxidation sequence, so there is no catabolism of fatty acids, and amino acid catabolism is very limited. Similarly there is no evidence of the co-fermentation of amino acids and carbohydrate or of fatty acids and carbohydrate (Barrett and Körting, 1977).

The anaerobic metabolism of parasitic helminths differs from that of free-living invertebrates in two important respects. First, in free-living invertebrates anaerobiosis is a transient state, the major response of free-living organisms to anoxia being to reduce their metabolic demand until aerobic conditions are restored. In contrast, the anaerobic pathways of parasitic helminths represent a steady state condition; carbohydrates are taken in and the end-products excreted. Unlike free-living organisms, helminths do not accumulate anaerobic end-products for later resynthesis into glycogen. Second, in parasitic helminths anaerobic metabolism persists in the presence of oxygen. Although anaerobic/aerobic transitions in helminths are accompanied by qualitative and quantitative changes in the end-products from carbohydrate breakdown, the anaerobic pathways are not inhibited by the presence of oxygen.

In general, parasitic helminths can be divided, on the basis of their end-products, into two groups. There are those which rely essentially on glycolysis alone and produce as end-products of carbohydrate breakdown lactate or some other reduced derivative of pyruvate and there are those which fix carbon dioxide and have what is often referred to as an '*Ascaris* type metabolism'. In the *Ascaris* type the primary end-products of carbohydrate catabolism are succinate and pyruvate, but these are usually further metabolised to propionate, acetate and volatile fatty acids such as 2-methylvalerate and 2-methylbutyrate. These two types of metabolism are convenient divisions, but are in no way absolute and there is in reality a continuous spectrum between the two. Most homolactic fermentors can fix carbon dioxide to a limited extent and most *Ascaris* type helminths produce some lactic acid.

Glycolytic Helminths

A number of parasitic helminths breakdown carbohydrate more or less quantitatively to lactate (Tables 8.1-8.4). In general, homolactic fermentation seems more characteristic of tissue or blood dwelling parasites than intestinal parasites. In the filarial worm *Litomosoides carinii* the amounts of lactate produced vary markedly with the oxygen tension. Under anaerobic conditions 80% of the carbohydrate can be accounted for as lactate, the remainder as acetate (Barrett, 1983). Aerobically lactate accounts for 30-40% of the end-products, acetate 25-35% and acetoin about 3%. Acetate is produced by the decarboxylation of pyruvate probably via the pyruvate dehydrogenase complex; acetoin is also probably produced by this enzyme complex.

A variant of the glycolytic scheme is found in the acanthocephalan *Moniliformis moniliformis* where the main end-product is ethanol, with small amounts of lactate, succinate and volatile fatty acids. The parasite has an NADP-linked alcohol dehydrogenase, but no pyruvate decarboxylase has been found. The most likely route for the decarboxylation of pyruvate to acetaldehyde in *M.moniliformis* is a partial reaction of the pyruvate dehydrogenase complex. Ethanol is an anaerobic end-product in some free-living and plant-parasitic nematodes, where again the pyruvate dehydrogenase complex is responsible for acetaldehyde formation (Barrett and Butterworth, 1984). Amongst animal parasitic nematodes alcohol dehydrogenase is extremely active in the larvae of *Anisakis* and is also present in *Strongyluris brevicaudata,* so possibly ethanol is an end-product in these parasites as well.

In free-living invertebrates such as annelids and arthropods alanine is often an important anaerobic end-product. The redox linked formation of alanine from pyruvate in these animals involves either a specific alanine dehydrogenase, or more likely, the combined activities of pyruvate transaminase and glutamate dehydrogenase. In parasitic helminths, although alanine formation has been noted in the cestode *Hymenolepis diminuta* and in the digenean *Schistosoma mansoni,* it is not a significant end-product. Similarly the reductive condensation of pyruvate and an amino acid to yield opines such as octopine or alanopine, again widespread in free-living invertebrates, has not been found in parasitic helminths.

Carbon Dioxide Fixation

Carbon dioxide fixation is characteristic of the large intestinal parasites such as *Moniezia expansa, Taenia taeniaeformis, Parascaris equorum* and *Ascaris lumbricoides* (Tables 8.1-8.4). The nematode *A. lumbricoides* is perhaps the best studied and can be taken as a model for the others (Figure 8.1). Glycogen is broken down in this nematode by a normal glycolytic sequence, as far as phosphoenolpyruvate. The levels of pyruvate kinase in *Ascaris* are low, and instead of forming pyruvate, carbon dioxide fixation takes place and the phosphoenolpyruvate is converted to oxaloacetate by the action of a cytoplasmic phosphoenolpyruvate carboxykinase (properly this enzyme should be called phosphopyruvate carboxylase, but the old name has become firmly entrenched in the parasitological literature). The phosphoenolpyruvate carboxykinase of *Ascaris* differs from the corresponding enzyme of vertebrates in that it is more active with IDP than GDP (ADP being inactive). Oxaloacetate is then reduced to malate by a cytoplasmic malate dehydrogenase, this reoxidises the NADH produced during glycolysis (from glyceraldehyde-3-phosphate dehydrogenase), fulfilling the function of lactate dehydrogenase in other tissues. Malate then enters the mitochondrion via a phosphate-dependent translocase where it undergoes a dismutation. Part of the malate is oxidatively decarboxylated to pyruvate via an NAD-linked malic enzyme. Malate is also in equilibrium with fumarate via fumarase (which in *A. lumbricoides* occurs in both the mitochondrion and the cytoplasm) and the fumarate is reduced to succinate by a fumarate reductase. So *A. lumbricoides* is using the second span of the tricarboxylic acid cycle, from succinate to oxaloacetate, but operating in the reverse direction and has what has been described as a partial, reversed tricarboxylic acid cycle.

In *A. lumbricoides* there are two systems for the reoxidation of NADH; in the cytoplasm, NADH formed during glycolysis is reoxidised by the reduction of oxaloacetate to malate; in the mitochondrion, the NADH formed by the malic enzyme is reoxidised by the reduction of fumarate to succinate. The breakdown of carbohydrate to succinate and pyruvate in *Ascaris* is, therefore, in redox balance and could proceed entirely anaerobically. A variation occurs in *Hymenolepis diminuta* and *H. microstomum,* where the malic enzyme is NADP-linked, whilst the fumarate reductase remains NAD-linked (in vertebrates the malic enzyme is also NADP-linked, but cytoplasmic). In the mitochondria of these cestodes there is a non-energy linked transhydrogenase which catalyses the reaction:

$$NADPH + NAD^+ + H^+ \rightarrow NADP^+ + NADH + H^+$$

enabling the malic enzyme to be coupled with fumarate reductase (Fioravanti, 1982; Fioravanti and Kim, 1983). Many of the small gastrointestinal nematodes of mammals may be like *Nippostrongylus brasiliensis.* Aerobically *N. brasiliensis* produces lactate, anaerobically lactate and succinate, succinate formation involving carbon dioxide fixation and the partial reversed tricarboxylic acid cycle.

Table 8.1 End-products of carbohydrate breakdown in nematodes.

species	aerobic	anaerobic
Ancylostoma caninum	acetate, propionate, traces of isobutyrate, 2-methyl butyrate	same
Angiostrongylus cantonensis	lactate	same
Ascaridia galli	lactate, acetate, propionate	same
Ascaris lumbricoides	2-methylvalerate, 2-methylbutyrate, traces of acetate, n-valerate, propionate, n-butyrate, 2-methylcrotonate, n-caproate, acetoin	same
Brugia pahangi		
(adult)	lactate, traces of alanine	same
(microfilaria)	lactate, acetate	lactate
Chandlerella hawkingi	lactate, traces of pyruvate, propionate, acetate	-
Dipetalonema viteae	lactate	-
Dirofilaria uniformis	lactate	-
Dirofilaria immitis	lactate	same
Dracunculus insignis	lactate	same
Haemonchus contortus	acetate, propionate, propanol, ethanol, traces of lactate, succinate	same
Heterakis gallinae	acetate, propionate, traces of lactate, pyruvate	same
Litomosoides carinii	lactate, acetate	more lactate, less acetate
Mecistocirrus digitatus	lactate, butyrate	lactate, acetate
Nippostrongylus brasiliensis	lactate, traces of pyruvate	lactate, succinate
Oesophagostomum radiatum		
(adult)	acetate, propionate, lactate, methylbutyrate	same
(infective larva)	acetate, propionate, lactate	same
Onchocerca gutturosa	lactate	same
Onchocerca lienalis	lactate	same
Parascaris equorum	lactate, propionate, 2-methylbutyrate	same
Setaria cervi	lactate	-
Trichinella spiralis		
(adult)	-	n-valerate, acetate, propionate, traces of caproate, butyrate, formate
(muscle larva)	n-valerate, acetate, propionate, traces of butyrate, caproate, lactate	same
Trichostrongylus colubriformis	-	acetate, propionate, traces of butyrate
Trichuris vulpis	aerobic lactate, propionate, succinate, traces of acetate, n-valerate, butyrate, formate	-

Table 8.2 End-products of carbohydrate breakdown in cestodes.

species	aerobic	anerobic
Cotugnia digonopora	lactate	-
Diphyllobothrium dendriticum		
(adult)	succinate, lactate	same
(plerocercoid)	succinate, lactate	same
Echinococcus granulosus		
(adult)	succinate, lactate, acetate, traces of alanine	same
(larva)	succinate, lactate, acetate, traces of pyruvate, ethanol	similar, with more succinate and pyruvate
Hymenolepis diminuta	lactate, acetate, succinate	similar, with more succinate
Hymenolepis microstomum	lactate, succinate, acetate, propionate	more succinate, less lactate
Ligula intestinalis		
(adult)	lactate, some succinate, propionate, acetate, malate, pyruvate	similar, less propionate
(plerocercoid)	lactate, succinate, acetate, propionate, pyruvate, malate	similar
Mesocestoides corti		
(tetrathyridia)	lactate, succinate, acetate	similar, more succinate
Moniezia expansa	lactate, succinate	similar, more succinate
Schistocephalus solidus		
(plerocercoid)	acetate, propionate	more propionate, less acetate
Spirometra mansonoides		
(adult)	-	acetate, propionate, traces of lactate, succinate
(larva)	-	acetate, lactate, traces of propionate, succinate
Taenia taeniaeformis		
(adult)	lactate, acetate, pyruvate, succinate, ethanol	similar, more succinate
(larva)	similar to adult	similar to adult

Table 8.3 End-products of carbohydrate breakdown in digeneans.

species	aerobic	anaerobic
Calicophoron ijimai	lactate, succinate, acetate, propionate, pyruvate, malate	same.
Clinostomum complanatum	lactate, pyruvate, traces of succinate	lactate
Dicrocoelium dendriticum	lactate	lactate acetate, propionate, succinate
Echinostoma liei	n-valerate, traces of n-caproate, propionate, butyrate, acetate, 2-methyl-butyrate, lactate, succinate	same
Fasciola gigantica	lactate, acetate	-
Fasciola hepatica	acetate, propionate, traces of lactate, isobutyrate, isovalerate, 2-methyl-butyrate, succinate	same
Paragonimus westermani	acetate, propionate, n-butyrate, 2-methylbutyrate, n-valerate, n-caproate	-
Schistosoma mansoni	lactate, traces of valerate	same
Schistosoma haematobium	lactate, traces of acetate	-
Schistosoma margrebowiei	lactate	-
Schistosoma douthitti	lactate	-

Table 8.4 End-products of carbohydrate breakdown in acanthocephalans.

species	aerobic	anerobic
Echinorhynchus gadi	lactate	lactate, succinate
Moniliformis dubius	ethanol, traces of lactate, succinate, acetate, butyrate, formate	same
Neoechinorhynchus emydis	lactate	-
Neoechinorhynchus pseudemydis	lactate	-
Polymorphus minutus	-	lactate, succinate

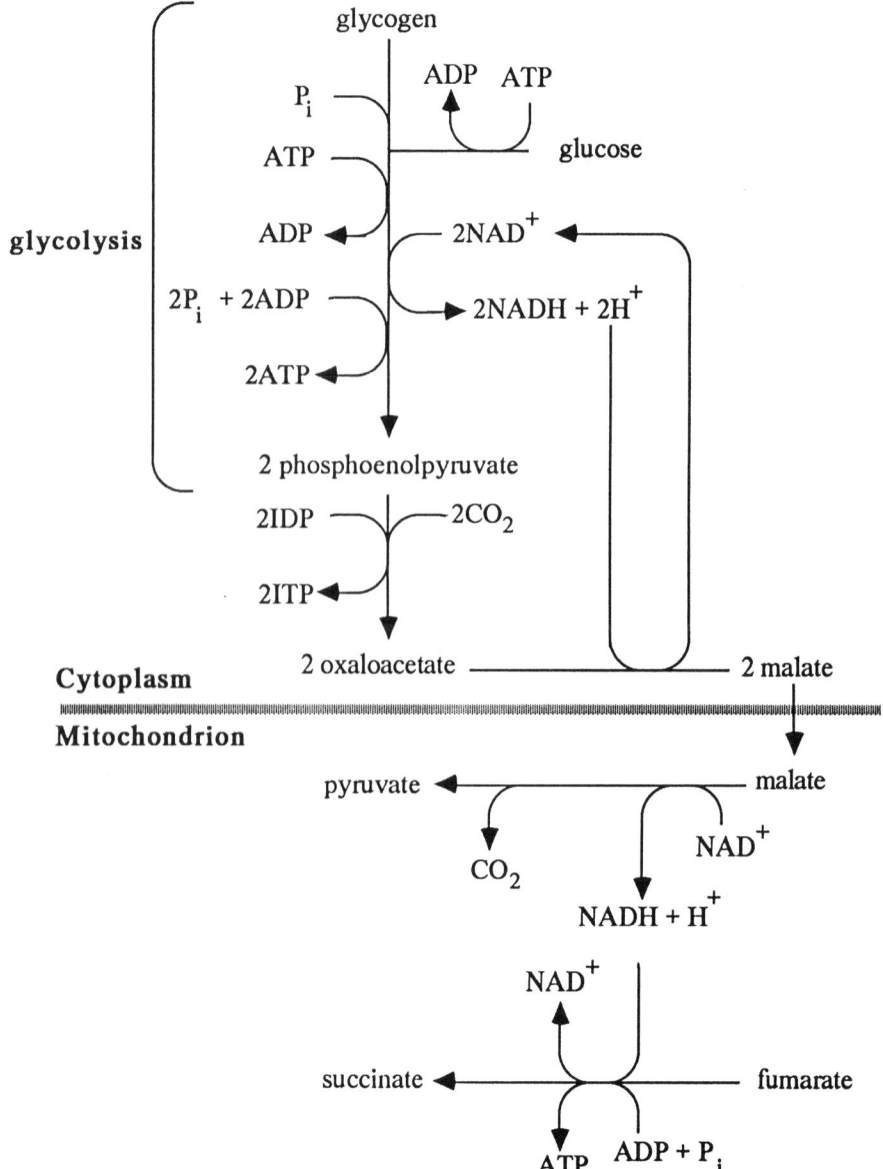

Figure 8.1 The pathway of carbohydrate catabolism in *Ascaris lumbricoides*.

The redox state of the free [NAD$^+$]/[NADH] couple in the cytoplasm of *A. lumbricoides* muscle has been estimated as between 725:1 and 2214:1, a value comparable with the ratio of 1000:1 found in normal rat liver (in ischaemic liver the ratio falls to 100:1). Coupling the reoxidation of cytoplasmic NADH in *Ascaris* muscle to malate dehydrogenase rather than lactate dehydrogenase allows the cytoplasmic [NAD$^+$]/[NADH] couple to remain in a relatively oxidised state despite metabolism being anaerobic. In the mitochondria of *Ascaris* the free [NAD$^+$]/[NADH] ratio is considerably lower than comparable estimates for the redox state of the couple in rat liver mitochondria, 0.1:1 as compared with 10:1. This may be correlated with the fact that in the *Ascaris* mitochondrion fumarate is being reduced to succinate, whilst in the mammalian mitochondrion succinate is being oxidised to fumarate.

The reduction of fumarate to succinate in *Ascaris* involves part of the cytochrome chain and results in a site 1 associated phosphorylation of ADP (ATP/malate ratio = 0.5). There is some debate as to whether the succinate dehydrogenase of helminths and the fumarate reductase are different functions of the same enzyme or two separate enzymes. Succinic dehydrogenases isolated from different sources can differ considerably in their properties. The enzyme from aerobic tissues (such as rat liver) is always membrane bound, has a high affinity for succinate, is only weakly inhibited by fumarate and is activated by substrate. The ratio of the forward to backward reaction of the aerobic enzyme (succinate oxidation:fumarate reduction) is high at about 60:1. In contrast the succinate dehydrogenases from obligate anaerobes, such as anaerobic bacteria, are usually cytoplasmic, they have a high affinity for fumarate and a low affinity for succinate, they are not activated by substrate, and the ratios of the forward to the backward reaction are a thousand times less than that of the aerobic enzyme (0.03:1 in *Micrococcus lactyliticus*).

The succinate dehydrogenases from helminths which reduce fumarate to succinate resemble the aerobic type enzyme in being membrane bound and, at least in *Fasciola hepatica* and *H. diminuta,* the enzyme can exist in active and inactive forms (Barrett, 1978). The succinate oxidation:fumarate reductase ratio of the heminth enzyme ranges from 0.3 to 5, mid-way between the classical aerobic and anaerobic enzymes. Work with *A. lumbricoides* and *A. galli* suggests that they have a single reversible enzyme. However, in *N. brasiliensis* there may be two succinic dehydrogenases, one reversible, like the *Ascaris* enzyme, the other similar to the typical aerobic mammalian enzyme (Fry and Brazeley, 1984).

The proposed scheme for carbohydrate breakdown in *Ascaris* gives four moles of nucleoside triphosphate for every C6 unit of glycogen catabolised. Of these four moles, two are produced by the phosphoenolpyruvate carboxykinase reaction and will be in the form of ITP or GTP, not ATP. However, Ascaris has high levels of nucleoside diphosphate kinase activity, which rapidly catalyses the transfer of high energy phosphate from GTP or ITP to ADP.

$$NTP + ADP \rightarrow NDP + ATP \quad K_{eq} = 0.91$$

The levels of inosine and guanosine nucleotides in *Ascaris* are not particularly high and are comparable to the nucleotide levels found in vertebrate smooth muscle.

Anaerobic End-Products

The initial end-products of carbohydrate breakdown in helminths which fix carbon dioxide are pyruvate and succinate. Succinate is often excreted unchanged although it is frequently metabolised to propionate and other short chain volatile fatty acids. Pyruvate is rarely excreted by parasitic helminths and is almost always metabolised to acetate or some other derivative. In *Ascaris* for example the main end-products of carbohydrate catabolism are 2-methylvalerate and 2-methylbutyrate with small amounts of acetate, propionate, butyrate, valerate, caproate and 2-methylcrotonate, whilst in the liver fluke *F. hepatica* the main end-products are propionate and acetate.

One might have expected evolution to have led to the optimisation of biochemical pathways for the best use of metabolic resources, with energy producing pathways being highly conserved. Parasitic helminths, however, show great variability in their catabolic pathways, suggesting that optimising ATP production is not the only consideration and that there must be competing selective demands on the pathways or their end-products (Barrett, 1984).

In an anaerobic system each oxidative step must be balanced by a reductive one. Anaerobic coupling is not possible in a cyclic system, so anaerobic pathways must, of necessity, be linear. Producing a range of different end-products gives a parasitic helminth a degree of flexibility in balancing its redox couples. Different end-products may be produced in different cellular compartments, for example lactate and ethanol are typical cytoplasmic end-products. The various tissues of the parasite may also produce different end-products, depending on their metabolic roles. The end-product of one tissue may serve as substrate for another and, as in mammals, certain end-products, for example malate, may be tissue specific substrates.

The different end-products produced by helminths vary in their redox relationships. Some such as lactate are in redox balance, other end-products such as acetate are net producers of NADH whilst fatty acids are net users of NADH and may act as electron sinks. Overall redox balance can be maintained by combining end-products which are net producers of NADH with those which are net users, in stoichiometric amounts. The combination of acetate (a net producer of NADH) with propionate (a net user) in the ratio of 1:2 would maintain redox balance. An acetate:propionate ratio of approximately 1:2 has been reported for a variety of helminth parasites. Mammalian blood contains significant amounts of pyruvate (up to 0.2 mM), if blood dwelling parasites can use this source of pyruvate for their lactate dehydrogenase reaction they will have an external electron sink.

As well as differing in their redox relationships, the various end-products differ in terms of the number of moles of ATP produced/mole of glucose catabolised. Lactate production results in the formation of 2ATP/mole of glucose catabolised, the production of other fatty acids offers additional sites for ATP production. These include the reduction of fumarate to succinate, the decarboxylation of pyruvate and succinate to acetate and propionate and the reduction of 2- methylcrotonate (tiglate) and 2-methyl-2-pentenoate to 2-methylbutyrate and 2-methylvalerate. In the reduction of fumarate to succinate, fumarate acts as an alternative electron acceptor for the helminth cytochrome chain. Anaerobically electron flow proceeds from NADH, through complex I to the quinone, then to the succinic dehydrogenase complex; so there is a site 1 associated phosphorylation of ADP. The decarboxylation of succinate to propionate in helminths follows a similar pathway to that found in mammalian liver, succinate being metabolised via succinic thiokinase, methylmalonyl-CoA mutase, methylmalonyl-CoA racemase and propionyl-CoA carboxylase to propionyl-CoA. An ATP is produced during the decarboxylation of methylmalonyl-CoA, but one is also required in the initial conversion of succinate to succinyl-CoA so there is no net production of ATP. However, propionyl-CoA is a high energy compound and there is evidence in helminths that the bond energy is conserved via CoA transferase reactions.

$$\text{Propionyl-CoA} + \text{Succinate} \rightarrow \text{Succinyl-CoA} + \text{Propionate}$$

The CoA transferase may also be linked to GTP formation via the reversal of succinic thiokinase.

$$\text{Succinyl-CoA} + \text{GDP} + P_i \rightarrow \text{Succinate} + \text{GTP} + \text{CoA}$$

Propionate is a major product of ruminant microorganisms and many of the helminth parasites of ruminants produce propionate as an end-product *in vitro*. The pathway from succinate to propionate is freely reversible and it is possible that *in vivo* the parasites could use this pathway for glyconeogenesis from propionate (Ward, 1982). The nematode *Haemonchus contortus* excretes propanol as a major end-product. The reduction of propionate to propanol is energetically most unlikely; a possible route would be the direct reduction of propionyl-CoA.

The decarboxylation of pyruvate to acetate in helminths involves the pyruvate dehydrogenase complex which yields acetyl-CoA, this is then cleaved to acetate via CoA transferase systems as described for propionate. The pyruvate dehydrogenase complex is also the source of acetoin, as well as of acetaldehyde in those species which produce ethanol.

The formation of 2-methylvalerate and 2-methylbutyrate from acetate and propionate (or more properly acetyl-CoA and propionyl-CoA) in *Ascaris* mitochondria involves a series of condensation and reduction reactions reminiscent of some of the steps of β-oxidation (Komuniecki, Komuniecki and Saz, 1981). The penultimate step in branched chain fatty acid production, the reduction of 2-methylcrotonyl-CoA to 2-methylbutyryl-CoA and of 2-methyl-2-pentenoyl-CoA to

2-methylvaleryl-CoA may be linked to the electron transport chain in a reaction analogous to fumarate reductase (Komuniecki, Rioux and Thissen, 1984). So the formation of these acids (Figure 8.2), like the formation of succinate may be associated with an electron chain mediated site 1 phosphorylation of ADP (Rioux and Komuniecki, 1984).

Equally important as ATP/mole glucose catabolised may be the power output of a pathway, that is how fast can it generate ATP. There is an inverse relationship between power output and efficiency (Barrett, 1984); the propionate pathway has a low rate of working, but a high ATP yield, acetate production has a high potential power output, but a low ATP production/mole glucose. The end-products selected by helminths must represent a compromise between substrate conservation and rate of working.

In free-living organisms, tissue compatibility and the need to resynthesise carbohydrate from accumulated end-products at the end of anaerobiosis restrict the choice of end-products. In helminths anaerobiosis is a steady state condition, end-products are not stored in the tissues, nor are they used as substrates for glyconeogenesis. Helminths can, therefore, afford to produce 'dead end' compounds such as acetoin and volatile fatty acids, which cannot be readily remetabolised.

The Role of Aerobic Processes

Despite the essentially anaerobic nature of carbohydrate catabolism in parasitic

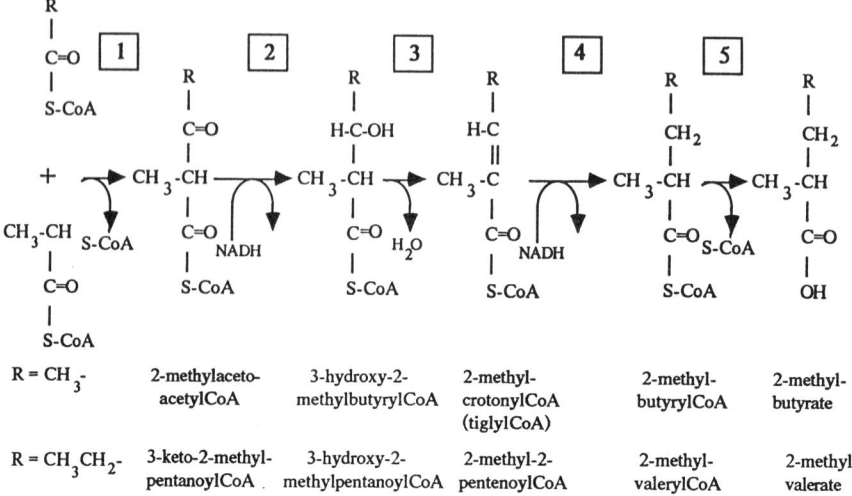

Figure 8.2 Possible pathways of 2-methylbutyrate and 2-methylvalerate formation in *Ascaris lumbricoides*.

helminths, they all use oxygen when it is available. That is, at least in air, they have a measurable oxygen consumption. The rates of oxygen uptake in parasitic helminths are comparable to those found in free-living animals; for example the oxygen consumption (ml O_2/gm fresh weight/h) of *A. lumbricoides* is 0.5, *N. brasiliensis* 1.3 and *H. diminuta* 0.3. Comparable values for free-living invertebrates are *Arenicola* 0.03, resting locust 1.7 and *Mytilus* 0.06. Where investigated in detail parasitic helminths have all been shown to possess cytochromes and to be capable of oxidative phosphorylation.

Cytochromes corresponding to b, c, c_1 and a/a_3 have been demonstrated in a wide variety of helminths. In addition, a number of helminths, including *A. lumbricoides, M. expansa* and *F. hepatica* contain a carbon monoxide-reactive haemoprotein, with an absorption spectrum similar to cytochrome o (Cheah, 1975; Cheah and Prichard, 1975). In both *M. expansa* and *A. lumbricoides* the quinone component of the cytochrome chain is not ubiquinone, as in mammals, but rhodoquinone. Rhodoquinone has been isolated from a whole range of helminths, occurring either on its own or in conjunction with ubiquinone (Allen, 1973).

Oxygen uptake by parasitic helminths such as *A. lumbricoides* or *F. hepatica* shows a number of peculiar characteristics. First, they are all oxygen conformers, that is, the rate of oxygen uptake varies with the partial pressure. Second, they show cyanide insensitive respiration and in some helminths cyanide may actually stimulate oxygen uptake. Finally, when mitochondria from these helminths oxidise substrates *in vitro,* hydrogen peroxide accumulates. This suggests that in addition to cytochrome a/a_3, these parasitic helminths possess an alternative oxidase with properties similar to the alternative oxidase described in plants (Laties, 1982). In *A. lumbricoides* the alternative pathway accounts for some 70% of the total oxidase activity, there is no evidence for oxidative phosphorylation associated with the alternative pathway and the oxidase is specifically inhibited by radical scavenging agents such as salicylhydroxamic acid. There is also general agreement that branching occurs at or near the level of the quinone.

On the basis of carbon monoxide binding, it is tempting to equate the alternative terminal oxidase in helminths with cytochrome o. But carbon monoxide binding alone does not show conclusively that a particular cytochrome is acting as a terminal oxidase. The photochemical action spectrum of carbon monoxide-inhibited respiration in *Ascaris* mitochondria reveals only cytochrome a_3 (Cheah and Chance, 1970), the cytochrome o-carbon monoxide complex presumably not being light reversible. The role of the o-type cytochrome as the alternative terminal oxidase in helminths is, therefore, by no means proven. Cytochrome o in helminths is certainly substrate reducible; it might have the same relationship to cytochrome b as do b_T and b_K in the mammalian chain or it could be specifically associated with the fumarate reductase complex.

In plants lipoxygenase activity is often mistaken for a cyanide insensitive terminal oxidase; another possibility is a quinone oxidase. Most helminths live in a micro-aerobic environment and when incubations are carried out under air, the increased oxygen tension may lead to the quinone component of the cytochrome chain being oxidised directly by molecular oxygen rather than passing electrons on to the b cytochromes. This could be non-enzymatic auto-oxidation or involve a quinone oxidase, therefore either of these would give an apparently cyanide insensitive, oxygen dependent terminal oxidase which was inhibited by radical scavenging agents.

As well as a cytochrome chain, all parasitic helminths seem to have a complete sequence of tricarboxylic acid cycle enzymes. However, the enzymes at the beginning of the tricarboxylic acid cycle in parasitic helminths often have very low activities. In particular aconitase is usually extremely difficult to detect and the NAD-linked isocitrate dehydrogenase is often absent. The activities of NADP-linked isocitrate dehydrogenase and 2-oxoglutarate dehydrogenase are also both relatively low in parasitic helminths compared with mammalian tissues. In contrast to the other enzymes in the first span of the tricarboxylic acid cycle, the activity of citrate synthase is usually high. In free-living organisms, the ratio of the activities of citrate synthase to glyceraldehyde-3-phosphate dehydrogenase is often taken as an indicator of the relative contribution of aerobic and anaerobic processes. In parasitic helminths, citrate synthase may be part of a shuttle involved in the transport of acetyl-CoA across the mitochondrial membrane (Stryer, 1981), this would explain its high activity in an anaerobic tissue, particularly if acetate was being produced as an end-product.

Carbon balance studies suggest that, in the majority of helminths, the classical tricarboxylic acid cycle accounts for less than 10% of the total carbohydrate catabolised. The tricarboxylic acid cycle would seem to be making an insignificant contribution to energy metabolism and its main role may be in the interconversion of carbon skeletons to provide synthetic intermediates. However, the complete aerobic catabolism of one mole of glucose yields 36 moles of ATP, compared with 2 to 6 moles of ATP for anaerobic catabolism. The complete oxidation of 10% of the total carbohydrate catabolised could, at least in theory, yield more ATP than the anaerobic catabolism of the remaining 90%. So potentially, the tricarboxylic acid cycle in parasitic helminths could make a significant contribution to energy metabolism, despite its relatively low activity.

Different species of helminths show considerable differences in the *in vitro* effects of oxygen on such things as survival, motility, reproduction and development. Some parasites show a marked Pasteur effect, others do not (Table 8.5). Similarly the ability of parasitic helminths to accumulate an oxygen debt varies from species to species and is in no way correlated with the presence of a Pasteur effect. The widespread variation in the effects of oxygen on parasitic helminths may reflect differences in underlying metabolic pathways. Parasitic helminths can all survive anaerobically, often for considerable periods of time, but most, if not all of them probably require at least some oxygen for normal sustained growth and reproduction.

Synthetic Reactions

There are a number of enzyme reactions which have an absolute dependency on

Table 8.5 The occurrence of a Pasteur effect in parasitic helminths.

Pasteur Effect	No Pasteur Effect
Nematoda	
Ancylostoma caninum	*Ascaris lumbricoides* (adult)[a]
Ascaridia galli	*Dracunculus insignis*
Ascaris lumbricoides (juvenile)	*Trichuris vulpis*
Dirofilaria immitis	
Eustrongylides ignotus (juvenile)[a]	
Haemonchus contortus	
Litomosoides carinii[b]	
Nippostrongylus brasiliensis	
Digenea	
Fasciola hepatica (juvenile)	*Fasciola hepatica* (adult)
Cestoda	
Hymenolepis diminuta[a]	*Echinococcus granulosus*
Schistocephalus solidus (plerocercoid)	
Taenia taeniaeformis	
Acanthocephala	
Macracanthorhynchus hirudinaceus	

[a] show oxygen debt; [b] no oxygen debt

molecular oxygen, that is oxygen is the physiological oxidant in the reaction, rather than an intermediate electron carrier such as NAD or flavoprotein. These enzymes are the oxidases, hydroxylases and oxygenases; a number of these enzymes are involved in important synthetic reactions such as quinone tanning, steroid synthesis, fatty acid desaturation and detoxification reactions including aromatic hydroxylations and O-demethylase reactions.

Quinone tanning occurs in the eggs of monogeneans, digeneans and pseudophyllidean cestodes, but this occurs after the eggs have been laid and there is no evidence for quinone tanning in the adult. Helminths, both free-living and parasitic are unable to synthesise steroids *de novo*. The step in steroid synthesis which has an absolute requirement for molecular oxygen is the cyclisation of squalene via squalene epoxide to give lanosterol. The evidence from helminths suggests that the block to steroid synthesis occurs before the cyclisation step, probably in the formation of squalene. However, in helminths, as in insects, there may be multiple lesions in the steroid pathway. Parasitic helminths appear to be unable to desaturate preformed fatty acids (a reaction which proceeds via an epoxide intermediate) nor can they synthesise the porphyrin nucleus of haem. In the latter pathway the conversion of coproporphyrinogen to protoporphyrin IX involves the conversion of a propionate side chain to a vinyl side chain in a reaction analogous to fatty acid desaturase. Only very low levels of the enzyme responsible, coproporphyrinogen oxidase have been detected in *A. lumbricoides* mitochondria, but it is not clear if this is the primary lesion in the pathway (Cain, 1976). Oxidative detoxification reactions seem to be totally absent from parasitic helminths, or at least from cestodes and nematodes (Douch and Blair, 1975; Munir and Barrett, 1985; Precious and Barrett, 1989).

The apparent absence from parasitic helminths of many of the known oxygen requiring synthetic reactions raises the question as to whether helminths can function as obligate anaerobes. However, there is at least one other important synthetic reaction requiring molecular oxygen which does occur unambiguously in helminths, that is proline hydroxylase (Fujimoto and Prockop, 1969). This enzyme converts proline to hydroxyproline, a key step in collagen synthesis, collagen being a major structural protein in helminths. So parasitic helminths carry out at least one reaction which requires molecular oxygen and so they must, strictly speaking, be considered obligate aerobes.

In free-living organisms the aerobic catabolism of carbohydrate via glycolysis and the tricarboxylic acid cycle provides, as well as energy, some ten key intermediates for synthetic reactions. Of these, four are produced by glycolysis (hexose-phosphate, triose-phosphate, phosphoenolpyruvate and pyruvate) and so are available from all anaerobic pathways. Three other intermediates, acetyl-CoA, succinyl-CoA and oxaloacetate, although not produced by homolactic fermentors, are intermediates in several anaerobic sequences. The remaining synthetic intermediates, 2-oxoglutarate, tetrose-phosphate and pentose-phosphate are not intermediates in any of the anaerobic pathways and must be produced by other reactions. The pentose-phosphate pathway is a source of pentose and tetrose-phosphates, whilst 2-oxoglutarate can be formed by amino acid catabolism or from the first span of the tricarboxylic acid cycle. These additional pathways are, however, all net producers of reducing power and so in some way must be coupled to a suitable electron sink.

Evolution of Helminth Pathways

The metabolic adaptations found in adult helminths do not necessarily apply to their free-living and intermediate stages. These stages normally have an aerobic metabolism with an active tricarboxylic acid cycle and β-oxidation sequence and a mammalian type cytochrome chain. Adult parasites must, therefore, possess the information required to synthesise aerobic pathways, but it is not being expressed.

The end-products produced by parasitic helminths and the pathways they use are very similar to the anaerobic metabolism of free-living invertebrates, the difference being that in parasites there are changes in the regulatory mechanisms, such that anaerobic processes predominate, even in aerobic environments. Many sites within the vertebrate host have low redox potentials or fluctuating oxygen tensions, both of which would favour anaerobic metabolism. However, given the much greater efficiency of aerobic catabolism, why do parasites not make full use of oxygen when it is available? One possibility is that there are environmental factors, other than just oxygen tension which favour the use of anaerobic pathways by helminths. A number of suggestions have been put forward, but none is wholly convincing. The high ambient carbon dioxide levels in the tissues of terrestrial vertebrates may present parasites with the problem of tissue acidification. The excretion of organic acids and the presence of pathways which involve carbon dioxide fixation may help to combat the problem (Podesta, Mustafa, Moon, Hulbert and Mettrick, 1976). So the anaerobic pathways of parasitic helminths could be more a response to high carbon dioxide levels rather than low oxygen levels. Carbon dioxide fixation, however, also occurs in free-living organisms and in the parasites of aquatic vertebrates where carbon dioxide tensions are low, whilst many parasites of terrestrial vertebrates are homolactic fermentors and their metabolism is not affected by carbon dioxide levels (Barrett, 1984). In free-living invertebrates, one of the adaptations to increase in temperature is an increase in aerobic glycolysis. The helminth parasites of birds and mammals are essentially ectotherms in an endothermic environment. The increase in anaerobic metabolism could be a response to temperature stress, but anaerobic pathways also predominate in the parasites of ectotherms.

Alternatively parasitic helminths may have arisen from forms that were initially anaerobic (Bryant, 1982), the earliest parasites inhabiting anaerobic sites within the body and migrating to more aerobic sites later. But one has then to explain how parasitic helminths later acquired aerobic metabolism in their free-living stages.

In free-living organisms the transition from aerobic to anaerobic metabolism is controlled primarily by the levels of adenine nucleotides. Both glycolysis and the respiratory chain compete for available ADP and inorganic phosphate; when glycolysis wins, the result is a Crabtree effect; when the respiratory chain wins, a Pasteur effect. The Crabtree effect is the suppression of oxygen uptake in tissues by the presence of exogenous glucose, the Pasteur effect is the reduction in the rate of carbohydrate catabolism in the presence of oxygen.

Cytochrome systems generally have a high affinity for ADP and when oxygen is available will phosphorylate ADP to give high ATP/ADP ratios. This inhibits glycolysis, first by limiting the availability of ADP for the phosphoglycerate kinase and pyruvate kinase reactions and secondly by allosteric inhibition of the regulatory enzymes, phosphorylase, hexokinase, phosphofructokinase and pyruvate kinase. These same enzymes are also regulatory in helminth pathways and are modulated by similar effectors. No new regulatory enzymes have yet been discovered in parasitic helminths nor have any unique modulators been described.

In parasitic helminths it is the anaerobic pathways which are dominating the cytochrome system, resulting in a permanent Crabtree effect. Glycolysis and the cytochrome chain are, of course, located in different cellular compartments. The extremely high glycolytic capacity of parasitic helminths could effectively sequester ADP in the cytoplasm, thus inhibiting the mitochondrial systems. The dominance of the anaerobic pathways in parasitic helminths may be further increased by the lack of side reactions removing glycolytic intermediates and by the absence of significant glyconeogenesis from lactate. The failure to achieve true state III respiration in helminth mitochondria may also be exacerbated by the limitations of the helminth translocase systems (Barrett, 1976).

Summary and Conclusions

Parasitic helminths belong to three separate phyla and certainly within the Nematoda and Platyhelminthes parasitism has arisen independently on several separate occasions. The anaerobic pathways found in parasitic helminths show considerable diversity, but in general they resemble the anaerobic pathways found in free-living invertebrates. There are, however, two major differences; first anaerobic metabolism is a steady state condition in parasitic helminths and secondly anaerobic metabolism persists in the presence of air. This poses a paradox; if parasitic helminths are steady state anaerobes why do they have functional cytochrome chains and require at least some oxygen for normal sustained growth and survival? Alternatively, if parasitic helminths are capable of oxidative metabolism, why do anaerobic processes persist in the presence of oxygen? Possibly only certain of the parasite's tissues are aerobic, for example the nervous tissue. This would make only a minor contribution to the overall energy budget, yet be of vital importance for survival. There is evidence in parasitic helminths for mitochondrial heterogeneity and tissue specific metabolism; the end-products produced in one tissue acting as substrates for another.

Oxygen may be required for certain specific synthetic reactions such as the synthesis of hydroxyproline, other key enzymic reactions may be directly linked to the cytochrome chain thus ensuring its retention in anaerobes. These might include the fumarate and volatile fatty acid reductases of *Ascaris*.

There is no evidence that control of catabolic pathways in parasitic helminths is fundamentally different from that of free-living organisms. The dominance of anaerobic cytoplasmic pathways over aerobic mitochondrial systems may be due to competition for ADP coupled with peculiarities in the mitochondrial translocase systems.

References

Allen, P.C. (1973) 'Helminths: Comparison of their rhodoquinone.' *Experimental Parasitology* **34**, 211-219

Barrett, J. (1976) 'Bioenergetics in helminths', in H. Van den Bossche (ed.), *Biochemistry of Parasites and Host-Parasite Relationships* (Elsevier, North Holland, Amsterdam), pp. 67-80

Barrett, J. (1978) 'Activation of succinate dehydrogenase from adult *Fasciola hepatica* (Trematoda)', *Parasitology* **76**, 269-275

Barrett, J. (1981) *Biochemistry of Parasitic Helminths*, (MacMillan, London)

References 163

Barrett, J. (1983) 'Biochemistry of filarial worms', *Helminthological Abstracts A* **52**, 1-18
Barrett, J. (1984) 'The anaerobic end-products of helminths', *Parasitology* **88**, 179-198
Barrett, J. and Butterworth, P.E. (1984) 'Acetaldehyde formation by mitochondria from the free-living nematode *Panagrellus redivivus,*' *Biochemical Journal* **221**, 535-540
Barrett, J. and Körting, W. (1977) 'Lipid catabolism in the plerocercoids of *Schistocephalus solidus* (Cestoda:Pseudophyllidea)', *International Journal for Parasitology* **7**, 419-422
Bryant, C. (1982) 'The biochemical origins of helminth parasitism', in L.E.A. Symons, A.D. Donald and J.K. Dineen (eds), *Biology and Control of Endoparasites*, (Academic Press, Sydney), pp. 29-52
Cain, G.D. (1976) '*Ascaris lumbricoides*: Coproporphyrinogen oxidase activity in eggs and muscle', *Experimental Parasitology* **40**, 112-115
Cheah, K.S. (1975) 'Properties of *Ascaris* muscle mitochondria 1. Cytochromes', *Biochimica et Biophysica Acta* **387**, 107-114
Cheah, K.S. and Chance, B. (1970) 'The oxidase systems of *Ascaris*- muscle mitochondria', *Biochimica et Biophysica Acta* **223**, 56-60
Cheah, K.S. and Prichard, R.K. (1975) 'The electron transport systems of *Fasciola hepatica* mitochondria', *International Journal for Parasitology* **5**, 183-186
Douch, P.G.C. and Blair, S.S.B. (1975) 'The metabolism of foreign compounds in the cestode, *Moniezia expansa* and the nematode *Ascaris lumbricoides* var *suum*', *Xenobiotica* **5**, 279-292
Fioravanti, C. (1982) 'Mitochondrial malate dehydrogenase, decarboxylating (malic enzyme) and transhydrogenase activities of adult *Hymenolepis microstoma* (Cestoda)', *Journal of Parasitology* **68**, 213-220
Fioravanti, C.F. and Kim, Y. (1983) 'Phospholipid dependence of the *Hymenolepis diminuta* mitochondrial NADPH:NAD transhydrogenase', *Journal of Parasitology* **69**, 1048-1054
Fry, M. and Brazeley, E.P. (1984) 'NADH-fumarate reductase and succinate dehydrogenase activities in mitochondria of *Ascaridia galli* and *Nippostrongylus brasiliensis*', *Comparative Biochemistry and Physiology* **77B**, 143-150
Fujimoto, D. and Prockop, D.J. (1969) 'Protocollagen proline hydroxylase from *Ascaris lumbricoides*', *Journal of Biological Chemistry* **244**, 205-210
Komuniecki, R., Komuniecki, P.R., and Saz, H.J. (1981) 'Pathway of formation of branched-chain volatile fatty acids in *Ascaris* mitochondria', *Journal of Parasitology* **67**, 841-846
Komuniecki, R., Rioux, A., and Thissen, J. (1984) 'NADH-dependent tiglyl-CoA reduction in disrupted mitochondria of *Ascaris suum*', *Molecular and Biochemical Parasitology* **10**, 1-10
Laties, G.G. (1982) 'The cyanide-resistant alternative path in higher plant respiration', *Annual Review of Plant Physiology* **33**, 519-555
Munir, W.A. and Barrett, J. (1985) 'The metabolism of xenobiotic compounds by *Hymenolepis diminuta* (Cestoda: Cyclophyllidea)', *Parasitology* **91**, 145-156
Podesta, R.B., Mustafa, T., Moon, T.W., Hulbert, W.C. and Mettrick, D.F. (1976) 'Anaerobes in an aerobic environment: Role of CO_2 in energy metabolism of *Hymenolepis diminuta*', in H. Van den Bossche (ed.), *Biochemistry of Parasites and Host-Parasite Relationships* (North Holland, Amsterdam), pp. 81-88
Precious, W.Y and Barrett, J. (1989) 'The possible absence of cytochrome p-450 linked xenobiotic metabolism in helminths,' *Biochimica et Biophysica Acta* **992**, 215-222

Rioux, A. and Komuniecki, R. (1984) '2-Methylvalerate formation in mitochondria of *Ascaris suum* and its relationship to anaerobic energy generation', *Journal of Comparative Physiology B* **154**, 349-354

Stryer, L. (1981) *Biochemistry*, 2nd edit., Freeman, San Francisco

Ward, P.F.V. (1982) 'Aspects of helminth metabolism', *Parasitology* **84**, 177-194

Chapter 9

Annelids

Udo Schöttler and E.M. Bennet

Introduction

The three classes of annelids, Polychaeta, Oligochaeta and Hirudinea, are predominantly aquatic. Most polychaetes are marine, whereas oligochaetes and the hirudines occur mainly in fresh water. Relatively few species, the majority of which are oligochaetes, have accomplished the transition to terrestrial life and, since they possess no effective protection against desiccation, they are limited to damp environments such as terrestrial soils, moist jungle or grasslands. Within aquatic environments, annelids are also found where hypoxic or even anoxic situations can be encountered. Examples are the stagnant mudbanks of inland waters and the intertidal zones of the coastline which become dry at low tide.

Early in this century, the abilities of several annelids to tolerate fluctuations in ambient oxygen concentrations were investigated. The majority of the investigations were limited to the determination of survival under anoxic conditions (Packard, 1905; von Brand, 1927; Dausend, 1931; Jacubowa and Malm, 1931; Hecht, 1932). Survival times measured ranged from one day for *Amphitrite* to 25 days for *Owenia*. In addition, studies of *Tubifex* showed that resistance to anoxia is temperature-dependent. The animals survived for longer at 2°C than at 18°C. Von Brand was the only worker who attempted to correlate biochemical data with resistance to anoxia. He showed that *Owenia* had large stores of glycogen (5% of the wet weight) and suggested that the catabolism of glycogen enabled it to survive for long periods of time.

Work on anaerobic metabolism of annelids was not resumed until the end of the 1950s. Dales (1958) showed that the lugworm *Arenicola marina,* a marine polychaete, mobilised glycogen under anaerobic conditions, but did not accumulate lactate. Dales concluded from these results 'It is now quite clear that in *A. marina* metabolism of glycogen under anaerobic conditions leads to products other than lactic acid'.

The first evidence to indicate which metabolic pathways are utilised by euryoxic annelids to produce energy under anaerobic conditions was furnished by experiments with *Alma emini,* an oligochaete from the tropical swamps of central Africa (Coles,

1970). This work showed quantitatively that propionate and acetate were produced under anaerobic conditions. Coles' investigations provided the basis for a body of experimental work on anaerobic metabolism of free-living, euryoxic annelids.

It is necessary to differentiate between environmentally dependent and functionally dependent anaerobiosis. Environmentally dependent anaerobiosis occurs when animals inhabiting an area where oxygen becomes depleted are either unable to move out of it or are unable to leave it quickly enough. Environmentally dependent anaerobiosis is thus predominantly found in animals with limited motility. Functionally dependent anaerobiosis arises when, as a result of excessive muscular activity, such as hunting of prey or escape from predators, the requirement for oxygen is greater than its availability. In contrast to environmentally dependent anaerobiosis, which always affects the whole animal, functionally dependent anaerobiosis is limited to individual tissues.

Anaerobic Metabolism during Environmentally Dependent Anaerobiosis

Most of the experiments on environmentally dependent anaerobiosis in annelids have been carried out on *A. marina*. The advantage of using this polychaete is that it is relatively large. It is thus possible to conduct separate experiments on individual tissues and track the metabolism of specific radioisotopes to trace pathways. The metabolic capacity for anaerobic energy production shown by *A. marina* is also found in other euryoxic annelids. The following discussion will describe, in the first instance, the findings from *A. marina*. Where generalisations are not possible, variations among different species are discussed.

The End-products of Anaerobic Energy Metabolism

A. marina inhabits the sandy and sand-mud sediments of the tidal flats. The animals live in U-shaped burrows that may be 30 cm deep and are seldom vacated. During high tide the animals pump surface water, rich in oxygen, through their burrows. During this time energy is predominantly produced via the respiratory chain. At low tide *A. marina* can be exposed to several hours of anoxia. Under laboratory conditions it survives 5-6 days in oxygen-free medium at a temperature of 12°C (Toulmond, 1975).

Unlike aerobic metabolism, the energy required for maintaining life in the absence of oxygen has to be provided mainly by substrate-linked phosphorylation. Carbohydrate forms the most important, if not the only, energy source. Relatively energy-rich compounds accumulate as end-products of metabolism because complete oxidation to CO_2 and H_2O is not possible. It has been shown that, apart from glycogen, *A. marina* can also utilise aspartate and the phosphagen phosphotaurocyamine as energy sources. The end-products of anaerobic metabolism which accumulate are D- and L-alanine, strombine, succinate and the volatile fatty acids, acetate and propionate; some of the acids are excreted (Zebe, 1975; Surholt, 1977; Felbeck and Grieshaber, 1979; Felbeck, 1980; Siegmund and Grieshaber, 1983; Siegmund, Grieshaber, Reitze and Zebe, 1985). Strombine is an end-product of anaerobic glycolysis formed by the condensation of pyruvate and glycine. As with lactate production, the NADH formed in glycolysis is reoxidised in the production of strombine (Fields, Eng, Ramsden, Hochachka and Weinstein, 1980).

Table 9.1 gives the concentrations of end-products measured at various times during anaerobiosis. *A. marina* does not produce all the identified end-products continuously. D- and L-alanine, succinate and strombine accumulate only in the first 12 hours. Propionate and acetate play only a minor role during this time, but form the sole end-products during long-term anaerobiosis.

Subsequent investigations, mainly on body-wall musculature, which is by far the largest tissue, showed that the switch from aerobic to optimal anaerobic metabolism occurs in two phases (Schöttler, Wienhausen and Westermann, 1984). These phases are called the transition period and the switching period

The Transition Period

During the transition period (Figure 9.1), which takes about three hours in *A. marina* the switch from aerobic to anaerobic energy production occurs. *A. marina* has only limited endogenous oxygen reserves; as anoxic conditions increase, the animal has to switch to anaerobic energy production. At that stage, metabolic pathways that are immediately available have to be used. They may not necessarily provide the most efficient form of anaerobic energy production.

Some of the energy produced during this period comes from the mobilisation of the phosphagen stores. Increases in the concentrations of glycolytic metabolites indicate that, at the same time or with only a slight delay, the rate of glycogen

Table 9.1 Anaerobic metabolism in the lugworm, *Arenicola marina*. Concentrations of endproducts and the phosphagen after different periods of incubation (in µmoles/g dry weight).

Time (h)	D,L-alanine	strombine*	succinate	propionate		acetate		p-taurocyamine
				animal	water	animal	water	
0	106	1.5	0.8	0.2	-	2.4	-	39
3	149	6.5	12.9	2.7	-	6.3	-	17
12	157	11.1	17.4	9.5	6.8	7.4	4.2	12
24	151	12.8	19.6	13.3	28.6	10.8	9.6	10
48	159	9.8	18.2	13.7	72.4	12.2	25.9	11

(From Siegmund *et al.*, 1985)

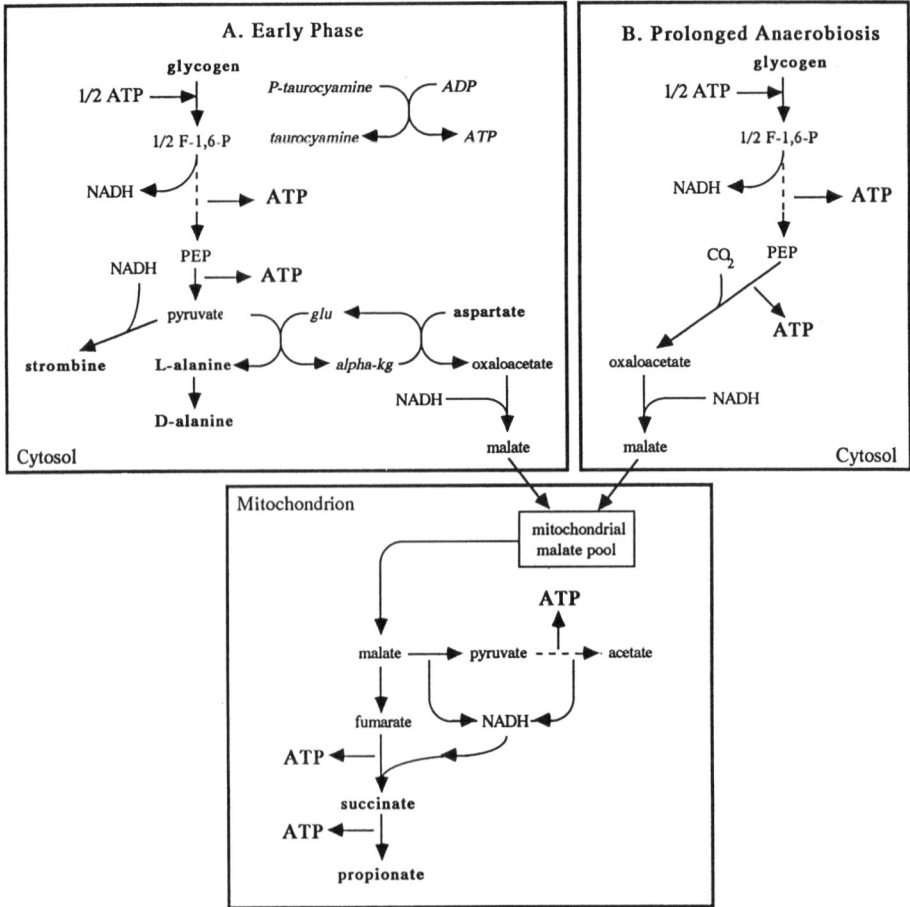

Figure 9.1 Metabolic pathways involved in anaerobic energy production in *Arenicola marina*.

catabolism increases (Schöttler, 1980). Activation of glycogen phosphorylase is, in contrast to that of skeletal muscle of vertebrates, not dependent on phosphorylation. Available data (Kamp, 1986) suggest that the activity of the enzyme is controlled by the concentration of inorganic phosphate (P_i). The rapid mobilisation of phosphagen in the initial phase results in a distinct increase in P_i concentration and consequent activation of glycogen phosphorylase. Glycogen is metabolised via the Embden-Meyerhof pathway to pyruvate and is further converted to strombine or L- or D-alanine. D-alanine is formed from L-alanine by a reaction catalysed by an alanine racemase (Sticher, 1985). During the transition period, aspartate and glycogen are catabolised simultaneously. The amino group of aspartate is transferred to pyruvate. Aspartate and alanine aminotransferase are involved in this process (Felbeck, 1980) (Figure 9.1A).

In the cytoplasm, oxaloacetate arising from aspartate is converted to malate by the action of malate dehydrogenase. At the same time, NADH, originating from the glyceraldehyde phosphate dehydrogenase step of the Embden Meyerhof pathway, is reoxidised, a necessary process during the transition period to facilitate the extensive mobilisation of glycogen. Malate is further metabolised in the mitochondria to succinate, acetate and propionate. These reactions are discussed in more detail below.

Despite the use of various energy sources, *A. marina* cannot match energy production with its requirements. This is indicated by a large drop in the energy charge, from 0.89 to 0.82, during this phase. The drop occurs despite the fact that the rate of substrate turnover during this time is much higher than during prolonged anaerobiosis. If, subsequently, the rate of substrate utilisation were not reduced, the glycogen reserves would be exhausted after little more than one day (Schöttler *et al.* 1984).

The Switching Period

After 2-3 hours substrate utilisation is considerably decreased. At the same time, the drop in energy charge that occurs between 3 and 24 hours is small (0.82 to 0.78) in comparison with that in the first three hours, indicating that energy utilisation is also reduced to a considerable extent (Schöttler *et al.*, 1984). This reduction in metabolic activity represents an important adaptation of a euryoxic organism to anaerobic conditions. The more economical it is in the utilisation of its energy stores, the more likely it is to survive long-term anoxia.

In addition to the reduction in metabolic rate there is an important change in the way glycogen is utilised. After about three hours, an increasing proportion of glycogen, which had previously been catabolised almost quantitatively to alanine and strombine, is converted to succinate, propionate and acetate (Schöttler *et al.*, 1984).

The metabolism of glycogen to succinate requires the carboxylation of a glycolytic metabolite. This occurs at the level of phosphoenolpyruvate and is catalysed by phosphoenolpyruvate carboxykinase (Schöttler and Wienhausen, 1981). Since a metabolic switch occurs at the level of phosphoenolpyruvate, this point is termed the 'phosphoenolpyruvate branchpoint' (Hochachka and Mustafa, 1972). The switch depends on regulation of at least one of the enzymes competing for phosphoenolpyruvate, either pyruvate kinase or phosphoenolpyruvate carboxykinase. Energetically the pyruvate kinase and PEPCK reactions are of equal value. In both cases an energy-rich nucleotide is gained. The controlling factor seems to be that, during the switching period, the affinity of pyruvate kinase for the substrate phosphoenopyruvate is reduced (Schöttler, 1980; Englisch, 1989). During 24 h anaerobiosis, the K_m for phosphoenolpyruvate $_{0.5}$ increases from 1.4×10^{-4} to 2.9×10^{-4} mol/l at pH 7.3. Even though, in contrast to *Mytilus*, it was not possible to isolate two different forms of pyruvate kinase in *Arenicola marina* (Holwerda, Kruitwagen and de Bont, 1981), it seems certain that phosphorylation of the enzyme is involved in the reduction of affinity during anoxia. Of course, it is doubtful whether similar conditions prevail in *Arenicola marina* and *Mytilus*.

Analytical isoelectric focusing and staining for enzyme activity identified at least five discrete bands under both aerobic and anaerobic conditions (Englisch, 1989). This suggests that the phosphorylation and dephosphorylation of pyruvate kinase is not achieved by an 'all or nothing' principle but occurs sequentially and that transitional stages exist in *Arenicola marina*. In the former case only two variants occur - 'phosphorylated' or 'dephosphorylated'- these being independent of the maximum number of phosphate groups binding to the enzyme molecule. In the latter case, the number of transition states will depend on the maximum degree of phosphorylation attainable. As the separate transitional steps differ only slightly in their chemical properties, it is understandable that, despite intensive efforts, it has not been possible quantitatively to isolate different forms of the enzyme.

Affinity of pyruvate kinase for phosphoenolpyruvate is increased as intracellular pH falls. Recent NMR data have shown that the intracellular pH falls within a few hours, from about 7.3 to less than 6.8, during anoxia (Kamp and Juretschke, 1989). As a result, the phosphoenolopyruvate$_{0.5}$ value changes from 1.4×10^{-4} to 4.7×10^{-4} mol/l within 24 hours of anaerobiosis. In contrast, pyruvate kinase from aerobic animals shows a less remarkable change in affinity for phosphoenolpyruvate from 1.4×10^{-4} to 2.1×10^{-4} mol/l between pH 7.3 and 6.8.

It is possible that the processes of regulation are more complex. There is evidence that aggregation and dissociation reactions could be involved in control of activity (Englisch, 1989). In addition, the pyruvate kinase from the cuticulomuscular tube of *Arenicola marina* is associated with cellular structures under aerobic conditions. Under anaerobic conditions the degree of association is diminished possibily as a result of the drop in pH values within 6-12 hours (Englisch, 1989). Available data suggests it is unlikely that allosteric effectors, such as the activators fructose-1,6-bisphosphate and AMP and the inhibitors L-alanine, ATP and phosphotaurocyamine play a role in the anaerobic regulation of pyruvate kinase.

The oxaloacetate that is derived from the carboxylation of phosphoenolpyruvate is reduced to malate by the action of the cytoplasmic malate dehydrogenase, and takes on the same function as the oxaloacetate derived from the catabolism of aspartate during the transition period.

The switch at the phosphoenolpyruvate branchpoint is complete by about 9-12 hours. After that time, glycogen is metabolised only via oxaloacetate (Figure 9.1B). At the same time, aspartate catabolism ceases and the phosphagen reserves are depleted to such an extent that they play no further role in the course of anaerobiosis. Thus, glycogen has become the only energy source. Consistent with its physiological capabilities, *A. marina* has adjusted optimally to anoxia.

The Role of Mitochondria during Anaerobic Energy Metabolism

The metabolism of glycogen or aspartate beyond malate, takes place in the mitochondria. The enzymes involved are located in the matrix or the inner membrane. Of particular significance are, first, the existence of two possibilities for the chemical conversion of malate - it can either be oxidised or reduced - and second, that fumarate assumes the role of electron acceptor in the absence of oxygen. Fumarate is formed from malate in a reaction catalysed by fumarase and is further converted to succinate by the action of fumarate reductase; at this step NADH is oxidised. Fumarate reductase is not a single enzyme but a complex enzyme system, connecting segments I and II of the respiratory chain, NADH-quinone-oxidoreductase and succinate-quinone-oxidoreductase. In contrast to aerobic conditions, where both oxidoreductases transport electrons in the direction of quinone, under anaerobic conditions, hydrogen is transported via the NADH-quinone-reductase to quinone and from there in the opposite direction via succinate-quinone-oxidoreductase to fumarate.

As segment I of the respiratory chain is involved in this reaction, the formation of succinate from fumarate is coupled to an equimolar yield of ATP. However, there must be a difference from the respiratory chain known from higher animals. The redox potential of ubiquinone ($E'=+0.10$ mV) does not permit the transfer of electrons to fumarate ($E'=-0.03$ mV). What type of quinone is present in *A. marina* and other free-living euryoxic invertebrates - and whether ubiquinone is also present - has not yet been investigated.

The electrons necessary for reducing fumarate originate from the oxidative catabolism of malate (Figure 9.2). Anaerobically, malate is not only oxidised to oxaloacetate, as under aerobic conditions, but is also oxidatively decarboxylated to pyruvate by the action of malic enyzme. Pyruvate is further converted to acetyl CoA *via* pyruvate dehydrogenase. Two possibilities exist for further anaerobic metabolism. The end-product acetate can be formed from acetyl CoA. Acetyl CoA can also condense with oxaloacetate to form citrate. At present it is not known which are the deciding factors for the fate of acetyl CoA. From results to hand it can be argued that, in the body wall musculature of *A. marina*, a greater portion proceeds in the direction of acetate. Citrate can be converted via the TCA cycle to succinyl-CoA and possibly to succinate. The latter can only take place during the initial phase of anaerobiosis when succinate concentrations are still low. Oxidative catabolism to succinyl CoA or succinate involves five dehydrogenases: malic enzyme and mitochondrial malate, pyruvate, isocitrate and 2-oxoglutarate dehydrogenases. Since intramitochondrial metabolism of malate can only proceed if the redox state is balanced, the following stoichiometric equations, which have been confirmed experimentally with isolated mitochondria (Schroff and Schöttler, 1977; Schöttler 1977), apply:

172 Annelids

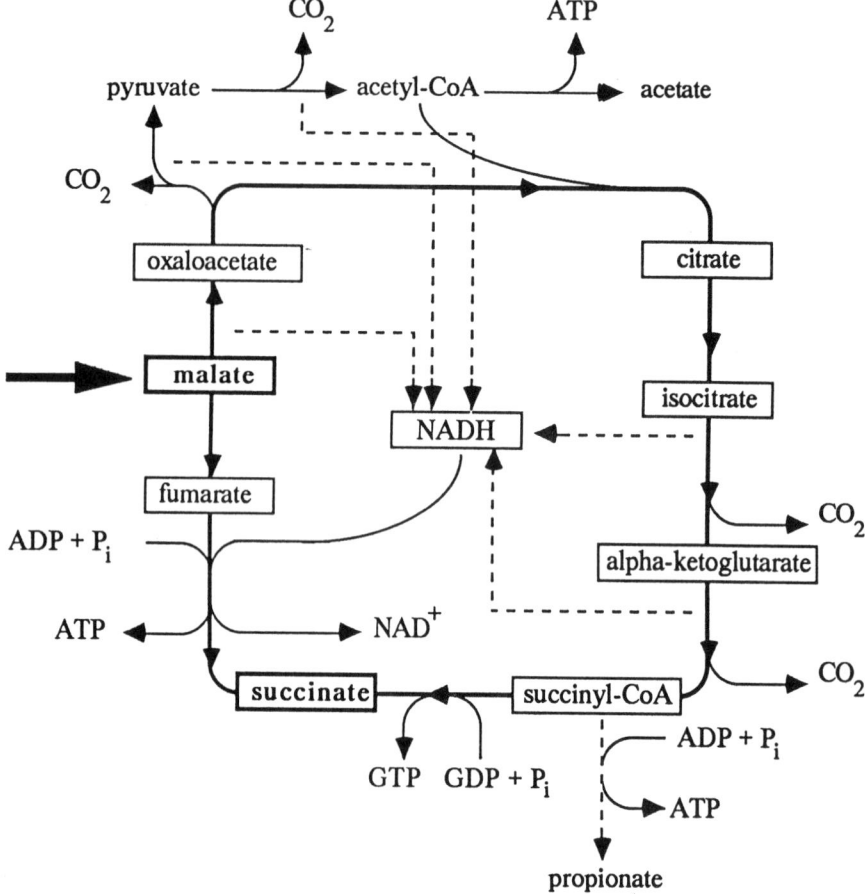

Figure 9.2 Anaerobic metabolism of malate in mitochondria in the body-wall musculature from *Arenicola marina*. (Modified after Schroff and Schöttler, 1977.)

$$2 \text{ malate} + 5\text{NAD}^+ \rightarrow \text{succinate} + 5\text{NADH} + 4\text{CO}_2$$
$$5 \text{ malate} + 5\text{NADH} \rightarrow 5 \text{ succinate} + 5\text{NAD}^+$$
$$\overline{7 \text{ malate} \rightarrow 6 \text{ succinate} + 4\text{CO}_2}$$

However, as can be seen from Table 9.1, succinate is an end-product of energy metabolism only during the initial phase of anaerobiosis. Under prolonged anaerobiosis it is further converted to propionate (Zebe, 1975), which then becomes the major end-product. In isolated mitochondria from the body wall musculature of

A. marina propionate is formed intramitochondrially from succinate. The formation of propionate is greatly increased by the addition of malate (Schroff and Zebe, 1980).

The metabolic pathways from succinate to propionate have been elucidated by Schöttler (1983). The initial substrate for propionate formation is succinyl CoA which, under anaerobic conditions, is also derived from 2-oxoglutarate in the TCA cycle as described above. However, since far more propionate is produced than can be formed from 2-oxoglutarate under anaerobic conditions, there must be another way to convert succinate to succinyl CoA. *In vitro* this is possible through the action of succinyl CoA synthetase. However, this step requires an energy-rich nucleotide and energy in particular is in low supply during anaerobic conditions. It is thus questionable whether this reaction can occur *in vivo*. Succinyl CoA is converted to D-methylmalonyl-CoA, *via* L-methylmalonyl-CoA, by the action of methylmalonyl-CoA-mutase and methylmalonyl-CoA-racemase. D-methylmalonyl-CoA is decarboxylated to propionyl-CoA by the action of propionyl CoA-carboxylase. ATP is formed in this reaction (Figure 9.3).

The metabolic pathway to this point is a reversal of the one occurring in the digestive tract of mammals, where propionate is conveyed in the direction of glycogen synthesis (Flavin, Ortiz and Ochoa, 1955; Flavin and Ochoa, 1957). Whereas, in mammals, propionate is activated to propionyl-CoA by the action of a kinase with concomitant dephosphorylation of ATP to AMP, the formation of propionate from propionyl CoA during anaerobic metabolism in *A. marina* is catalysed by a CoA transferase. This reaction, which has also been shown to occur in endoparasites and molluscs (Köhler, Bryant and Behm, 1978; van Vugt, van der

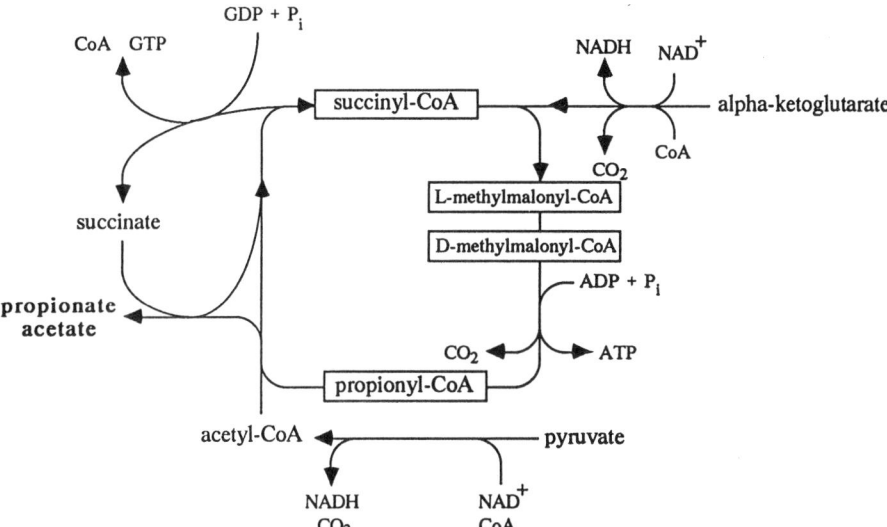

Figure 9.3 Anaerobic formation of propionate and acetate in mitochondria of the body-wall musculature from *Arenicola marina*.

Meer and van den Burgh, 1979; Saz and Pietrzak, 1980; Pietrzak and Saz, 1981; Schulz and Kluytmans, 1983), does not result directly in the production of energy in the form of an energy-rich nucleotide. However, CoA is transferred directly to succinate without the consumption of an energy rich CoA-substrate. The action of the CoA-transferase thus results in 'recycling' of energy-rich thioester bonds, forming a 'succinate-propionate' cycle where one ATP is synthesised for every propionate produced (Figure 9.3).

The formation of succinyl CoA by the action of 2-oxoglutarate dehydrogenase or, *in vitro*, by succinyl-CoA-synthetase thus functions only as a primary step. Once set in motion, the cycle can run as long as succinate is available (Table 9.2).

In these experiments, succinyl CoA was formed by the action of succinyl-CoA synthetase. The reaction was started with the addition of succinate and was linear for 60 min. Inhibition of succinyl-CoA synthetase by addition of 3-mercaptopicolinic acid caused a reduction in propionate synthesis of about 80%. Inhibition of succinyl-CoA synthetase 15 min after the start of the reaction affected the formation of propionate only minimally. This shows that the cycle, once set in motion, requires no additional energy.

The succinate-propionate cycle could also function as a trap for coenzyme A, into which it becomes locked by 2-oxoglutarate dehydrogenase, leading to depletion of the free coenzyme. The scheme for the mitochondrial portion of anaerobic energy production developed so far requires that a permanent and sufficient concentration of free coenzyme A be available. Therefore some of the coenzyme A must be continuously removed from the succinate-propionate cycle. This is made possible via succinyl-CoA, upon which both methylmalonyl-CoA mutase and succinyl-CoA synthetase act.

Experiments with isolated mitochondria from the body wall musculature of *A.*

Table 9.2 Inhibition of the formation of propionate by 3-mercaptopicolinic acid.

3-MPA (μg/mg protein)	formation of propionate (μg/mg protein)	% inhibition
0	198 ± 24	0
30	110 ± 17	45
100	39 ± 8	80
addition after 15 min		
100	164 ± 27	17

Mitochondria prepared from body-wall musculature of *Arenicola marina* were sonicated and centrifuged for 1 h at 150,000 x g. The supernatant was dialysed and then incubated for 1 h at 25° with: 70 mMol/l HEPES ph 7.2; P_i 8 mMol/l; Mg^{2+} 2 mMol/l; ADP 1 mMol/l; GTP 2 mMol/l CoA 0.25 mMol/l. The reaction was started with succinate 10 mMol/l *marina* have shown that, even under anaerobic conditions, succinyl-CoA synthetase catalyses the splitting and not the formation of succinyl-CoA, thus liberating the necessary free coenzyme A. The energy yielded as a result of the cleavage is stored in the form of an energy rich nucleotide (Schöttler, 1986). It is not yet clear how anaerobic metabolism is regulated at the succinyl-CoA step. This will require determination of intramitochondrial concentrations of the intermediary metabolites and examination of the kinetic properties of methylmalonyl-CoA mutase and succinyl-CoA-synthetase.

In addition to propionate, another volatile fatty acid, acetate, is formed as an end-product of anaerobic metabolism. Wienhausen (1979, 1981) demonstrated that acetate is also formed intramitochondrially, in association with this equimolar synthesis of ATP. The details remain obscure.

More recent findings, obtained with submitochondrial fractions, indicate that the release of acetate from acetyl-CoA and the formation of ATP occur in two separate, not necessarily coupled, reactions (Schöttler, unpublished), and that the formation of acetate and propionate is catalysed by the same enzyme.

Investigations into the specificity of CoA transferase have shown that, although the enzyme transfers coenzyme A specifically to succinate, it reacts less specifically with respect to the substrate donating coenzyme A. Hence, butyryl-CoA, acetoacetyl-CoA, malonyl-CoA and acetyl-CoA may react, as well as propionyl CoA. These compounds can therefore be used as primers for the succinate-propionate cycle *in vitro.* (Table 9.3). The consequences for the living organism, is that ATP does not arise directly by the release of acetate from acetyl-CoA but is produced subsequently *via* a thiokinase reaction (succinyl-CoA synthetase) coupled to CoA-transferase.

Table 9.3 Formation of propionate after starting with different CoA-thioesters.

Starter Substance	Formation of Propionate (nmol/mg protein/h)
priopionyl-CoA	242 ± 14
succinyl-CoA	228 ± 9
acetyl-CoA	224 ± 16
butyryl-CoA	246 ± 23
malonyl-CoA	214 ± 19
acetoacetyl-CoA	227 ± 12

A mitochondrial supernatant fraction was prepared as described in Table 9.2, and incubated with: 70 nMol/l HEPES pH 7.0; P_i 8 mMol/l; Mg^{2+} 2 mMol/l; ADP 1 mMol?l; succinate 10 mMol/l. The reaction was started by the addition of the different CoA-thioesters (0.25 mMol/l)

Tissue-specific Differences

This discussion of anaerobic metabolism in *A. marina* has concentrated on results obtained with the whole animal (Table 9.1) or body wall musculature and they agree well. However, the chloragogen tissue associated with the gut has also been investigated (Schöttler *et al.,* 1984). During long-term anaerobiosis the same pathways employed for energy production in body wall musculature are used in this tissue, that is, glycogen is catabolised to propionate and acetate. Distinct differences are evident in the transition phase, however. Aspartate and phosphotaurocyamine stores are considerably lower than in the body wall musculature and are depleted within two hours. Also, much less alanine is accumulated during this time. As it has not yet been possible to measure strombine it remains to be seen if glycolysis leading to strombine has greater significance in this organ complex than in the body wall musculature. The drastic drop in energy charge from 0.84 to 0.68 after two hours is, however, an indication that anaerobiosis during the transition phase is a heavy burden for the gut/chloragogen tissue. There is some indication that the switch at the phosphoenolpyruvate branchpoint is accomplished more quickly in the gut/chloragogen tissue than in the body wall musculature. The energy charge stabilises after three hours and remains constant. At the same time more succinate and propionate is accumulated than aspartate is catabolised.

The Role of Coelomic Fluid in Anaerobic Metabolism of Arenicola marina

The coelomic fluid comprises more than 50% of the body weight and contains only a small number of cells, except when it harbours developing gonadal products. Metabolites that have been found to increase in the coelomic fluid during anaerobiosis thus originate predominantly, if not totally, from the tissues. The metabolites that have so far been identified are: glucose, succinate, acetate and propionate. Substances that decrease in concentration have not yet been found. Glucose and succinate concentrations rise during the first three hours and reach a plateau after six to nine hours. Glucose increases from 0.2 to 0.7 mmol/ml; succinate from 0.1 to 0.8 mmol/ml. It is not known from where the glucose originates, nor whether increased concentrations are an indication of enhanced transport between various organs. The same applies to succinate. It also remains to be seen whether the change in concentration is associated with adjustment to a new equilibrium between cells and body fluid or to increased transport between various organs of the body. Output to the surrounding medium could not be shown for either substance.

In contrast to succinate and glucose, the concentrations of the volatile fatty acids reach higher values (2-2.5 mmol/ml). In this case, the increase occurs between 3 and 24 hours. As a result, the coelomic fluid, to a limited degree, takes on the function of a reservoir for end-products of anaerobic metabolism.

The Excretion of Volatile Fatty Acids

A characteristic of anaerobic energy metabolism after 24 hours is the quantitative excretion of acetate and propionate (Table 9.1). Even though substances with relatively high energy content are thereby lost by the animal, this process has specific advantages. Acidosis, which could develop from the accumulation of larger quantities of acid end-products, is avoided, as well as a disturbance in osmotic balance which could occur where anoxia lasts several days.

By what means the acids get into the medium had for some time been an open and uninvestigated question. Excretion via the nephridia or the gills seemed possible. Even loss via the gut was not excluded. Recently it has been possible to demonstrate that gills and gut are not involved in excretion. Volatile fatty acids mainly pass through the undifferentiated body surface. The effective mechanisms in this case are diffusion for acetate, and diffusion plus facilitated diffusion for propionate (Holst and Zebe, 1986).

Unresolved Problems

Investigations in the last 10 years have given us some insight into the process of anaerobic metabolism in *A. marina*. Yet, apart from the ones already mentioned, there remain many unanswered questions, particularly concerning the transition and switching phases.

The first problem concerns the role of amino acids in anaerobic metabolism. The catabolism of aspartate and the accumulation of alanine are linked, as described above, because the amino group of aspartate is transferred to pyruvate. However, this is not in metabolic balance. *A. marina* accumulates about twice as much alanine as it catabolises aspartate. Therefore there must exist at least one other mechanism by which amino groups are made available for the reaction. The possibility that other amino acids such as asparagine or glutamine serve as donors of amino groups can be excluded because the concentration of these substances is too low. The formation of alanine from pyruvate through fixation of ammonia is also not possible, as *A. marina* does not have the appropriate alanine dehydrogenase (Sticher, 1985).

The catabolism of aspartate is also out of balance with respect to the carbon skeleton. Only a little more than half the total aspartate carbon is fixed into succinate, propionate or acetate. At the same time, the concentrations of other metabolites participating in this pathway do not increase to account for the discrepancy.

A second problem which should be mentioned concerns the drastic reduction in the metabolic rate that occurs towards the end of the transition period. In part, this reduction can be traced to the drop in cellular pH values. At this stage, however, the possibility that superimposed hormonal or nervous regulation may be involved cannot be excluded.

Anaerobic Metabolism in Other Annelids

Anaerobic metabolism has also been investigated in a number of other annelids: the marine polychaetes *Euzonus mucronata* (Ruby and Fox, 1976), *Nephtys hombergii* (Schöttler, 1982), various *Nereis* species (Schöttler, 1979), *Scoloplos armiger* (Schöttler and Grieshaber, 1988), the aquatic oligochaetes *Tubifex* spp. (Schöttler and Schroff, 1976; Hoffman, 1981; Seuss *et al.*, 1983; Famme and Knudsen, 1984, 1985) and *Lumbriculus variegatus* (Putzer, Gnaiger and Lackner, 1984; Putzer, 1985), the terrestrial oligochaetes *Lumbricus terrestris, Lumbricus rubellus* (Gruner and Zebe, 1977) and *Eisenia foetida* (Zebe and Heiden, 1983), as well as the leech *Hirudo medicinalis* (Zebe, Salge, Wiemann and Wilps, 1981). With the exception of *Tubifex* and *Nereis virens* the investigations cited above were limited largely to the measurement of various metabolites important for anaerobic energy metabolism.

All species that are able to survive anaerobically for more than 24 hours form and excrete propionate as either the sole or, quantitatively, the most important end-product during anaerobiosis lasting more than 12 hours. In addition, acetate has been shown to be an end-product, the amount varying between species. Propionate/acetate ratios varied from 2.5 in *Tubifex* and *Lumbriculus* to >25 in the medicinal leech *Hirudo*. It was not possible to detect acetate from *E. mucronata*.

It can thus be concluded that anaerobic glycogen catabolism leading to propionate and acetate, as described in detail for *A. marina*, is widely distributed and probably the rule among the euryoxic annelids. The nereids are exceptions, as they accumulate lactate as well as propionate and acetate during anaerobiosis of several days' duration. In contrast to metabolism during long-term anaerobiosis, species-specific, differences occur in the transition phase. All annelids that have been studied mobilise their phosphagen reserves during this phase. On the other hand, the parallel catabolism of glycogen and aspartate is limited largely to marine polychaetes. In these organisms high intracellular concentrations of free amino acids, which include aspartate, are involved in the maintenance of osmotic equilibrium (Schoffeniels and Gilles, 1972).

In fresh water and terrestrial species, however, the concentration of aspartate is too low for its catabolism to provide energy to any significant extent. Yet *Tubifex* and *Lumbriculus*, as well as the terrestrial species *Lumbricus terrestris, Lumbricus rubellus* and *Eisenia foetida*, accumulate alanine in the first hours of anoxia. In *Tubifex*, the carbon skeleton of the accumulated alanine originates from the catabolism of glycogen (Schöttler and Schroff, 1976). The origin of the amino group is not known. In two species, *Lumbriculus variegatus* and *Hirudo medicinalis*, unusually high malate concentrations have been demonstrated. For *Lumbriculus* maintained aerobically a value of 9 mmol/g fresh weight is quoted (Putzer, 1985); in *Hirudo,* 8 mmol/g fresh weight was measured in the body wall musculature and 9 mmol/g in the blood (Zebe et al., 1981). Similarly, concentrations of malate in *Tubifex* (Seuss et al., 1983) and *Eisenia foetida* (Zebe and Heiden 1983) of 4 and 3 mmol/g fresh weight respectively are much higher than those in *Arenicola marina* (0.2 mmol/g fresh weight in body wall musculature). In all four species malate is almost completely catabolised in the first hours of anoxia and, at the same time, similar quantities of succinate accumulate.

It appears that in *Hirudo* and *Lumbriculus* only very little, if any, glycogen is mobilised in this phase. It is possible, although it has yet to be proved experimentally, that in these animals malate is catabolised by the pathway depicted in Figure 9.2.

Lumbricus terrestris, Lumbricus rubellus and *Eisenia foetida* accumulate lactate as an additional end-product. In *Eisenia* and *L. terrestris* lactate formation is limited to the first hours; in *L. rubellus* it extends over at least the first 24 hours. The reduction in energy requirements noted in *A. marina* under anaerobic conditions also appears to occur in other euryoxic annelids. Finally, it has to be noted that in almost all investigations, the phosphagens were either incompletely or not determined and that a detailed investigation of the transition phase has been carried out in only a few cases (*Tubifex* and *Lumbriculus*).

Conclusions

Anaerobic energy production is always less efficient than its aerobic counterpart. The energy yield from even the most favourable pathway in euryoxic annelids, that is, the anaerobic catabolism of glycogen to propionate, is more than double that of anaerobic glycolysis, but only about 15% of that gained from aerobic catabolism of the same substrate. The maintenance 'normal' life activities under anaerobic conditions includes not only locomotion and feeding, but also growth and, probably, the formation of gonadal products - an animal that lives under anaerobic conditions must mobilise much greater quantities of energy-yielding substrate than one that lives under aerobic conditions. This is possible if the energy-yielding substrate is directly accessible and if simple precursors for *de novo* synthesis of macromolecules are available in sufficiently high concentrations in the surrounding medium. Such a situation exists for intestinal endoparasites of higher vertebrates. But it is questionable whether it is possible to sustain such 'normal' life functions under anaerobic conditions in situations where complex macromolecular food needs to be taken up and digested as is the case for free-living euryoxic invertebrates.

Unlike intestinal endoparasites, free-living euryoxic invertebrates are primarily adapted to produce energy aerobically. Their limited resistance to anoxia, at least under laboratory conditions, does not appear to permit a permanent life of anaerobiosis. It is more probable that it is an adaptation that allows the animals to survive transient periods when the environmental pO_2 drops below the level at which they can utilise oxygen. An added advantage for survival is that at the same time energy requirements are drastically reduced.

Euryoxic invertebrates 'withstand' periods of anoxia in a state of quiescence. The possible danger of becoming easier prey for predators does not exist because the predators, being fast moving animals, must either avoid anoxic environments or, reduce their own metabolic activities.

As a final *caveat*, one must note that, until now, experiments carried out under anaerobic conditions have been performed almost exclusively in closed systems and without sources of food. It is very possible that these artificial conditions influence the responses of the animals. As far as the authors are aware, an alternative experimental approach has so far been tried only by Famme and Knudsen (1984, 1985), who incubated *Tubifex* in an anaerobic perfusion system in artificial sediment, which included a food source. The worms not only survived for about 300 days, but they also multiplied successfully. As a result of these observations, the authors classed *Tubifex* as an obligatory anaerobe, in the sense of Wieser, Ott, Schiemer and Gnaiger (1974). These authors define an obligate anaerobe as an organism whose life processes function better in the absence of oxygen than in its presence. The observation by Famme and Knudsen (1984, 1985) that *Tubifex* reproduced under the above anaerobic experimental conditions is particularly surprising. Even endoparasites, which live in intestines of higher vertebrates, and which are forced into anaerobic energy production by the absence or low activity of cytochrome oxidase, enzymes of the citrate cycle or beta-oxidation, seem to require oxygen in certain stages of the embryonic or larval development (von Brand, 1972). Thus, there is some doubt as to whether the system of Famme and Knudsen (1984, 1985) was truly anoxic. This aside, future anaerobic experiments with other euryoxic annelids should be performed under the most natural conditions possible, that is, in sediment with food provided.

Until about 1985, experiments on anaerobic metabolism of euryoxic annelids were limited, on the whole, to laboratory experiments. The significance of the ability to produce energy anaerobically in the natural habitat was not considered. Recently, there have been attempts to make good this deficiency; that is, to correlate ecological and metabolic/physiological questions. For example, recent studies tried to determine whether polychaetes that inhabited the sediment of tidal flats made use of their apparent euryoxic capacity during low tide exposure lasting several hours, whether these conditions constrained them to anaerobic energy production

The results so far are inconclusive and it seems certain that a simple answer cannot be given. In *Scoloplos armiger*, a worm 10 cm long but only about 2mm wide, anaerobic energy production could not be shown, even after more than 6 h. These worms, capable of maintaining aerobic metabolism even at very low oxygen tensions (< 20 torr), ascend into the upper, oxidative layer during low tide and utilise the residual low oxygen concentrations present there (Schöttler and Grieshaber, 1988).

In contrast, adult *Arenicola marina*, capable of producing energy almost totally anaerobically at an oxygen partial pressure of 20 torr, is forced to anaerobic energy production during dry spells lasting several hours. There are also distinct seasonal differences (1989). During the colder winter months (December - March), when the basal metabolic rate of the animals is clearly reduced, anaerobic energy production cannot be demonstrated, even at low tide after more than 6 h. On the other hand, during the summer months, notably August and early September, breakdown of phosphagen and aspartate is significant, as is the accumulation of succinate and free fatty acids, due to anaerobic energy production. At this time the coelom contains large quantities of eggs and sperm. They are produced from April until June and released into the coelomic fluid, where they grow and mature. During August and September the gametes are the second largest utiliser of O_2, after the cuticular musculature. This, and the higher basal metabolic rate observed during summer, are the reasons for the pronounced anaerobic energy production during dry periods.

Further, the growth of the eggs of *Arenicola marina* in the upper intertidal zone (with regular 6-8 hour dry periods) occurs significantly more slowly than in the lower intertidal zone (2-4 hour dry periods) (unpublished). This effect is of no biological consequence, as the larval shedding phase is preceded by a switch to a synchronous phase to ensure that all animals shed larvae within a few days (Olive, 1984). During the synchronising period, those animals inhabiting the upper intertidal zone could compensate for the initial delay.

But even within one species the results cannot unreservedly be applied to all age groups, as experiments with juvenile *Arenicola marina* (O-generation) have shown (Schiedek, 1989). These often occur in dense colonies, the so-called nursery beds of Reise (1981), in the upper intertidal zone near the high tide line, which regularly lies dry for up to 10 hours during the diurnal cycle and in which adult *Arenicola marina* does not occur. As in *Scoloplos armiger*, juvenile *A. marina* are also capable of almost total aerobic energy production during dry spells. In this case, it is also due to the ability to maintain aerobic energy production, though limited, even at low partial pressure of oxygen (15 torr). The more efficient utilisation of low oxygen concentrations by the juveniles may be just a question of size. The juveniles have a far more favourable surface area/volume than adults - even in adults, about 50% of the oxygen is taken up through the cuticle and not the gills (Mangum, 1976). In addition the cuticle of the juveniles is considerably thinner and therefore presents less of a diffusion barrier.

References

Brand, T. von (1927) 'Stoffbestand und Ernährung einiger Polychaeten und anderer mariner Würmer', *Zeitschrift Vergleichende Physiologie* 5, 643-698

Brand, T. von (1972) *Parasitenphysiologie* (Gustav Fischer Verlag, Stuttgart)

Coles, C.G. (1970) 'Some Biochemical Adaptations of the Swamp Worm *Alma emini* to Low Oxygen Levels in Tropical Swamps', *Comparative Biochemistry and Physiology* **34**, 481-489

Dales, R.P. (1958) 'Survival of Anaerobic Periods by Two Intertidal Polychaetes, *Arenicola marina* (L.) and *Owenia fusiformis* Delle Chiaje', *Journal of the Marine Biological Association of the UK* **37**, 522-529

Dausend, K. (1931) 'Über die Atmung der Tubificiden', *Zeitschrift für Vergleichende Physiologie* **14**, 557-608

Englisch, H. (1989) ' Die Regulation des Energiestoffwechsels bei Fakultative Anaeroben Evertbraten: Untersuchungen an *Arenicola marina* L. unterspezieller Berücksichtigung von Eigenschaften und Bedeutung der Pyruvat-Kinase.' Thesis (University of Münster, FRG)

Famme, P. and Knudsen, J. (1984) 'Total Heat Balance Study of Anaerobiosis in *Tubifex tubifex* (Müller)', *Journal of Comparative Physiology* **154**, 587-591

Famme, P. and Knudsen, J. (1985) 'Anoxic Survival, Growth and Reproduction by the Fresh water Annelid, *Tubifex* sp., Demonstrated Using a New Simple Anoxic Chemostat', *Comparative Biochemistry and Physiology* **81A**, 251-253

Felbeck, H. (1980) 'Investigations on the Role of the Amino Acids in Anaerobic Metabolism of the Lugworm *Arenicola marina* L', *Journal of Comparative Physiology* **137**, 183-192

Felbeck, H. and Grieshaber, M.K. (1979) 'Investigations on Some Enzymes Involved in the Anaerobic Metabolism of Amino Acids of *Arenicola marina* L', *Comparative Biochemistry and Physiology* **66B**, 205-213

Fields, J.H.A., Eng, A.K., Ramsden, W.D., Hochachka, P.W., and Weinstein, B. (1980) 'Alanopine and Strombine are Novel Amino Acids Produced by a Dehydrogenase Found in the Adductor Muscle of the Oyster, *Crassostrea gigas*', *Archives of Biochemistry and Biophysics* **20**, 110-114

Flavin, M. and Ochoa, S. (1957) 'Metabolism of Propionic Acid in Animal Tissues I. Enzymatic Conversion of Propionate to Succinate', *Journal of Biological Chemistry* **229**, 965-979

Flavin, M., Ortiz, P.J. and Ochoa, S. (1955) 'Metabolism of Propionic Acid in Animal Tissues', *Nature (London)* **176**, 823-826

Gruner, B. and Zebe, E. (1977) 'Studies on the Anaerobic Metabolism of Earthworms', *Comparative Biochemistry and Physiology* **60B**, 441-445

Hecht, F. (1932) 'Der chemische Einfluß organischer Zersetzungstoffe auf das Benthos, dargelegt an Untersuchungen mit marinen Polychaeten, insbesondere *Arenicola marina* L.', *Seckenbergiana*, 14, 199-200

Hochachka, P.W. and Mustafa, T. (1972) 'Invertebrate Facultative Anaerobiosis', *Science* **178**, 1056-1060

Hoffman, K.H. (1981) 'Phosphagens and Phosphokinases in *Tubifex* sp.', *Journal of Comparative Physiology* **143**, 237-243

Holst, H. and Zebe, E. (1986) 'Volatile Fatty Acid Excretion During Anaerobiosis in the Lugworm *Arenicola marina*', *Comparative Biochemistry and Physiology* **83A**, 189-196

Holwerda, D.A., Kruitwagen, E.C.J. and de Bont, A.M.T. (1981) 'Regulation of Pyruvate Kinase and Phosphoenolpyruvate Carboxykinase Activity During Anaerobiosis in *Mytilus edulis* L.', *Molecular Biology* **1**, 165-171

Jacubowa, L. and Malm, E. (1931) 'Die Beziehungen einiger Benthos-Formen des Schwarzen Meeres zum Medium', *Biologisches Zentralblatt*, **51**, 105-116

Kamp, G. (1986) 'The Mode of Glycogen Phosphorylase Activation During Work and Hypoxia.' *Zoologische Beitrage Neue Folge.*. **30**, 171-186

References

Kamp, G. and Juretschke, H. P. (1989) "Hypercapnic and Hypocapnic Hypoxia in the Lugworm, Arenicola marina: A ^{31}P-NMR Study'. *Journal of Experimental Zoology* 252, 219-227

Köhler, P., Bryant, C. and Behm, C.A. (1978) 'ATP Synthesis in a Succinate Decarboxylase System from Fasciola hepatica mitochondria', *International Journal for Parasitology* 8, 399-404

Mangum, C.P. (1976) 'The Oxygenation of Hemoglobin in Lugworms.' *Physiological Zoology* 49, 85-99

Olive, P.J.W. (1984) 'Environmental Control of Reproduction in Polychaeta.' *Fortschritte der Zoologie* 29 17-38

Packard, W.H. (1905) 'On Resistance to Lack of Oxygen and a Method of Increasing this Resistance', *American Journal of Physiology* 15, 30-41

Pietrzak, S.M. and Saz, H.J. (1981) 'Succinate Decarboxylation to Propionate and the Associated Phosphorylation in *Fasciola hepatica* and *Spirometra mansonoides*', *Molecular and Biochemical Parasitology* 3, 61-70

Putzer, V. (1985) 'Der anaerobe Stoffwechsel des Glanzwurmes *Lumbriculus variegatus*', Thesis (University of Innsbruck, Austria)

Putzer, V., Gnaiger, E. and Lackner, R. (1984) 'Flexibility of Anaerobic Metabolism in Aquatic Oligochaetes (*Tubifex* sp.). Biochemical and Calorimetric Changes Induced by Deproteinized Hydrolysate of Bovine Blood', *Comparative Biochemistry and Physiology* 82A, 965-970

Reise, K. (1981) 'Ökologische Experimente zur Dynamik und Vielfalt der Bodenfauna der Nordseewatten.' *Verhandlung der Deutschen Zoologischen Gesellschaft* 1-15

Ruby, E.G. and Fox, D.L. (1976) 'Anaerobic Respiration in the Polychaete *Euzonus (Thoracophelia) mucronata*', *Marine Biology* 35, 149-153

Saz, H.J. and Pietrzak, S.M. (1980) 'Phosphorylation Associated with Succinate Decarboxylation to Propionate in *Ascaris* mitochondria', *Archives of Biochemistry and Biophysics*. 202, 388-395

Schoffeniels, E. and Gilles, R. (1972) 'Ionoregulation and Osmoregulation in Mollusca', in M. Florkin and B.T. Scheer (eds.), *Chemical Zoology, Vol. 7 Mollusca*, (Academic Press, New York), pp. 393-418

Schöttler, U. (1977) 'NADH Generating Reactions in Anaerobic *Tubifex* mitochondria', *Comparative Biochemistry and Physiology* 58B, 261-265

Schöttler, U. (1979) 'On the Anaerobic Metabolism of Three Species of *Nereis* (Annelida)', *Marine Ecology Progress Series* 1, 249-254

Schöttler, U. (1980) 'Der Energiestoffwechsel bei biotop-bedingter Anaerobiose: Untersuchungen an Anneliden', *Verhandlung der Deutschen Zoologischen Gesellschaft*, 228-240

Schöttler, U. (1982) 'Vergleichende Untersuchungen zum Anaerobiose-stoffwechsel von Polychaeten', Habilitationsschrift (Universität Münster, FRG)

Schöttler, U. (1982) 'An Investigation on the Anaerobic Metabolism of *Nephtys hombergii* (Annelida: Polychaeta)', *Marine Biology* 71, 265-269

Schöttler, U. (1983) 'Untersuchungen zur Bildung von Propionat aus Succinat in Submitochondrialen Fraktionen aus dem Hautmuskelschlauch von *Arenicola marina*', *Verhandlung der Deutschen Zoologischen Gesellschaft*, 321

Schöttler, U. (1986) 'Weitere Untersuchungen zum Anaeroben Energiestoffwechsel des Polychaeten *Arenicola marina* L.' *Zoologische Beitraege Neue Folge* 30, 141-152

Schöttler, U. and Schroff, G. (1976) 'Untersuchungen zum anaeroben Glykogenabbau bei *Tubifex tubifex* M.', *Journal of Comparative Physiology* 108, 243-253

Schöttler, U. and Grieshaber, M. (1988) 'Adaptation of the Polychaete Worm *Scoloplos armiger* to Hypoxic Conditions.' *Marine Biology* 99, 215-222

Schöttler, U. and Wienhausen, G. (1981) 'The Importance of the Phosphoenolpyruvate Carboxykinase in the Anaerobic Metabolism of Two Marine Polychaetes', *Comparative Biochemistry and Physiology* **68B**, 41-48

Schöttler, U., Wienhausen G. and Westermann, J. (1984) 'Anaerobic Metabolism in the Lugworm *Arenicola marina* L.: The Transition from Aerobic to Anaerobic Metabolism', *Comparative Biochemistry and Physiology* **79B**, 93-103

Schroff, G. and Schöttler, U. (1977) 'Anaerobic Reduction of Fumarate in the Body Wall Musculature of *Arenicola marina* (Polychaeta)', *Journal of Comparative Physiology* **116**, 325-336

Schroff, G. and Zebe, E. (1980) 'The Anaerobic Formation of Propionic Acid in the Mitochondria of the Lugworm, *Arenicola marina*', *Journal of Comparative Physiology* **138**, 35-41

Schulz, T.K.F. and Kluytmans, J.H. (1983) 'Pathway of Propionate Synthesis in the Sea Mussel *Mytilus edulis* L.', *Comparative Biochemistry and Physiology* **75B**, 365-372

Schiedek, D (1989) 'Euryhalinität und Resistenz gegen Hypoxia als Mechanismen Ökologischer Anpassung bei Juvenilen *Arenicola marina*.' Thesis (University of Münster, FRG)

Seuss, J., Hipp, E. and Hoffmann, K.H. (1983) 'Oxygen Consumption, Glycogen Content and the Accumulation of Metabolites in *Tubifex* During Aerobic-Anaerobic Shift and Under Progressing Anoxia', *Comparative Biochemistry and Physiology* **75A**, 557-562

Siegmund, B. and Grieshaber, M.K. (1983) 'Determination of Meso-alanopine and D-strombine by High Pressure Liquid Chromatography in Extracts from Marine Invertebrates', *Hoppe-Seyler's Zeitschrift fur Physiologische Chemie* **364**, 807-812

Siegmund, B., Grieshaber, M.K., Reitze, M. and Zebe, E. (1985) 'Alanopine and Strombine are End-products of Anaerobic Glycolysis in the Lugworm *Arenicola marina* (Annelida, Polychaeta)', *Comparative Biochemistry and Physiology* **82B**, 337-345

Sticher, U. (1985) 'Untersuchungen zum Stoffwechsel des D-Alanin in marinen Anneliden', Diplom-Arbeit, (Universität Münster, FRG)

Surholt, B. (1977) 'Production of Volatile Fatty Acids in the Anaerobic Carbohydrate Catabolism of *Arenicola marina*', *Comparative Biochemistry and Physiology* **58B**, 147-150

Toulmond, A. (1975) 'Blood Oxygen Transport and Metabolism in the Confined Lugworm *Arenicola marina* (L.)', *Journal of Experimental Biology* **63**, 647-660

van Vugt, F., van der Meer, P. and van den Bergh, S.G. (1979) 'The Formation of Propionate and Acetate as Terminal Processes in the Energy Metabolism of the Adult Liver Fluke *Fasciola hepatica*', *International Journal of Biochemistry* **10**, 11-18

Wienhausen, G. (1979) 'Der Anaerobiosestoffwechsel von *Arenicola marina*: Untersuchungen über die alternativen Wege der Umsetzung von Phosphoenolpyruvat und die Bildung von Acetat', Thesis (Universitat Münster, FRG)

Wienhausen, G. (1981) 'Anaerobic Formation of Acetate in the Lugworm *Arenicola marina*', *Naturwissenschaften* **68**, 206

Wieser, W., Ott, J., Schiemer, F. and Gnaiger, E. (1974) 'An Ecophysiological Study of Some Meiofauna Species Inhabiting a Sandy Beach at Bermuda', *Marine Biology* **26**, 235-248

Zebe, E. (1975) 'In vivo-Untersuchungen über den Glucose-Abbau bei *Arenicola marina* (Annelida, Polychaeta)', *Journal of Comparative Physiology* **101**, 133-145

Zebe, E., Salge, U., Wiemann, C. and Wilps, H. (1981) 'The Energy Metabolism of the Leech *Hirudo medicinalis* in Anoxia and Muscular Work', *Journal of Experimental Zoology* **218**, 157-163

Zebe, E. and Heiden, T. (1983) 'Anaerobiosis in Earthworms: Investigation of *Eisenia foetida*', *Verhandlung der Deutschen Zoologischen Gesellschaft*, 221.

Chapter 10

Molluscs

Albertus de Zwaan

Introduction

Many molluscs tolerate hypoxic or anoxic conditions for days. Anoxic survival of over one month at moderate temperature is common especially among bivalve molluscs. Since aquatic biotopes do not reliably supply continuous and sufficient oxygen, many aquatic invertebrates have developed biochemical strategies to cope with environmental anoxia. The solubility and rate of diffusion of oxygen in water are low compared with air. While atmospheric oxygen guarantees a constant supply of oxygen, oxygen availability in water varies with a number of physical and biotic factors such as temperature, the concentration of solutes and total biomass. As a consequence, benthic inhabitants of the littoral water and shallow limnic biotopes face regular periods of oxygen lack.

The Origins of Anoxia

Oxygen lack is more rarely encountered by pelagic species. For example, cephalopods, unlike most other molluscs, are stenoxic. The Pectinidae, the bivalved scallops, can survive anoxic conditions only for a short time. In littoral species, uptake of oxygen from the water by the gills ceases on exposure to air at low tide. This does not always evoke anaerobic metabolism as some species are able to switch to air breathing. In a study of four intertidal molluscs it was observed that *Mytilus edulis* and *Scrobicularia plana* became anaerobic within their habitats whereas this was not the case with *Cardium edule* and *Patella vulgata*. When kept under artificial anoxic conditions, the metabolisms of all species responded identically, exploiting the same anaerobic pathways (Brinkhoff, Stockmann and Grieshaber, 1983; Widdows, Bayne, Livingstone, Newell and Donkin, 1979). In experiments with different *Littorina* species no anaerobic end-products were formed during aerial exposure (Kooyman, van Zoonen, Zurburg and Kluytmans, 1982).

In shallow waters, solar radiation may also provide oxygen by way of photosynthesis. Snow and ice coverings, as well as algal blooms, may block solar penetration resulting in low oxygen tension or complete anoxia in the deeper layers. Benthic organisms may therefore face seasonally long-term periods of severe hypoxia or anoxia. Also, fresh water snails, which usually live at the water surface, may occasionally inhabit bottom layers poor in oxygen, for example, when they expel air from their lung cavities and sink as a response to danger.

Algal blooms occur frequently in large ponds and along coastal areas and the resulting unfavourable oxygen conditions may cause high mortalities. During the last decade, blooms of dinoflagellates and diatoms developed during the summer along the coast of the Northern Adriatic Sea, following eutrophication due to effluents of waste products from agriculture and industry. In late summer, when the photoperiod becomes shorter, these floating blooms utilise more oxygen than they produce and, at the same time, prevent penetration of light. This may cause complete anoxia for benthic organisms for a couple of days and even longer periods of severe hypoxia. In this area live two large populations of bivalves, *Venus gallina* and *Scapharca inequivalvis*. Both species have been subjected to experimental anoxia and it appeared that the latter was able to survive about four times longer than *V. gallina*. After six days these figures were 50 and 100 per cent, respectively. Field observations confirm that the algal blooms cause high mortality in the populations of *V. gallina*, whereas *S. inequivalvis* is apparently unaffected (Cortesi, unpublished results).

Of the six classes of molluscs only the gastropods have colonised land, especially the snails and the slugs of the order Stylommatophora (subclass Pulmonata). Although land molluscs live in high ambient oxygen concentrations, survival of experimental anoxia for months has been reported for a few terrestrial pulmonates during aestivation at low temperatures (Meenakshi, 1964; van der Horst, 1974). *Helix pomatia* is thought to be aerobic in spring and summer, but anoxic in winter when surviving in waterlogged soil (Wieser, 1981).

So far I have dealt with anaerobiosis as a consequence of the environmental limitation of oxygen supply and therefore it concerns the whole animal body. The condition may be imposed for several months, and is termed 'environmental anaerobiosis'. This type of anaerobiosis, both for invertebrates and vertebrates, is accompanied by a drastic reduction in metabolic rate, of one to two orders of magnitude. Environmental anaerobiosis is therefore characterised by a low ATP turnover rate which requires a low ATP generation rate. It is for this reason also called the 'low power output mode' of anaerobiosis.

Avoidance of Anoxia at the Cellular Level

Some gastropods and bivalves show aerobic respiration when submerged. Avoidance of anoxia at the cellular level is a widely developed strategy to survive environmental anoxia in both plants and animals. For instance, the roots of some waterlogged plants have anatomical adaptations that permit oxygen uptake from the atmosphere; in diving mammals, redistribution of the circulatory system enables the conservation of oxygen for the heart and brain during underwater excursions. As a rule, molluscan cells switch to anaerobic processes when environmental changes prevent the normal respiratory system from functioning. They are able to exploit fermentative processes that provide a higher ATP yield per substrate equivalent than lactate fermentation, the only type known from hypoxic vertebrate tissues.

In marine species, mitochondrial pyruvate fermentation is connected with the conversion of aspartate into succinate and/or propionate. This is not unique for molluscs, but is shared by 'lower and middle' invertebrate phyla, excluding the adult arthropods and echinoderms. The same mechanisms are exploited by the adult stages of many parasitic helminths in the vertebrate host. Usually, several anaerobic end-products accumulate simultaneously or in succession. Succinate, propionate, acetate and alanine (especially in marine species) are ubiquitous end-products. Pathways resulting in these types of products are poorly developed in scallops and cephalopods, as might be expected from the low anaerobic capacity of both groups. For this reason, these types of fermentation are usually considered to be powerful adaptive strategies for survival without oxygen. Stoichiometric equations, however, show that ATP yield, compared with lactate fermentation per equivalent carbohydrate, hardly exceeds a factor of two. Therefore, the impact on the energetic budget is of much less importance than the marked decrease induced in energy expenditure. A more important advantage is that CO_2 and volatile fatty acids can easily be released into the environment, thus delaying the development of metabolic acidosis, a serious threat during long-term anaerobiosis. Functional anaerobiosis, as opposed to environmental anaerobiosis, usually involves local sites within organisms, especially muscular tissues. It is a consequence of a high energy expenditure and serves to augment energy production by respiration. The resulting high glycolytic rate is unfavourable, because it causes a fast depletion of energy reserves and/or causes self-poisoning by organic acid accumulation - metabolic acidosis. The latter greatly limits the time of operation.

Apart from explosive locomotory activity, a high energy demand met by both respiration and fermentation also occurs in animals which recover their aerobic cellular homeostasis after long periods of environmental anaerobiosis (de Zwaan, De Bont, Zarburg, Bayne and Livingstone, 1983a; Schick, Gnaiger, Widdows, Bayne and de Zwaan, 1986). Both recovery and functional (exercise) anaerobiosis are high power output modes of anaerobic metabolism. In all metazoan cells the fermentative pathway is fundamentally identical with classical glycolysis in terms of energy yield and in the way redox balance is maintained. This is achieved by coupling the glyceraldehyde 3-phosphate dehydrogenase step with a pyruvate reductase step. The latter results in the accumulation of lactate or an iminodicarboxylic acid (opine).

Low Power Output Mode of Anaerobic Metabolism

Over a period of more than 15 years I have focused my research on the anaeroic metabolism of the blue mussel, *Mytilus edulis*. One of the difficulties I encountered was intra-specific biological variation; in separate experiments, animals differed considerably in the types and amounts of end-products accumulating. Unlike vertebrates, where lactate is the sole accumulating compound, anaerobic pathways in invertebrates are obviously not fixed and it has indeed been shown that various factors affect the type and ratio of accumulating end-products. Moreover, different tissues do not respond in a similar manner. Results vary with experimental conditions (aerial exposure or oxygen-free sea water), the season and the duration of the anaerobic incubation.

Recently, we have caged mussels from one location and, after one month of acclimation at the location site in which the animals could settle by byssus threads, the cages were transferred to different sites in the North Sea and the Western Scheldt of The Netherlands. After two months of exposure, the mussels were recovered and subjected to anaerobic stress for 48 hours. Pronounced differences were found, especially in the ratio of the sum of pyruvate derivatives to the sum of succinate plus propionate. This shows that even the natural collection site (in our experiment there were differences in salinity and in the content of various industrial pollutants) may produce different results which make it difficult to compare results obtained by different research groups. It is therefore dangerous to conclude from the literature that there will be consistent differences between one species and another. In spite of these difficulties, general characteristics of anaerobic metabolism, that are shared by a great number of free living invertebrates, can be recognised

Some years ago, I designed a metabolic map based on data gathered over many years in which I tried to express the flexibility of anaerobic metabolism in the sea mussel. This was first published in 1983 (de Zwaan, 1983) and is here published in a slightly modified form (Figure 10.1). Different metabolic panels represent metabolic transitions in time as they may occur in the intertidal habitat of this sessile animal (and other marine molluscs) over the tidal cycle and depending on their position on the shore.

From the moment all tissues of the mussel are exposed to air at ebbing tide, there is an increase of the steady state levels of alanine and succinate. In adductor muscle, and probably in the ventricles of bivalves and gastropods, there is, initially, also the formation of an opine from pyruvate, but in small amounts relative to alanine. Radiolabel studies with ^{14}C-glucose and aspartate, in combination with specific inhibitors, have revealed that carbon from glycogen mainly appears as alanine, whereas the carbon skeleton of aspartate is converted to succinate via the citric acid cycle between succinate and oxaloacetate in an anticlockwise (reductive) direction. Transaminase inhibitors and iodoacetate applied in this initial or in the transition stage of anaerobiosis lessen markedly the concentration changes in alanine, aspartate and succinate, suggesting a direct connection between the glycogen-alanine (plus

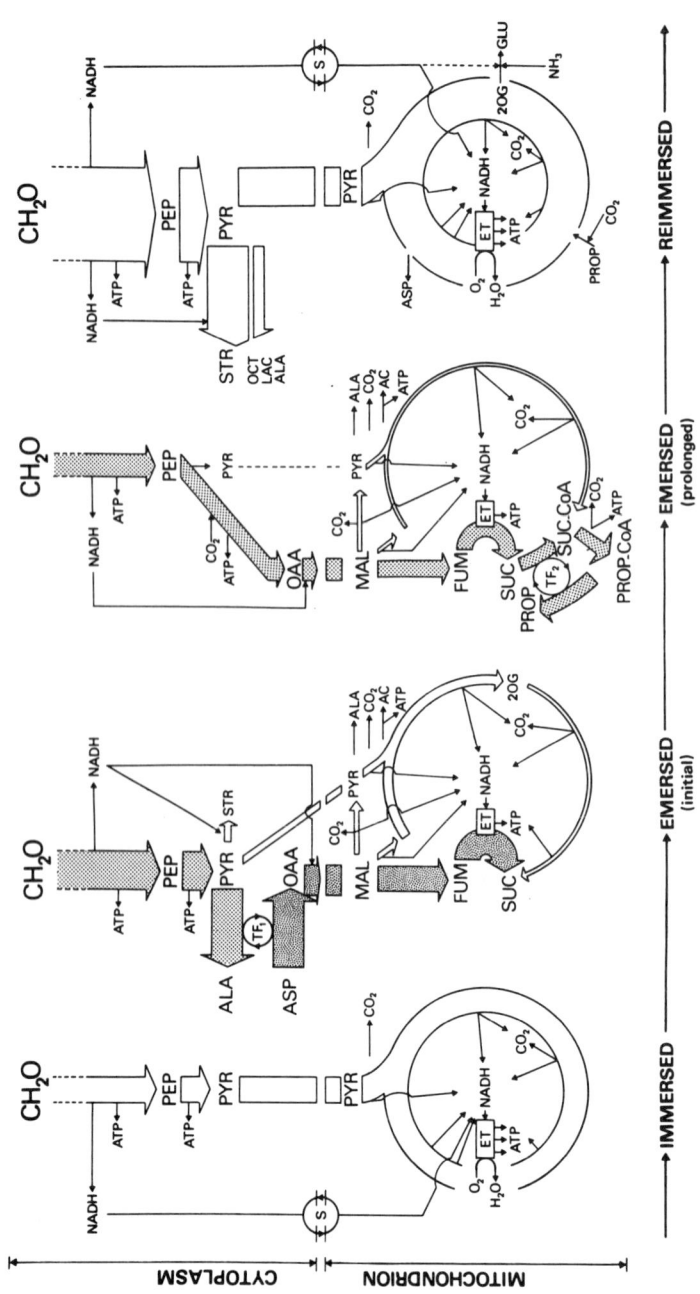

some opine) pathway and the aspartate-succinate pathway by transaminase reactions (Collicutt and Hochachka, 1977; Foreman and Ellington, 1983; de Zwaan, De Bont and Verhoeven, 1982; de Zwaan, De Bont and Hemelraad, 1983b; see also Figure 10.2). Only glutamate can serve directly as an amino-group donor for oxaloacetate and pyruvate (de Zwaan and Dando, 1984) and it is therefore assumed that alanine formation depends upon the glutamate-pyruvate transamination with aspartate being the amino group donor for 2-oxoglutarate. The coupling between the simultaneous carbohydrate and aspartate fermentation is thus based on nitrogen transfer via the 2-oxoglutarate/glutamate couple and involves alanine aminotransferase (ALAT) and aspartate aminotransferase (AAT). In this manner, equimolar amounts of alanine and oxaloacetate are formed at the expense of the reduction of an equimolar amount of NAD. Since glyceraldehyde 3-phosphate dehydrogenase, which catalyses the oxidative step of the glycolytic chain, is located in the cytosol and both oxaloacetate and NADH cannot directly penetrate the mitochondrial inner membrane, it is assumed that further conversion of oxaloacetate involves cytosolic reduction by malate dehydrogenase. Malate so produced would maintain cytosolic redox balance, but this assumes that coupling of alanine formation to aspartate metabolism occurs within the cytosol (later, I will argue that, for certain tissues, this is not very likely).

Malate can be taken up by mitochondria and converted to succinate by two different routes in order to maintain mitochondrial redox balance. This occurs when two out of seven malate molecules are used to form one citrate molecule, which subsequently is oxidised to succinate via the citric acid cycle running in the conventional (oxidative) way and five malate molecules are reduced to succinate after an initial dehydration via fumarate (see de Zwaan, 1983). The overall equation is:

$$\text{malate} \rightarrow 0.86 \text{ succinate} + 0.57 \, CO_2$$

Indeed, we have been able to show that isolated intact mantle mitochondria can take up exogenous malate and transform it in a manner which is close to this

---

Figure 10.1 A metabolic map to account for the degradation of glycogen (and aspartate) with aerobiosis (immersed), environmental anaerobiosis (emersed, initial, prolonged) and postanaerobic recovery (reimmersed). From the left to the right, the pathways represent gradual transitions in the carbon flow, which occur in the course of long-term anaerobiosis. The width of the bars are indications of the relative carbon flux through those parts of the pathway (redrawn after De Zwaan, 1983). Abbreviations: CH_2O, glycogen; ET, electron transfer chain; OCT, octopine; PROP, propionate; AC, acetate; STR, strombine; S, malate aspartate shuttle; TF, transfer of NH_2 from aspartate to alanine by amino transferase (ALAT and AAT) reactions; TF2, transfer of CoA by acyl-CoA transferase.

equation (de Zwaan, Holwerda and Veenhof, 1981). By measuring changes in steady state concentrations of all citric acid cycle intermediates, amino acids and volatile fatty acids we designed a carbon and hydrogen flow scheme for the anaerobic conversion of malate. The overall reaction appeared to be:

$$\text{malate} (+ 0.06 \text{ NH}_3) \rightarrow 0.06 \text{ pyruvate} + 0.06 \text{ alanine} + 0.11 \text{ acetate} + 0.04 \text{ fumarate} +$$
$$0.08 \text{ propionate} + 0.61 \text{ succinate} + 0.61 \text{ CO}_2$$

The reduction of fumarate to succinate was via NADH, generated by the oxidative steps catalysed by malic enzyme, malate dehydrogenase, pyruvate dehydrogenase, isocitrate dehydrogenase and 2-oxoglutarate dehydrogenase, contributing 44, 8, 26, 9 and 13 per cent respectively. The anaerobic operation of the citric acid cycle could further be confirmed by monofluoroacetate inhibition and the involvement of pyruvate dehydrogenase by arsenite inhibition (de Zwaan *et al.*, 1981).

We have examined the NADH-fumarate reductase reaction, using intact mitochondria and submitochondrial particles from mantle tissue of *M. edulis*. On incubating intact mitochondria anaerobically with malate at 25°C, an enzyme activity of 1.8 mmole succinate h^{-1}, g^{-1} fresh weight was found (Holwerda and de Zwaan, 1979). This value is well above the *in vitro* succinate production of the sea mussel. The enzyme activity could be stimulated by a factor of 1.44 with ADP plus inorganic phosphate; ATP was formed with a P/2e$^-$ ratio of 0.46. Oligomycin and several uncouplers, including dicoumarol and 2,4 dinitrophenol inhibited the synthesis of ATP, indicating that it is coupled to the transport of electrons from NADH to fumarate.

Kinetics of the NADH-fumarate reductase reaction were studied with cyanide-poisoned sub-mitochondrial particles (Holwerda and de Zwaan, 1980). The pH optimum was 8.0, whereas a pH optimum of 7.7 was found for succinate formation from malate by intact mitochondria. The K_m values for NADH and for fumarate were 0.04 mM and 0.06 mM, respectively. These low values seem to offer a good adaptation for the oxidation of low concentrations of NADH by low concentration of fumarate. Inhibitors (rotenone, amytal, malonate and ethanol) established the presence of two flavoproteins, NADH-quinone reductase and quinone-fumarate reductase, or the so-called complexes I and II of the electron transport chain. Antimycin A was only inhibitory at high concentrations. This probably means that the cytochromes of complex III are not involved in electron transport between NADH and fumarate, as, in other systems, antimycin A inhibits at much lower concentrations. NADPH appeared to be a poor substrate for the fumarate reductase reaction. However, when added together with NAD$^+$, maximum reductase activity was obtained, indicating the presence of a transhydrogenase associated with the submitochondrial particles and explaining why 'malic enzyme', which is NADP-dependent in the two bivalve molluscs so far tested (de Zwaan, 1977), can generate reducing equivalents for the fumarate reductase system.

During prolonged anoxia succinate is converted into propionate (Figure 10.1, 3rd panel). When *M. edulis* is exposed to air at 12°C, it takes about 16 hours before the accumulation of propionate begins (Kluytmans, De Bont, Janus and Wijsman, 1977). The pathway of propionate production has been studied by Schulz, Kluytmans and Zandee (1982), Schulz, van Duin and Zandee (1983) and Schulz and Kluytmans (1983) in mantle mitochondria of the sea mussel. Propionate is formed in a cyclic manner in a pathway through which succinyl-CoA is converted to succinate in four steps catalysed by methylmalonyl-CoA isomerase, methylmalonyl-CoA racemase, propionyl-CoA carboxylase and acyl (propionyl)-CoA, thus closing the cycle. In the propionyl-CoA carboxylase step ATP is formed by substrate linked phosphorylation of ADP. Mitochondrial participation can therefore add two extra sites of ATP generation to the cytosolic substrate linked phosphorylations. By comparing overall equations for fermentations it appears that the maximum energetic advantage, in terms of extra ATP yield, is 2.14 times higher than that of simple lactate fermentation (see de Zwaan, 1983).

$$\text{glycosyl U} + 6 O_2 + 39 \text{ ADP} \rightarrow 6 CO_2 + 6 H_2O + 39 \text{ ATP}$$
$$\text{glycosyl U} + 3 \text{ ADP} \rightarrow 2 \text{ lactate} + 3 \text{ ATP}$$
$$\text{glycosyl U} + 2 \text{ ASP} + 4{,}71 \text{ ADP} \rightarrow 1.71 \text{ succinate} + 1.14 CO_2 + 2 \text{ ALA} + 4.71 \text{ ATP}$$
$$\text{glycosyl U} + 2 \text{ ASP} + 6.43 \text{ ADP} \rightarrow 1.71 \text{ propionate} + 2.8 CO_2 + 2 \text{ ALA} + 6.43 \text{ ATP}$$
$$\text{glycosyl U} + 0.86 CO_2 + 4.71 \text{ ADP} \rightarrow 1.71 \text{ succinate} + 4.71 \text{ ATP}$$
$$\text{glycosyl U} + 6.43 \text{ ADP} \rightarrow 1.71 \text{ propionate} + 0.86 CO_2 + 6.43 \text{ ATP}$$

The reduction in metabolic rate, however, compensates for or even overrides the reduction in fermentative yield of ATP as compared with respiration (Storey, 1985). A Pasteur effect (increased carbon flow through the glycolytic pathway) is therefore avoided during environmental anaerobiosis. The flux may even be reduced.

During the initial stage, three sources contribute to the anaerobic power output: endogenous stores of ATP, phosphagen and catabolism (Table 10.1, 0-4 h). A drop in the steady state level of ATP only contributes marginally to the ATP consumption (0-9 per cent), but transphosphorylation may be as important as catabolism (Table 10.1; an average contribution of 44 percent to ATP provision during the first four hours in the eight gastropods and bivalves listed). In this phase the glycolytic flux has not yet reached its lower limit, which causes opine as well as alanine production. The ATP turnover rate may be five times greater than in the following prolonged stage (compare ATP values at 0-4 h and > 10 h in Table 10.1). When the conversion of aspartate to succinate is blocked by aminooxyacetate the loss of mitochondrial ATP production is compensated for by two mechanisms, (1) hydrolysis of phosphoarginine is strongly enhanced and (2) there is an increased opine production (Foreman and Ellington, 1983; de Zwaan *et al.*, 1982).

Table 10.1 ATP turnover rates at rest (unstressed aerobic animals), environmental and exercise (functional) anaerobiosis.

species	tissue	M ATP		A/R	percentage			conditions
		resting (R)	anaerobic (A)		ATP	phospho-arginine	anaerobic catabolism	
environmental 0-4h								
1. *Nassa mutabilis*	F		>0.036		nd	48	52	
2. *Cardium tuberculatum*	F		>0.30		nd	67	23	
3. *Cardium edule*	F		>0.035		nd	46	54	
4. *Mytilus edulis*	AM	0.339[a]	0.032		9	60	31	
5. *Modiolus squamosus*	PA		0.049		2	3	95	
	CA		0.020		7	42	51	
5. *Geukensia sp*	TA		>0.021		nd	20	80	
6. *Lima hians*	AM		0.121		4	65	31	
7. *Placopecten magellanicus*	CA	0.43[b]	0.115	0.26	8	25	67	
>10h								
8. *Lymnaea stagnalis*	TA	1.07[c]	>0.011	0.01	nd	nd	100	
8. *Cardium edule*								
4. *Mytilus edulis*	AM	0.339[a]	0.008	0.023	0	0	100	20-24h anoxia
		0.339	>0.072	0.21	nd	nd	100	6h forced valve closure
9. *Geukensia sp*	TA		>0.008		nd	nd	100	
6. *Lima hians*	AM		0.021		14	24	52	

For both anaerobic conditions a comparison is made of ATP equivalents derived from ATP, when determined, phosphagen and anaerobic catabolism (given as a percentage of total ATP turnover rate).

M ATP = μmoles ATP/g wetweight/min.

Resting M ATP is calculated from respiration at rest, assuming that 1 mol O_2 is equivalent to 6 mol ATP. VO_2, taken from [a]Prosser, 1973; [b]Thompson *et al*, 1980; [c]Prosser, 1973; and [d]Baldwin and Lee, 1979.

Table 10.1 (continued)

species	tissue	M ATP resting (R)	M ATP anaerobic (A)	A/R	percentage ATP	percentage phospho-arginine	percentage anaerobic catabolism	conditions
exercise								
10. Nassarius coronatus	F		1.4		nd	36	64	8s, 125 lashing movements
1. Nassa mutabilis	F		10.6		1.5	51	47.5	73s, 37 escape movements
			5.7		2	26	72	132s, 35 escape movements, previously 4h anoxia
11. Buccinum undatum	F		0.76		4	32	64	15min, 30 escape movements
12. Busycon contrarium	RRM		>0.13		nd	23	77	KCl contractions *in vitro*
14. Strombus luhanus	F		2.93		nd	41	59	2min, 31 leaps
2. Cardium tuberculatum	F		5.6		0	72	28	53s, 7-11 leaps
3. Cardium edule	F		0.77		nd	78	22	10min, electrical stimulation
14. Lima hians	AM		11.6		2	59	39	3min, 65 valve snaps
15. Chlamys apercularis	AM		15.42		22	22	0	90s, 45-55 snaps
16. Pecten jacobaeus	AM		2.38		13	57	30	5min, 23 snaps
			6.50		7	38	55	4min, 42 snaps in air
17. Agropecten irradiens	A		9.66		7	50	43	3.5min, 54 snaps
Placopecten magellanicus	PA	0.43[b]	12.65	29.4	12.5	69	18.5	2min, 30 snaps
18. Limaria fragikis ($-O_2$)	AM	0.155[d]	5,07	32.7	10	30	69	2min, 30 snaps
19. Hapalochlaena maculosa	M		6.39		14	31	56	2.5, 7s swimming movements
	TN		15.97		9	44	47	2.5, 7s swimming movements

M, mantle; AM, adductor muscle; PA, phasic adductor muscle; CA, catch adductor muscle; F, foot; RRM, radula retractor muscle: TA, total animal.

References: 1, Gäde et al., 1984; 2, Gäde, 1980; 3, Meinardus and Gäde, 1981; 4, Ebberink et al., 1979; 5, Nicchita and Ellington, 1983; 6, Gäde, 1983; 7, De Zwaan et al., 1980; 8, De Zwaan et al., 1976; 9, De Zwaan et al., 1983b; 10, Baldwin et al., 1981; 11, Koorman and Grieshaber, 1980; 12 Ellington, 1982; 13, Baldwin and England, 1982; 14, Gäde, 1981; 15, Grieshaber, 1978; 16 Grieshaber and Gäde, 1977; 17, Chih and Ellington, 1983; 18, Baldwin and Morris, 1983; 19, Baldwin and England, 1980.

The transition from the initial stage to the prolonged stage is indicated by deviation of the main flow of carbon from glycogen at the level of phosphoenolpyruvate (PEP), away from pyruvate and towards oxaloacetate. It means that from this moment the catalytic potential of PEP-carboxykinase exceeds that of pyruvate kinase. This transition will be gradual and carbon from oxaloacetate may temporarily be derived from both aspartate and glycogen. In Figure 10.1 the transition at the PEP branchpoint (initial versus prolonged) coincides with the start of the succinate conversion to propionate plus CO_2. It is presented in this manner in order to emphasise that both events usually take time to develop during anaerobiosis but need not be synchronised.

The latency period between the onset of anoxia and the first appearance of propionate is possibly related to the phylogeny and the habitat of the species. Land gastropods (Stylommatophora) seem not to produce any volatile fatty acid, whereas marine gastropods show extremely large latency period (in general over 24 hours at 14°C). Fresh water bivalves may not show latency periods and, in limnic gastropods, latency periods appear to be short. Marine bivalves generally produce propionate within one day of anoxia, although exceptions exist. Two species of the genus *Venus* did not produce any propionate within 48 h (Kluytmans and Zandee, 1983).

As well as the volatile fatty acids mentioned so far, others have been reported to accumulate in fresh water gastropods and in marine and fresh water bivalves. Among these are butyric, iso-butyric, valeric and iso-valeric acids. However, their rate of formation is one to two orders of magnitude lower. When anoxic conditions are encountered in water these volatile fatty acids are excreted, probably by way of diffusion, since they have low molecular weights and amphiphatic properties (Klutymans, Veenhof and de Zwaan, 1975; de Zwaan, Mohamed and Geraerts, 1976; Van den Thillart and De Vries, 1985).

Succinate is a major, ubiquitous anaerobic end-product of environmental anaerobiosis, except in the Pectinidae and the cephalopods, where the succinate pathway is poorly developed, in comparison with the octopine pathway. During shell valve closure (environmental hypoxia) octopine appeared as the main end-product in the 'catch' adductor muscle of the scallop, *Placopecten magellanicus* (de Zwaan, Thompson and Livingstone, 1980). In these groups, succinate levels can be higher under anaerobic conditions, but the increase is not of quantitative importance. Even in mammalian tissues some succinate is produced under anaerobic stress, especially by perfused anaerobic hearts.

The reason for low succinate production is apparently the same in all these groups, namely the virtual absence of PEP-carboxykinase and the low aspartate levels. Moreover, cephalopods store small amounts of glycogen in heart and mantle muscle (about 20 mmole glycosyl units g^{-1} wet tissue in both organs in *Sepia officinalis*; Storey and Storey, 1983; Hoeger and Mommsen, 1985). Scallops do not differ from good molluscan anaerobes in the amount of glycogen they contain. Gastropods and bivalves with a high capacity for anaerobic survival store between

100 and 250 mmol glycosyl units g^{-1} wet weight in both ventricles (Ellington, 1981; 1982a,b,c) and other muscular tissue (de Zwaan and Putzer, 1985). In the scallop *Placopecten magellanicus* concentrations were 340 and 140 glycosyl units g^{-1} wet weight for phasic and catch adductor muscle, respectively (de Zwaan *et al.*, 1980). Zammit and Newsholme (1976) found that the activity of alanine aminotransferase is much lower in adductor muscle of scallops (*Pecten maximus* and *Chlamys varius*) than in muscle from other bivalves. This, together with the low aspartate levels found in scallops (about one tenth that in other marine bivalves; de Zwaan *et al.*, 1980) explains why coupling between carbohydrate and aspartate turnover cannot be important in these animals.

Low levels of aspartate are also found in fresh water and terrestrial molluscs, although succinate production starts as soon as anaerobic conditions are initiated. Unlike marine species, addition of aminooxyacetate to *in vitro* incubations of foot tissue of the fresh water gastropod *Marisa cornaurietis* had no effect on the rate of production of succinate (Livingstone and de Zwaan, 1983). The same was true for gill and adductor muscle in *Anodonta cygnea* (Figure 10.2). The PEPCK inhibitor, 3-mercaptopicolinate had a strong inhibitory effect on the formation of succinate. Fresh water molluscs thus may not have a clearly distinct transition stage, but change almost instantaneously to a catabolic pattern as depicted in Figure 10.1 under 'prolonged' aerial exposure.

Formation of alanine is also ubiquitous among molluscs, although its quantitative importance seems inversely related to the importance of aspartate as carbon donor for succinate. In fresh water gastropods and bivalves, pyruvate is, anaerobically, converted mainly to other compound such as acetate and lactate. In (terrestrial) pulmonates, lactate is the main fermentation product. When produced in substantial amounts it may even be excreted. This has been observed for *Lymnaea stagnalis* and the bivalve *Tellina planata* (de Zwaan *et al.*, 1976; Kluytmans, De Bont, Kruitwagen, Ravestein and Veenhof, 1983). In two gastropods, *Helix pomatia* and *L. stagnalis,* it has been observed that, during anaerobiosis, the haemolymph concentrations of lactate and succinate were two or three times higher than in the tissues, which indicates active transport from the tissue into the haemolymph (Wieser, 1981; Wijsman, Van der Lugt and Hoogland, 1985).

The High Power Output Mode of Anaerobic Metabolism

When carbohydrates are catabolised by aerobic respiration, all the NADH formed is oxidised by the electron transport chain located in the mitochondrion. However, the inner mitochondrial membrane is a permeability barrier for pyridine nucleotides and the electrons derived from extramitochondrial NADH enter the electron transport chain by so-called shuttles. In molluscs, two shuttles may be present; the malate-aspartate shuttle and the glycerol phosphate shuttle. The former operates in non-muscular tissue, in ventricles and in muscles of gastropods and bivalves (de Zwaan,

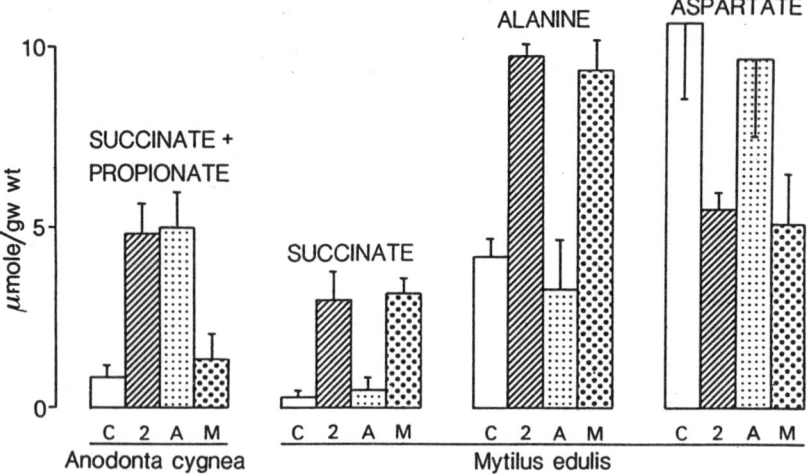

Figure 10.2 Levels of alanine, aspartate, succinate and succinate plus propionate in excised adductor muscles. The posterior adductor muscle from *M. edulis* and the anterior plus posterior adductor muscles from *A. cygnaea* were incubated in O_2 free sea water (*M. edulis*) and tap water (*A. cygnaea*) adjusted to pH 7 with or without metabolic inhibitors. Open bars (C), levels 15 min after the muscles were excised and immediately placed in ice-cold incubation medium without inhibitor; other (diagonal lines or dotted) bars, levels after 2 h incubation at 20°. Code for the inhibitors is as follows; 2, no inhibitor added; A, aminooxyacetate added (2mM); M, 3-mercaptopicolinate added (5mM). Data for *M. edulis* after De Zwaan *et al.*, 1982, with new results for M added; data for *A. cygnaea* after De Zwaan *et al.*, 1984.

1977; Ellington, 1982c), whereas the latter operates in the mantle muscle of cephalopods (Storey and Storey, 1983). Both shuttles contain a cytosolic dehydrogenase which forms a redox couple with glyceraldehyde 3-phosphate dehydrogenase (GAPDH); they are, respectively, malate dehydrogenase and glycerol phosphate dehydrogenase (GPDH). Alternatively, pyruvate reductase may form a redox couple with GAPDH, in which case pyruvate derivatives accumulate in the cytosol. There can be up to four pyruvate oxidoreductases in one molluscan tissue (de Zwaan and Dando, 1984; Gäde and Grieshaber, 1986). Cytosolic GPDH and malate dehydrogenase activities are considered to be indices of aerobic capacity, and the sum of the separate pyruvate oxidoreductase activities (PRH) indices of the anaerobic capacity. The ratio PRH/GPDH has been determined for a number of cephalopods, both for different tissues within one organ and for different organs of a single species in order to establish the relative potentials for anaerobic glycolysis versus aerobic muscle work (Storey and Storey, 1983). The ratio is highest in the more sluggish species such as *Nautilus* and *Octopus* which swim with short bursts of jet propolusion. The pelagic squid *Symphlectoteuthis_oualaniensis* showed much lower values, and has a high capacity for sustained swimming which is obviously powered by aerobic energy supply.

In the presence of an adequate oxygen supply, the shuttles can exclude the pyruvate reductases from cytosolic NADH which, for instance, explains why, in the resting vertebrate muscle, no lactate accumulates. As the work load progressively increases, a stage is reached where the rate of NAD reduction exceeds the shuttle capacity for intercompartmental hydrogen transport. There will then be synchronous reoxidation of NADH in the cytosol by the pyruvate reductases. At which level of glycolytic flux this anaerobic threshold is reached depends on aerobic capacity, determined by the O_2 binding capacity, the transport efficiency of the circulatory system and the presence of myoglobin for facilitating oxygen diffusion in the cell.

Cephalopods are at an advantage in these respects, as they comprise the only molluscan class to possess closed circulatory systems and blood circulation is further improved by a pair of branchial hearts to support the function of the systemic heart. Pelagic squids are outstanding high speed swimmers, capable of long migrations. Swimming depends on the mantle cavity acting as a pump powered by the mantle musculature. The mantle muscle consists of muscle fibres sandwiched between outer and inner tunics which contain collagen fibres. The mantle muscles are circular fibres lying in the transverse plane of the mantle. There are three zones within the circular fibres, inner, outer and central zones.

The peripheral zones, which are relatively richly vascularised, areas, possess high succinic dehydrogenase and citrate synthetase activity and have high mitochondrial densities (Bone, Pulsford and Chubb, 1981; Chantler, 1983). They are classified as helical smooth muscle types, and are primarily involved in sustained aerobic work, reflected in a high GPDH/PRH ratio and a high aerobic capacity. The aerobic metabolism of the cephalopod mantle muscle involves coupling the utilisation of carbohydrate to the oxidation of proline and, to a lesser extent ornithine and arginine, which are the main fuels for oxidative metabolism (Hoeger and Mommsen, 1985). Short bursts of muscular work to catch prey or evade predators relies on the central mantle muscle zone which, accordingly, posesses a low GDH/PRH ratio, low mitochondrial density and low activities of 'aerobic' enzymes. The same is true for the mantle muscle of more sluggish cephalopods (eg. *Octopus*) which lack the capacity for sustained swimming. Cellular adaptations to take advantage of a high oxygen supply are not present and these species indeed show relatively low respiratory rates (Storey and Storey, 1983).

Relatively low respiratory rates are also the case for non-cephalopod molluscs. They have open circulatory systems; heart rates and pressure development are low compared with cephalopods (Smith, 1985). Often there is a lack of oxygen binding pigments in the haemolymph and the cell, especially in bivalves. The gas transport function is poorly developed, and, in a number of bivalves, it has been demonstrated that oxygen consumption completely ceases when oxygen pressure of the haemolymph is still relatively high (Mangum and Van Winkle, 1973). Most of the oxygen seems to be absorbed directly by the tissues via their exposed surfaces (Famme, 1981). Zaba and Davies (1984) have argued that this mode of delivery cannot oxygenate cells which are more than 0.6 mm from the surface, which would preclude the use of oxygen by cells in the interior of tissues more than 1.2 mm thick. This agrees well with my own observations that when intact excised organs (gill, mantle, adductor muscle) are incubated in aerated sea water there is a considerable accumulation of succinate. Zaba and Davies (1984) refer to a number of studies concerning marine bivalves in which it was shown that CO_2 is by no means the sole carbon end-product of glycogen catabolism under aerobic conditions. Famme and Kofoed (1982) reported that organic carbon products were excreted at all oxygen tensions and Kluytmans and Zandee (1983) observed that high levels of acetate and propionate accumulated aerobically in all fresh water bivalves studied. They suggested that the propionate pathway was operating under aerobic conditions. From these observations we may conclude that in gastropods, and especially bivalves, the aerobic capacity is small (Baldwin and Lee, 1979) and that even small increases in energy demand exceeding maintenance metabolism will evoke anaerobic energy production.

Unlike environmental anaerobiosis, there appears to be uniformity in metazoan tissues concerning the anaerobic pathway exploited under aerobic conditions. ATP utilisation rates that exceed the aerobic capacity result in a rapid transphosphorylation of phosphoarginine and accumulation of one or more pyruvate derivatives (Table 10.1, exercise). In echinoderms, arthropods and vertebrates this is solely lactate and in all fresh water representatives of the other metazoan phyla it also appears to be solely lactate, although some fresh water bivalves possess the potential for octopine formation (Gäde and Grieshaber, 1986). In marine species, lactate dehydrogenase is usually replaced by enzymes which catalyse the reductive imination of pyruvate with an amino acid as co-substrate to form opines. To date, alternative pyruvate reductases, which use either arginine, alanine, glycine or taurine have been found in molluscs. The products formed by using the first three amino acids have now been detected in a number of different marine molluscs. The trivial names of the products are octopine, alanopine and strombine, respectively. An enzyme using taurine as a substrate has recently been reported to occur in muscle extracts of the abalone *Haliotus dissus hannai* (Sato and Gäde, 1986).

The enzymes that form opines are dominant in muscle tissue. Functionally they are analogous to LDH in that they keep the $NADH/NAD^+$ ratio low in the cytosol. The use of these enzymes is often found to be restricted to periods of contractile activity or periods of recovery following contractile activity or environmental anaerobiosis. Detailed overviews of our present knowledge of pyruvate metabolism in molluscs are given in two recent reviews (de Zwaan and Dando, 1984; Gäde and Grieshaber, 1986). In the latter, it is suggested that the presence of more than one pyruvate oxidoreductase is typical for species that have to cope with both functional and environmental hypoxia. A different enzyme operates during each condition thus avoiding the possibility that anaerobic glycolytic flux is diminished by end-product inhibition if the two hypoxic states follow each other consecutively.

Octopine is the preferred pyruvate derivative in mobile molluscs, and is formed during exercise and post exercise recovery. This appears to be related to high resting steady state levels of phosphoarginine, which in swimming bivalves (scallops, file shells) and cephalopods, range between 20 and 30 mmol per gram wet weight. *Cardium tuberculatum*, which can display strenuous escape movements with its foot, has comparable high resting values. It accumulates octopine during functional anaerobiosis but lactate during environmental anaerobiosis (Meinardus-Hager and Gäde, 1986). Whelks (*Buccinum undatum, Nassa mutabilis, Nassarius coronatus, Busycon contrarium*) and other gastropods, which exhibit less vigorous escape activities than scallops and pelagic cephalopods, have steady state levels of phosphoarginine which are lower (around 10 mmol) and in these animals usually not all pyruvate is transformed into octopine. In the gastropod *Strombus luhuanus*, for example, both octopine and strombine are formed at about equal amounts during exercise (Baldwin and England, 1982).

There has been much speculation about functional reasons for the replacement of lactate by octopine. Storey and Storey (1983) summarise six suggestions from literature, of which I will briefly discuss the one suggested by Zammit and Newsholme (1976). These authors suggested that octopine formation facilitates the breakdown of phosphoarginine to form ATP by removal of arginine. Tight coupling between glycolysis and arginine breakdown has been observed in the catch muscle of the scallop *P. magellanicus* (de Zwaan *et al.*, 1980) and in the mantle of the cuttle fish *Sepia officinalis* (Grieshaber and Gäde, 1976). In the phasic muscle of the scallops, however, the breakdown of arginine phosphate and the accumulation of octopine takes place essentially in series with no metabolic integration (Gäde, Weeda and Gabbott, 1978; Gäde, 1980; de Zwaan *et al.*, 1980). At least 75% of exercise in these mobile animals is sustained by transphosphorylation of phosphoarginine which leads to increasing arginine levels (Table 10.1). This, in turn, stimulates octopine dehydrogenase activity. Phosphoarginine breakdown precedes octopine formation and once glycolysis becomes the driving force, muscle performance appears to be strongly reduced. The consequence is fewer, less powerful, escape movements. This may explain the fact that in the slower moving whelks the contribution of glycolysis to ATP requirements is in general considerably higher (Table 10.1).

The predominance of transphosphorylation over catabolic ATP generation by the high power output mode of metabolism may be physiologically important, because hydrolysis of phosphoarginine does not lead to metabolic acidosis. In gastropods and bivalves, the sum of the concentrations of phosphoarginine, arginine and octopine does not change, neither during exercise nor during recovery. This and the virtual absence of octopine in the haemolymph indicates that there is no removal of octopine from muscular tissue and that octopine is probably oxidised *in situ*.

In cephalopods, the metabolic fate of octopine is assumed to be analogous to that of lactate in vertebrates. During sustained activity, there can be anaerobic depletion of glycogen in white vertebrate muscle without a corresponding increase in lactate in the blood. This is achieved by the simultaneous high lactate oxidation rate in the red muscle. A comparable cooperation between tissues is suggested by Bone *et al.* (1981) for the different fibre types in the cephalopod mantle muscle. The blood supply to the mantle first passes through the central zone of the mantle and then flows to the inner and outer zones. In this manner anaerobic metabolites (octopine) produced in the central zone during high speed jetting could be transferred to the more aerobic fibres of the peripheral zones for reoxidation.

An alternative hypothesis proposes release of octopine from the mantle muscle (Storey and Storey, 1983). These authors identified two major isozymes of ODH, present at different ratios in different tissues of *Sepia officinalis*; a 'mantle' enzyme and a 'brain' enzyme. Kinetic properties indicated that brain ODH is functionally analogous to the H-isozyme of LDH, and that mantle ODH is analogous to the M-type LDH. Gäde, Meinardus and Carlsson (1979) also found the ODH from the optic lobe of *Loligo vulgaris* to be analogous to vertebrate H-type and ODH from mantle to be an analogous to vertebrate M-type LDH. Further studies on *S. officinalis* with ^{14}C-octopine revealed that brain (and other non-muscular tissues) efficiently catalyses the oxidation of octopine *in vivo* and that the mantle muscle is primarily geared to the production of octopine and has a low capacity for its oxidation. A reversal of the ODH reaction in brain provides pyruvate for oxidation as an aerobic substrate, whereas arginine is recycled to the muscle. Other tissues may direct the pyruvate moiety into gluconeogenesis and cycle both glucose and arginine back to the muscle, as lactate is in the so-called Cori cycle in vertebrates.

Both hypotheses for the metabolism of octopine need further experimental evaluation and if the existence of the 'octopine Cori cycle' be unequivocally proven (and thus appears to be more than a nice 'storey'), I suggest that it bear the name of its discoverer and be called the 'Storey cycle'.

High power output anaerobiosis leads to a large flare-up of the carbon flux through the glycolytic pathway in spite of the fact that glycolysis may only contribute to a minor extent to ATP generation. In *P. magellanicus* we observed, for instance, a 58-fold increase of the glycolytic rate whereas the energetic contribution of catabolism was only about 10 per cent of the total ATP required (de Zwaan *et al.*, 1980, see Table 1).

Alternative Sources of Nitrogen for Anaerobic Alanine Formation

During the course of adaptation to a hyperosmotic environment the total amino acid pool increases due to large changes in alanine and moderate changes in other amino acids. ^{14}C-tracer studies with the bivalve *Modiolus demissus* in the presence of metabolic inhibitors, led to the conclusion that alanine is synthesised from amino acids derived from both endogenous pools and from protein degradation (Bishop, Greenwalt and Burcham, 1981). The use of aminooxyacetate further supported the suggestion that protein amino acids provide the required amino group via a number of amino transferase reactions. Aspartate levels of animals exposed to increased salinity did not differ from those of control animals, in sharp contrast to what occurs during anoxia stress. In the latter case, in marine molluscs, a decrease of aspartate level is accompanied by an increase in the level of alanine. It appears therefore that marine animals with high free aspartate levels have two mechanisms for alanine formation. However, in terrestrial and fresh water species that possess low steady state concentrations of aspartate, some alanine is produced during environmental anaerobiosis, although in lower quantities, relative to succinate, than in marine ones.

We have studied alanine formation in the fresh water bivalve *Anodonta cygnea* and obtained results which indicate a mechanism similar to that in marine species in response to hyperosmotic stress (de Zwaan, De Bont and Nilsson, 1984). Figure 10.3 shows that about 70 per cent free amino acid pool in the adductor muscle is due to approximately equal amounts of alanine and glutamate. After two hours of anaerobic incubation the total free amino acid pool increased by 180 per cent, 30 per cent being accounted for by the rise in glutamate and alanine. When aminooxyacetate was also present in the incubation medium, a comparable increase in pool size was observed, although levels of alanine and glutamate remained at control levels. These results are consistent with the suggestion that the increase in the free amino acid pool results from protein catabolism from which, via transamination, glutamate is formed (see insert in Figure 10.3). Subsequently, part of the glutamate transaminates with pyruvate resulting in *de novo* synthesis of alanine. In the adductor muscle of *M. edulis*, we have examined the ability of amino acids to transfer their amino groups to 2-oxoglutarate, pyruvate and oxaloacetate. Glutamate is formed by incubating muscle extract and 2-oxoglutarate with valine, isoleucine, leucine, asparagine, ornithine, glutamine, alanine and aspartate. The results with asparagine and glutamine may not be due to direct transamination but to conversion into aspartate and glutamate by the action of asparaginase and glutaminase, respectively. Pyruvate and oxaloacetate could only transaminate with glutamate (de Zwaan and Dando, 1984). The fate of the oxoacids which are simultaneously formed in amount equimolar to alanine (according to the scheme proposed in the inset of Figure 10.3 for *A. cygnea*) is not yet known, but they may be further transformed into succinate and volatile fatty acids with a possible gain of ATP.

Another possibility for the conservation of nitrogen for alanine formation is by fixation of ammonia via glutamate dehydrogenase. Ammoniogenesis does not come to a standstill during environmental anoxia, and NH_3 accumulates in the haemolymph during shell valve closure in bivalves (de Zwaan *et al.*, 1983a). We obtained evidence that regeneration of ammonia could be linked to the deamination of adenine nucleotides. After incubation of disrupted mitochondria from the mantle of the sea mussel with malate anaerobically for 2 hours, some 85 per cent was metabolised and 70 per cent appeared as succinate. About 10 per cent was recovered

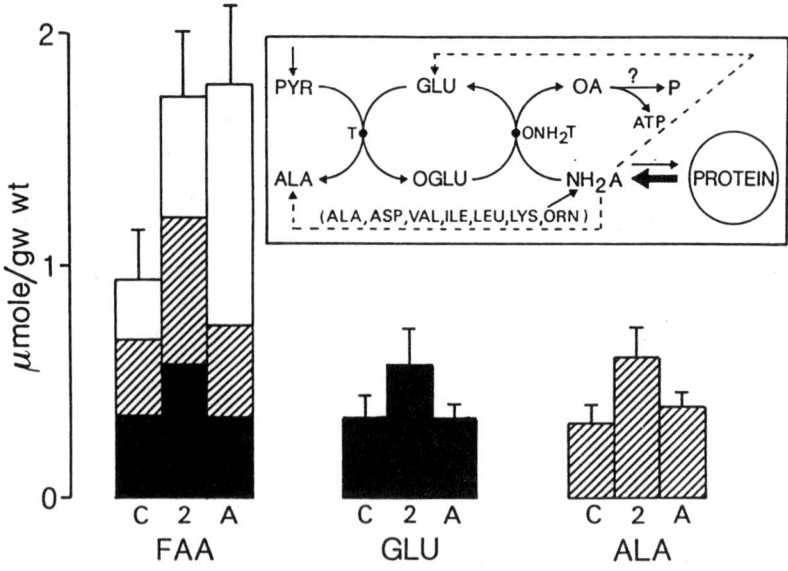

Figure 10.3 Levels of the free amino acid pool (FAA), glutamate (GLU, black bars) and alanine (ALA, diagonal lines) in excised anterior and posterior adductor muscle from *A. cygnaea*. Muscles incubated in tap-water adjusted to pH 7 for 2 h with 2mM aminooxyacetate (A) or without inhibitor (2) at 20°, levels at 15 min after the muscles were excised and immediately placed in ice cold tap water. The proposed reaction scheme for the transfer of nitrogen from various amino acids derived from protein catabolism to alanine by the concerted action of glutamate transferase (ONH_2T) and alanine transferase (T). Alanine transferase transfers NH_2 from glutamate to pyruvate, whereas glutamate transferase can transfer NH_2 from the amino acids listed between the brackets and 2-oxoglutarate. Abbreviations: OA, oxoacid; P, fermentation product; OGLU, 2-oxoglutarate; PYR, pyruvate.

as alanine, although there was a strong negative nitrogen balance when comparing changes in aspartate, glutamate and alanine. Ammonia levels increased five fold. When aminooxyacetate was present, alanine formation was blocked, whereas glutamate concentration increased markedly. In the absence or presence of the inhibitor, the sum of the adenine nucleotides dropped from 2.23 to 0.79-0.86 mmole per ml of mitochondrion suspension, respectively. In the presence of the inhibitor, the loss of nitrogen could be accounted for by the rise in ammonia plus alanine or glutamate. The decrease in adenine nucleotides was mainly caused by a fall in ATP concentration.

These results suggest that NH_3 is derived from nucleotides by the action of AMP-deaminase. We could indeed detect high activities of this enzyme in the adductor muscle and mantle (15.1 and 24.8 mmole NH_3 formed $h^{-1}.g^{-1}$ fresh weight, respectively) when activated by ATP. However, AMP-deaminase from the hepatopancreas of *Helix aspersa* was found to be mainly located in the cytosol (Campbell and Vorhaben, 1979). This enzyme is known to have a high affinity for enzyme complexes and, possibly, in our experiment, it was co-precipitated with the mitochondria. The general observation, however, that there is no substantial drop in the sum of the adenosine phosphates during anoxia in marine species does not support the idea of an important quantitive contribution to alanine formation in these molluscs. It could possibly contribute to the relatively low alanine formation in fresh water and terrestrial species (de Zwaan, De Bont and Verhoeven, unpublished).

Ammoniogenesis, and ammonia fixation by glutamate dehydrogenase, could also be shown by adding 2-oxoglutarate to intact mitochondria isolated from the mantle of the sea mussel. It appeared to be a moderate anaerobic substrate which, after three hours, was almost solely converted into succinate and glutamate (together accounting for about 20 per cent of the total). Redox balance was maintained by the 1:1 coupling of glutamate dehydrogenase (reductive amination) and 2-oxoglutarate dehydrogenase. When aspartate was also added, the utilisation rate of 2-oxoglutarate was enhanced by a factor of four. Under these circumstances, most 2-oxoglutarate appeared to be aminated by transamination with aspartate, whereas glutamate dehydrogenase appeared to be of minor importance, compared with fumarate reductase, in reoxidising NADH (de Zwaan, Veenhof and Holwerda, unpublished results). These observations show that glutamate dehydrogenase can contribute to the anaerobic reoxidation of NADH. In marine molluscs this may occur during periods in which insufficient fumarate is generated for the fumarate reductase complex and this may explain why small increases of glutamate are generally noticed in the transition stage of anaerobiosis (de Zwaan, 1977).

It has been known for some time that, in the lugworm *Arenicola marina*, both D- and L-alanine accumulate in about equal amounts during anaerobiosis (Schöttler, this book). Free D-alanine is extremely uncommon in nature and, until recent, had never been detected in molluscan tissues. Surprisingly, Matsushima, Katayama, Yamada and Kado (1984) found D- and L-alanine in about equal amounts in three out of five brackish or intertidal bivalves, namely *Corbicula japonica, Tapes philippinarum* and *Meretrix lamarckii*. *Mytilus edulis* and *Crassostrea gigas* possessed only L-alanine. The difference between the two groups appeared to be determined by the presence or absence of alanine racemase which catalyses racemisation between the two stereoisomers. Probably it is the L-form which is formed during anoxia or hypersalinity stress, about half of which will, in turn, be converted to the D-form when a racemase is present. The authors point to the fact that only the three infaunic bivalves possessed both stereoisomers in contrast to the two epifaunic species. For this reason they suggested that the distribution may be attributed to life-style. In this respect it is worth mentioning that *A. marina* is also an infaunic species.

Intracellular Location of Alanine Formation

I have discussed the idea that, in the transition phase of environmental anaerobiosis, the simultaneous utilisation of glycogen and aspartate, leading to the accumulation of alanine and succinate, is coupled by the reactions catalysed by ALAT, AAT and MDH. This results in the overall reaction:

$$\text{pyruvate} + \text{aspartate} + \text{NADH} + \text{H}^+ \rightarrow \text{alanine} + \text{malate} + \text{NAD} + \text{H}_2\text{O}.$$

Since a major part of malate is subsequently converted into succinate this would suggest a 1:1:1 stoichiometry between changes in the levels of alanine, aspartate and succinate. In Figures 10.1 and 10.4 it is assumed that the coupling leading to alanine and malate is located in the cytosol. Theoretically this is possible, as the whole enzyme machinery involved is generally present in the cytosol. Moreover, it enables a convenient way of maintaining redox balance in the cytosol.

Nonetheless there are several arguments in favour of a mitochondrial location for the nitrogen transfer from aspartate to alanine. These include the following: a) changes in the actual aspartate (or alanine)/succinate ratios are greater than unity; b) the cytosol and the mitochondrial compartments contain the complete enzyme set to catalyse the coupling; c) pyruvate is a good substrate for isolated mitochondria when added in combination with aspartate. Concerning the first argument, I have noticed in several independent experiments carried out with adductor muscle (*in vivo* and *in vitro*) that the ratio of changes in concentrations of alanine, aspartate and succinate is always close to 1:1:0.5 (de Zwaan *et al.*, 1982, 1983a and b). A similar ratio was recently established for adductor muscle of the bivalve *Scapharca inequivalvis* (de Zwaan and Cortesi, unpublished). Also, Ellington (1981; 1982a,b,c) reported that, in the ventricle of the whelk *Busycon contrarium*, although succinate varied inversely with aspartate this was without a 1:1 stoichiometry between the metabolites. On the contrary, there was a perfect 1:1 stoichiometric relationship between decreases in aspartate levels and increases in alanine levels. Gäde and Ellington (1983) have pointed to the fact that alanine+aspartate to succinate ratios in bivalves and gastropods can be as high as three. Also, Meinardus and Gäde (1981), from results obtained with the foot of *Cardium edule*, concluded that within the first ten hours of environmental anaerobiosis aspartate utilisation exceeds succinate accumulation.

In *in vitro* experiments with excised adductor muscle of the sea mussel, analysis of metabolites was carried out on tissue homogenised in incubation medium (de Zwaan *et al.*, 1982). The observed discrepancy between aspartate utilisation and succinate accumulation could therefore not be accounted for on the basis of release of malate or succinate by the tissue. Neither did conversion of succinate to propionate occur within the incubation period of three hours. This is also true for the ventricle of *B. contrarium* (Ellington, 1981; 1982a,b,c). Studies with isotopes (Collicutt and

Intracellular Location of Alanine Formation 207

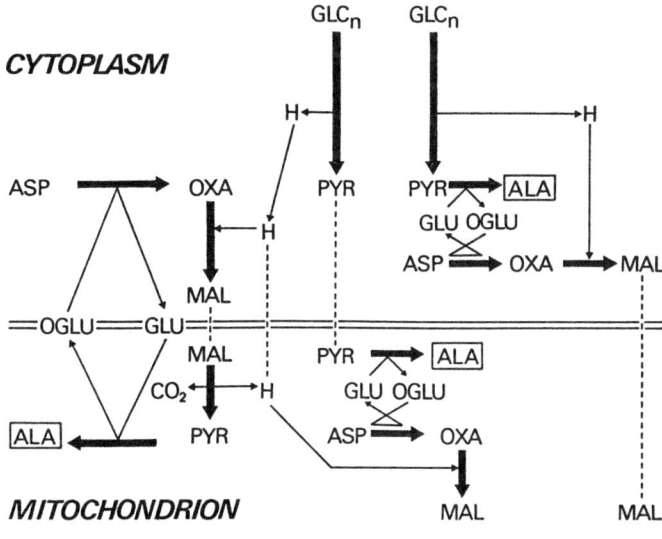

Figure 10.4 Metabolic maps showing two possibilities for the redox coupling of the glycogen-alanine pathway with the conversion of aspartate into malate in marine invertebrates. Alanine accumulates but malate is further metabolised within the mitochondrion. The left part is assumed to occur in (muscle) tissue which exploits 'classical glycolysis' during functional anaerobiosis (lactate and/or opine accumulation), but couples the conversion of pyruvate into alanine with the aspartate-succinate pathway during environmental anaerobiosis. In the latter case, the transport of hydrogen through the inner mitochondrial membrane is assumed to be dependent on the conversion of aspartate to CO_2 and alanine, analogous to the aspartate-malate shuttle when oxygen is available. The right part depicts the situation in tissues in which pyruvate reductases do not compete with alanine transferase for cytosolic pyruvate. The overall reactions at the bottom illustrate the stoichiometric differences with respect to the simultaneous mobilisation of glycogen and aspartate. Abbreviations: MAL, malate; OXA, oxaloacetate; GLC_n, glycogen. See also legend to Figure 10.3.

Hochachka, 1977), metabolic inhibitors (de Zwaan *et al.*, 1982) and the observation of malic enzyme activity in the mitochondria of bivalves and gastropods (de Zwaan, 1977; Ellington, 1981) have revealed that not only is the nitrogen of aspartate incorporated into alanine, but so also is a substantial part of its carbon.

Incorporation of aspartate carbon into alanine probably involves an initial, and then a final transamination step and the decarboxylation of malate to pyruvate resulting in the overall reaction:

$$\text{aspartate} \rightarrow CO_2 + \text{alanine (Figure 10.4)}.$$

As far as the second argument is concerned, it appears that ALAT has, in general, a higher activity in mitochondria relative to MDH than in cytosol. In a study of the subcellular distribution of aminotransferases in the gill tissue of four bivalves it was found that *Mercenaria mercenaria* has a trace activity of ALAT in the cytosol, whereas in *Modiolus demissus* it is strictly mitochondrial (Paynter, Ellis and Bishop, 1984; Paynter, Karam, Ellis and Bishop, 1985). Earlier studies on anaerobic metabolism of *M. demissus* have shown that its gill tissue behaves like that of other euryoxic bivalves, namely by accumulation of succinate and utilisation of aspartate, which leads to the conclusion that, at least in these species, alanine production and the coupling to aspartate turnover must be mitochondrial (Paynter *et al.*, 1984).

The third argument concerns direct evidence for the *in vitro* presence of a coupling between alanine and aspartate metabolism within intact mitochondria from the sea mussel. When pyruvate (6 mM) was added to a suspension of mantle mitochondria under anaerobic incubation at 20°C about 20 per cent was metabolised after two hours. Alanine accounted for 30 per cent of the converted pyruvate. When aspartate (6 mM) was added simultaneously, about 45 per cent of the added pyruvate was metabolised, which half appeared as alanine. A flow diagram to account for hydrogen and carbon flux is presented in Figure 10.5. Assuming coupling of pyruvate metabolism (giving alanine) and aspartate (giving succinate and propionate) by ALAT, AAT and MDH, it gives a reasonable balance for hydrogen, nitrogen. Glutamate levels did not change and NH_3 increased by 0.21 mmole.ml^{-1} of mitochondrial suspension.

Intracellular Translocation of Hydrogen for Maintenance of Redox Balance in the Cytosol

A major problem that arises when transferring the alanine and malate to the mitochondria is that the direct redox link between GADDH and MDH is lost, because the mitochondrial inner membrane is a barrier for hydrogen transport via the coenzyme NAD^+. In aerobically functioning tissues, the identical problem of transferring hydrogen from the cytosol to the mitochondrial respiratory chain is solved by the malate/aspartate shuttle. When anaerobic cellular conditions prevail, the mitochondrial malate dehydrogenase reverses its catalytic direction from oxaloacetate towards malate and, instead, catalyses the initial step of the conversion of oxaloacetate (from either aspartate and/or glycogen) into succinate. This prevents

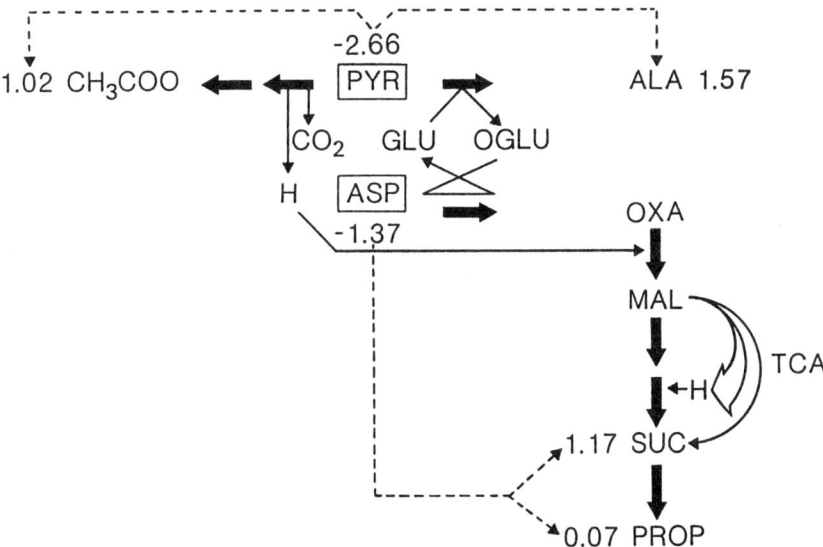

Figure 10.5 Carbon and hydrogen flow scheme for the conversion of pyruvate (5mM) plus aspartate (6mM) after 2 h of anaerobic metabolism with isolated intact mitochondria from the mantle of the sea mussel. The data represent the changes in the initial concentration of that particular metabolite. The dotted lines indicate the main carbon flow and the solid line the main hydrogen flow. Abbreviations: CM_3COO-acetate, SUC, succinate; PROP, propionate; TCA, tricarboxylic acid cycle. See also the legends to Figures 10.3 and 10.4.

its participation in the malate aspartate shuttle. This reversal of the catalytic direction of MDH is probably due to a rise in the $NADH/NAD^+$ ratio which, at the same time, reduces the activity of pyruvate dehydrogenase and hence increases the oxaloacetate available for malate dehydrogenase. Since, in bivalves, malic enzyme operates in the living cell in the oxidative decarboxylating direction (de Zwaan, 1977) it can take over the function displayed under aerobic conditions by MDH in the mitochondrial part of the malate-aspartate shuttle. Instead of a shuttle system in which aspartate is cycled, a net conversion of aspartate into alanine and CO_2 is obtained (Figure 10.4; de Zwaan *et al.*, 1983b).

As already summarised, several studies indicate that alanine formation from aspartate via decarboxylation of malate may be a general phenomenon, especially occurring in the muscle (adductor muscle, ventricle, foot) tissue of molluscs. The physiological relevance may be that this reaction sequence can functionally replace the malate/aspartate shuttle in transferring cytoplasmic hydrogen from the cytosol through the inner mitochondrial membrane in order to supply reducing equivalents for the mitochondrial production of malate from aspartate. Malate in turn can be transformed into succinate (and propionate) with the maintenance of redox balance by means of an initial dismutation reaction as already explained. The coupling of the aspartate-malate pathway with the glycogen-alanine pathway on one hand and the connection with the 'aspartate dependent hydrogen transfer system' is depicted in Figure 10.4. The scheme gives a good illustration as to why aspartate utilisation should be double succinate production. In the experiments described above, in which aspartate and pyruvate were together metabolised by anaerobic sea mussel mitochondria the ratios of changes in the levels of aspratate/alanine/ succinate were close to 1/1/1 (Figure 10.5). This is understandable because pyruvate was directly added and was not available as a consequence of an equimolar reduction of cytosolic NAD^+. No aspartate will therefore be needed for the transport of hydrogen from the cytosol to the mitochondrion. On the other hand the mitochondria are now themselves forced to generate reducing power for the reduction of oxaloacetate. The results shown in Figure 10.5 make it clear that this is obtained by the oxidation of part of the pyruvate into acetate, which *in vivo* is never a pronounced end-product in the sea mussel (Kluytmans *et al.*, 1977).

When, after a period of anaerobiosis, the aspartate pool is greatly diminished and succinate carbon is also delivered from carbohydrate (Figure 10.1), the discrepancy between the stoichiometry of aspartate and succinate will be less than two and at some stage the succinate production will exceed the aspartate utilisation. In *in vivo* studies with the sea mussel we observed that even after 18 hours following the onset of anaerobiosis (aerial exposure) the decrease in aspartate in adductor muscle exceeded the increase in succinate (no propionate formed) by a factor of two, whereas for the remaining part of the body, the discrepancy between the two metabolites was reversed within the first six hours. This could mean that the initial stage of Figure 10.1 is much more important and is maintained much longer in muscle tissue than in other tissues within a certain species.

In an extended comparative study including over 100 species of ten phyla, we have determined whether the presence of certain pathways may be correlated with the specific activities of certain enzymes involved in these pathways. The aspartate-malate shuttle predicts a relationship between MDH and GOT. This was observed for the vertebrates and most other phyla. The pathway depicted in Figure 10.4 also predicts a relationship between GOT and MDH, and also between GOT and GPT. This was observed for all the phyla except the vertebrates (Livingstone, de Zwaan, Leopold and Marteijn, 1983).

Reoxidation of Cytosolic NADH: Translocation of Hydrogen Versus Action of Pyruvate Oxidoreductases

The probability that the aspartate-alanine hydrogen translocase system is more pronounced and operates longer in muscular tissue may be linked with the fact that the pyruvate oxidoreductases are also dominant in muscle tissue. In contrast to most tissues, muscle is involved in both environmental and functional anaerobiosis. The first condition requires fuel conserving pathways (high ATP yield/unit fuel), the second conditions a high ATP output pathway (high ATP output/unit time). The pathways involved convert pyruvate into an end-product; the one operating under the first condition achieves this by transamination into alanine and the one operating under the second condition by reduction or reductive condensation with an amino acid into lactate or an opine, respectively. When aerobic conditions prevail, pyruvate is directed towards the citric acid cycle. Although the pyruvate oxidoreductases are always present in the cytosol they display their action only during functional anaerobiosis. It seems therefore that with aerobiosis and environmental anaerobiosis, cytosolic malate dehydrogenase suppresses the pyruvate oxidoreductases for oxidising glycolytic NADH. However, malate dehydrogenase is integrated in a system which serves to transfer hydrogen through the inner mitochondrial membrane and the capacity of the transport system (either the malate aspartate shuttle or the aspartate/alanine + CO_2 conversion) is therefore not determined *a priori* by the activity of malate dehydrogenase. During aerobic standard glycolytic flux, reoxidation of NADH generated by the GAPDH step will be entirely compensated for by the malate aspartate shuttle and this can be continued during environmental anaerobiosis because of the absence of a Pasteur effect and the possibility of modifying the shuttle for anaerobic functioning (by replacing MDH by malic enzyme). When, during high power output metabolism, the reduction rate of NAD^+ is much higher owing to a large flare-up of glycolysis, when not all NADH can be handled by the transport system, and the overshoot of NADH is taken up by the pyruvate oxidoreductases to form lactate or opines.

Support for the proposed higher affinity for cytosolic NADH by the aspartate coupled hydrogen transport system has been obtained from *in vitro* and *in vivo* studies with adductor muscle and ventricle in which fluxes through pathways were reduced by inhibitors (Foreman and Ellingtron, 1983; de Zwaan *et al.*, 1982, 1983b). The transaminase inhibitor aminooxyacetate appears to block both aspartate conversion into alanine and into succinate. Therefore ATP yield linked to the fumarate reductase complex is lost. However, aspartate conversion to alanine by decarboxylation also involves transaminase reactions which are also blocked. Under these conditions, the hydrogen transport system is unable to compete for cytosolic NADH with the pyruvate oxidoreductases. This results in extra opine formation and a higher transphosphorylation rate of phosphoarginine in order to compensate for the lost ATP formation by the mitochondrion. When iodoacetate, an inhibitor of

GAPDH, was applied no opine formation was observed, whereas accumulation of alanine and succinate was reduced. This is probably due to a reduced glycolytic flux with a concomitant reduction rate of NAD^+; consequently the capacity of the aspartate coupled hydrogen translocase system is large enough to deal with all generated NADH. Besides the absence of opine formation, it was noticed that the ratio of changes in the levels of alanine, aspartate and succinate was maintained at 1:1:0.5 which supports the assumption that the glycolytic flux must have been reduced (de Zwaan et al., 1982).

Metabolic suppression is an important and widely distributed means to survive long anoxic periods. Eels survive six hours of anoxia at 15°C, while trout die after about 12 minutes. Eels become quiescent, while trout become agitated at low pO_2. Similarly, when goldfish are transferred suddenly from normoxic into anoxic water, they die within two hours, which contrasts to the 12 hours survival of unstressed fish. We have also observed that stressed mussels in combination with anoxia are handicapped in suppressing metabolism and therefore in surviving anoxic conditions. In an experiment in which inhibitors were injected in the intrapallial cavity, the shell valves were held tightly together with rubber bands in order to avoid extrusion of the

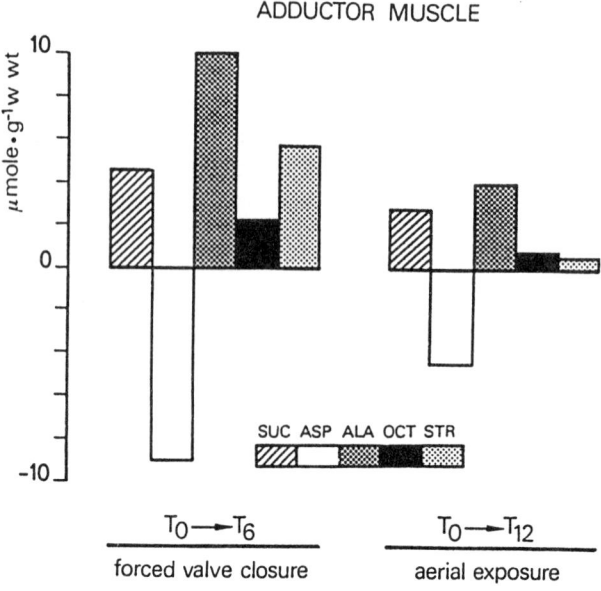

Figure 10.6 Changes in the initial concentrations of various metabolites in the adductor muscles of sea mussels out of water. The sea mussels were kept above the water surface for 6 h while closed by rubber bands (T_0 to T_6, forced valve closure). Others were normally exposed to air, without a rubber band for 12 h (T_0 to T_{12}, aerial exposure). For abbreviations, see legends to Figures 10.1 and 10.5.

injected liquid (de Zwaan et al., 1983b). In this state they were kept for 6 hours just above the water surface. The experiment was either carried out with mussels taken directly from the aquarium or with mussels kept above the water surface (without a rubber band) for 12 hours. It enabled us to compare changes in the levels of metabolites due to 12 hours natural aerial exposure without forced closure, by rubber band (environmental anaerobiosis) with those observed during the initial six hours of forced valve closure (stress). The results of this comparison are depicted in Figure 10.6. It appeared that, in the former group the rate at which the concentrations of aspartate and key end-products changed were much smaller, in spite of an incubation period which lasted twice as long. The forced closure thus resulted in a much higher energy expenditure, with roughly a ten times increase of catabolic ATP output, as determined from end-product formation (Table 10.1). Qualitatively, this caused a surprising difference. In the group subjected to forced closure, strombine and, to a lesser extent, octopine, accumulated in substantial amounts. This is an illustration that the ratio of the formation of succinate plus propionate/pyruvate derivatives is indeed determined by the glycolytic flux, which can be understood in the terms of the competition model of hydrogen translocation versus cytosolic pyruvate reduction.

References

Baldwin, J. and England, W.R. (1980) 'A comparison of anaerobic energy metabolism in mantle and tentacle muscles of the blue ringed octopus *Hapalochlaena maculosa* during swimming' *Australian Journal of Zoology* **281**, 407-412

Baldwin, J. and England, W.R. (1982) 'The properties and functions of alanopine dehydrogenase and octopine dehydrogenase from the pedal retractor muscle of Strombidae (Class Gastropoda)' *Pacific Science* **36**, 381-394

Baldwin, J. and Lee, A.K. (1979) 'Contribution of aerobic and anaerobic energy production during swimming in the bivalve mollusc *Limaria fragilis* (Family Limidae)' *Journal of Comparative Physiology* **129**, 361-364

Baldwin, J. and Morris, G.M. (1983) 'Re-examination of the contributions of aerobic and anaerobic energy production during swimming in the bivalve mollusc *Limaria fragilis* (Family Limidae)' *Australian Journal of Marine and Fresh water Research* **34**, 909-914

Baldwin, J., Lee, A.K. and England, W.R. (1981) 'The function of octopine dehydrogenase and D-lactate dehydrogenase in the pedal retractor muscle of the dog whelk *Nassarius coronatus*' (Gastropoda: Nassariidae). *Marine Biology* **62**, 235-238

Bishop, S.H., Greenwalt, D.E. and Burcham, J.M. (1981) 'Amino acid cycling in ribbed mussel tissue subjected to hyperosmotic shock' *Journal of Experimental Zoology* **215**, 277-287

Bone, Q., Pulsford, A. and Chubb, A.D. (1981) 'Squid muscle mantle' *Journal of the Marine Biological Association of the UK* **61**, 327-342

Brinkoff, W., Stockmann, K. and Grieshaber, M. (1983) 'Natural occurrence of anaerobiosis in molluscs from intertidal habitats' *Oecologia* **57**, 151-155

Campbell, J.W. and Vorhaben, J.E. (1979) 'The purine nucleotide cycle in *Helix* hepatopancreas' *Journal of Comparative Physiology* **129**, 137-144

Chantler, P.D. (1983) 'Biochemical and structural aspects of molluscan muscle' in S.M. Saleuddin and K.M. Wilbur (eds), *The Molluscs* Vol. 4, Part 1, pp. 78-143 (Academic Press, New York)

Chih, C.P. and Ellington, W.R. (1983) 'Energy metabolism during contractile activity and environmental hypoxia in the phasic adductor muscle of the baby scallop *Argopecten irradians* concentricus' *Physiological Zoology* **56**, 623-631

Collicutt, J.M. and Hochachka, P.W. (1977) 'The anaerobic oyster heart: coupling of glucose and aspartate fermentation' *Journal of Comparative Physiology* **115**, 147-157

de Zwaan, A. (1977) 'Anaerobic energy metabolism in bivalve molluscs' *Oceanography and Marine Biology* **15**, 103-187

de Zwaan, A. (1983) 'Carbohydrate catabolism in bivalves' in P.V. Hochachka (ed), *The Mollusca* Vol. 1, pp. 137-157 (Academic Press, New York)

de Zwaan, A. and Dando, P.R. (1984) 'Phosphoenolpyruvate-pyruvate metabolism in bivalve molluscs' *Molecular Physiology* **5**, 285-310

de Zwaan, A. and Putzer, V. (1985) 'Metabolic adaptations of intertidal invertebrates to environmental hypoxia (a comparison of environmental anoxia to exercise anoxia)' in M.S. Laverack (ed), *Physiological Adaptations of Marine Animals*, Vol. 39, pp. 33-62. Symposia of the Society for Experimental Biology (The Company of Biologists Limited, Dept of Zool., University of Cambridge)

de Zwaan, A. De Bont, A.M.T. and Verhoeven, A. (1982) 'Anaerobic energy metabolism in isolated adductor muscle of the sea mussel *Mytilus edulis* L' *Journal of Comparative Physiology* **149**, 137-143

de Zwaan, A., De Bont, A.M.T. and Hemelraad, J. (1983b) 'The role of phosphoenolpyruvate carboxykinase in the anaerobic metabolism of the sea mussel *Mytilus edulis* L' *Journal of Comparative Physiology* **153**, 267-274

de Zwaan, A., De Bont, A.M.T. and Nilsson, P. (1984) 'Anaerobic energy metabolism in two organs of fresh water mussel *Anodonta cygnea* L' Abstract in the First Int. Congress C.P.B. Liege Belgique

de Zwaan, A., De Bont, A.M.T., Zurburg, W., Bayne, B.L. and Livingstone, D.R. (1983a). 'On the role of strombine formation in the energy metabolism of adductor muscle of a sessile bivalve' *Journal of Comparative Physiology* **149**, 557-563

de Zwaan, A., Holwerda, D.A., and Veenhof, P.R. (1981) 'Anaerobic malate metabolism in mitochondria of the sea mussel *Mytilus edulis*' L. *Marine Biology Letters* **2**, 131-140

de Zwaan, A., Mohamed, A.M. and Geraerts, W.P.M. (1976) 'Glycogen degradation and the accumulation of compounds during anaerobiosis in the fresh water snail *Lymnaea stagnalis*' *Netherlands Journal of Zoology* **26**, 549-557

de Zwaan, A., Thompson, R.J. and Livingstone, D.R. (1980) 'Physiological and biochemical aspects of the valve snap and valve closure response in the giant scallop *Placopecten magellanicus*' *Journal of Comparative Physiology* **137**, 105-114

Ebberink, R.H.M., Zurburg, W.and Zandee, D.I. (1979) 'The energy demand of the posterior adductor muscle of *Mytilus edulis* in catch during exposure to air' *Marine Biology Letters* **1**, 23-31

Ellington, W.R. (1981) 'Energy metabolism during hypoxia in the isolated, perfused ventricle of the whelk, *Busycon contrarium*' *Journal of Comparative Physiology* **142**, 457-464

Ellington, W.R. (1982a) 'Metabolism at the pyruvate branch point in the radula retractor muscle of the whelk *Busycon contrarium*' *Canadian Journal of Zoology* **60**, 2973-2977

Ellington, W.R. (1982b) 'The recovery from anaerobic metabolism in invertebrates' *Journal of Experimental Zoology* **228**, 431-444

Ellington, W.R. (1982c) 'Cardiac energy metabolism in relation to work demand and habitat in bivalve and gastropod molluscs' in *Circulations. Respiration and Metabolism* (R. Gilles, ed.). pp. 356-366

Famme, P. (1981) 'Haemolymph circulation as a respiratory parameter in the mussel, *Mytilus edulis* L' *Comparative Biochemistry and Physiology* **69A**, 243-247

Famme, P. and Kofoed, L. H. (1982). 'Rates of carbon release and oxygen uptake by the mussel, *Mytilus edulis* L., in response to starvation and oxygen' *Marine Biology Letters* **3**, 241-256

Foreman, R.A. and Ellington, W.R. (1983). 'Effects of inhibitors and substrate supplementation on anaerobic energy metabolism in the ventricle of the oyster, *Crassostrea virginica*' *Comparative Biochemistry and Physiology* **74B**, 543-547

Gäde., G. (1980) 'The energy metabolism of the foot muscle of the jumping cockle, *Cardium tuberculatum* : Sustained anoxia versus muscular activity' *Journal of Comparative Physiology* **137**, 177-182

Gäde, G. (1981) 'Energy production during swimming in the adductor muscle of the bivalve, *Lima hians*: Comparisons with the date from other bivalve molluscs' *Physiological Zoology* **54**, 400-406

Gäde, G. (1983) 'Energy production during anoxia and recovery in the adductor muscle of the file shell, *Lima hians*' *Comparative Biochemistry and Physiology* **76B**, 73-78

Gäde, G. and Ellington, W.R. (1983) 'The anaerobic molluscan heart adaptation to environmental anoxia. Comparison with energy metabolism in vertebrate hearts' *Comparative Biochemistry and Physiology* **76A**, 615-620

Gäde, G. and Grieshaber, M.K. (1986) 'Pyruvate reductases catalyze the formation of lactate and opines in anaerobic invertebrates' *Comparative Biochemistry and Physiology* **38B**, 255-272

Gäde, G., Carlsson K.H. and Meinardus, G. (1984). 'Energy metabolism in the foot of the marine gastropod *Nassa mutabilis*, during environmental and functional anaerobiosis' *Marine Biology* **80**, 49-56

Gäde, G., Meinardus, G. and Carlsson, K-H. (1979) 'Tissue specific isozymes of octopine dehydrogenase from mantle muscle and optic lobe of *Loligo vulgaris* and octopine metabolism' *Animals and Environmental Fitness* (R. Gilles, ed.), vol. 2, pp. 53-54 (Pergamon Press, Oxford)

Gäde, G., Weeda, E. and Gabbott, P.A. (1978) 'Changes in the level of octopine during the escape response of the scallop *Pecten maximum* L.' *Journal of Comparative Physiology* **124**, 121-127

Grieshaber, M. (1978) 'Breakdown and formation of high-energy phosphates and octopine in the adductor muscle of the scallop, *Chlamys opercularis* (L.), during escape swimming and recovery' *Journal of Comparative Physiology* **126**, 269-276

Grieshaber, M. and Gäde, G. (1976). 'The biological role of octopine in the squid, *Loligo vulgaris* (Lamarck)' *Journal of Comparative Physiology* **126**, 269-276

Grieshaber, M. and Gäde, G. (1977) 'Energy supply and the formation of octopine in the adductor muscle of the scallop, *Pecten jacobaeus* (Lamarck)' *Comparative Biochemistry and Physiology* **58**, 249-252

Hoeger, U. and Mommsen, T.P. (1985) 'Role of free amino acids in the oxidative metabolism of cephalopod hearts' *Circulation, Respiration, and Metabolism* (R. Gilles, ed.). pp. 367-376. (Springer-Verlag, Berlin Heidelberg)

Holwerda, D.A. and de Zwaan, A. (1979) 'Fumarate reductase of *Mytilus edulis*' L. *Marine Biology Letters*. **1**, 33-40

Holwerda, D.A. and de Zwann, A. (1980) 'On the role of fumarate reductase in anaerobic carbohydrate catabolism of *Mytilus edulis* L' *Comparative Biochemistry and Physiology* **67B**, 447-453

Ho, M.S., and Zubkoff, P.L. (1982) 'Anaerobic metabolism of the ribbed mussel *Geukensia demissa*' *Comparative Biochemistry and Physiology* **73B**, 931-936

Kluytmans, J.H. and Zandee, D.I. (1983) 'Comparative study of the formation and excretion of anaerobic fermentation products in bivalves and gastropods' *Comparative Biochemistry and Physiology* **75B**, 729-732

Kluytmans, J.JH., De Bont, A.M.T., Janus, J. and Wijsman, T.C.M. (1977) 'Time dependent changes and tissue specificities in the accumulation of anaerobic fermentation products in the sea mussel *Mytilus edulis* ' L. *Comparative Biochemistry and Physiology* **58B**, 81-87

Kluytmans, J.H., De Bont, A.M.T., Kruitwagen, E.C.J., Ravestein, H.J.L. and Veenhof, P.R. (1983) 'Anaerobic capacities and anaerobic energy production of some mediterranean bivalves' *Comparative Biochemistry and Physiology* **75B**, 171-179

Kluytmans, J.H., Veenhof, P.R. and De Zwaan, A. (1975) 'Anaerobic production of volatile fatty acids in the sea mussel *Mytilus edulis* ' L. *Journal of Comparative Physiology* **104**, 71-78

Koorman, R. and Grieshaber, M. (1980) 'Investigations on the energy metabolism and on octopine formation of the common whelk, *Buccinum undatum* L. during escape and recovery' *Comparative Biochemistry and Physiology* **65B**, 543-547

Kooyman, D., van Zoonen, H., Zurburg, W., and Kluytmans, J. (1982) 'On the aerobic and anaerobic energy metabolism of *Littorina* species in relation to the pattern of intertidal zonation' in *Exogenous and Éndogenous Influences ɓn Metabolic and Neural Control* (A.D.F. Addink and N. Spronk, eds). Vol. 2 Abstr. Congr. Eur. soc. Comp. Physiol. Biochem. 3rd, pp. 134-135. (Noordwijkerhout, The Netherlands, Pergamon Press, Oxford)

Livingstone, D.R. and de Zwann, A. (1983) 'Carbohydrate metabolism of gastropods' *The Mollusca* (P.W. Hochachka, ed.). Vol. 1. pp. 177-242

Livingstone, D.R., de Zwaan, A., Leopold, M. and Marteijn, D. (1983). 'Studies on the phylogenetic distribution of pyruvate oxidoreductases' *Biochemical and Systematic Ecology* **11**, 415-425

Mangum, C. and Winkle, W. van (1973) 'Responses of aquatic invertebrates to declining oxygen conditions' *American Zoologist* **13**, 529-541

Matsushima, O., Katayama, H., Yamada, K. and Kado, Y. (1984) 'Occurrence of free D-alanine and alanine racemate activity in bivalve molluscs with special reference to intracellular osmoregulation' *Marine Biology Letters* **5**, 217-225

Meenashki, V.R. (1964) 'Aestivation in the Indian apple snail *Pila*. I. Adaption in natural and experimental conditions' *Comparative Biochemistry and Physiology* **11**, 379-386

Meinardus, G. and Gäde, G. (1981). 'Anaerobic metabolism of the common cockle, *Cardium edule* IV. Time dependent changes of metabolites in the foot and gill tissue induced by anoxia and electrical stimulation' *Comparative Biochemistry and Physiology* **70B**, 271-277

Meinardus-Hager, G. and Gäde, G. (1986) 'The pyruvate branchpoint in the anaerobic energy metabolism of the jumping cockle *Cardium tuberculatum* L.: D-lactate formation during environmental anaerobiosis versus octopine during exercise' *Experimental Biology* **45**, 91-110

Nicchita, C.V. and Ellington, W.R. (1983) 'Energy metabolism during air exposure and recovery in the high intertidal bivalve mollusc *Geukensia demissa granosissima* and the subtidal bivalve mollusc *Modiolus squamosus*' *Biological Bulletin* **165**, 708-722

Paynter, R.T., Ellis, L.L. and Bishop, S.H. (1984) 'Cellular location and partial characterization of the aline aminotransferase in ribbed mussel gill tissue' *Journal of Experimental Zoology* **232**, 51-58

Paynter, R.R., Karam, G.A., Ellis, L.L. and Bishop, S.H. (1985) 'Subcellular distribution of aminotransferases, and pyruvate branchpoint enzymes in gill tissue from four bivalves' *Comparative Biochemistry and Physiology* **82B**, 129-132

Sato, M. and Gäde, G. (1986) 'Rhodoic acid dehydrogenase: a novel into amino acid-linked dehydrogenase from muscle tissue of *Haliotis* species' *Naturwissenschaften* **73**, 207

Schultz, T.K.F., and Kluytmans, J.H. (1983) 'Pathway of propionate synthesis in the sea mussel *Mytilus edulis* L' *Comparative Biochemistry and Physiology* **75B**, 365-372

Schultz, T.K.F., Kluytmans, J.H., and Zandee, D.I. (1982) '*In vitro* production of propionate by mantle mitochondria of the sea mussle *Mytilus edulis* L.: overall mechanism' *Comparative Biochemistry and Physiology* **73B**, 673-680

Schultz, T.K.F., van Duin, M. and Zandee, D.I. (1983) 'Propionyl-CoA carboxykinase from the sea mussel *Mytilus edulis* L. Some properties and its role in the anaerobic energy metabolism' *Molecular Physiology* **4**, 216-230

Shick, J.M., Gnaiger, E., Widdows, J., Bayne, B.L. and A. de Zwaan. (1986) 'Activity and metabolism in the mussel *Mytilus edulis* L. during intertidal hypoxia and aerobic recovery' *Physiological Zoology* (in press)

Smith, P.J.S. (1985) 'Molluscan circulation: haemodynamics of the heart' *Circulation, Respiration, and Metabolism* (R. Gilles, ed.). pp.344-355 (Springer-Verlag, Berlin, Heidelberg)

Storey, K.B. (1985) 'A re-evaluation of the Pasteur effect: New mechanisms in anaerobic metabolism' *Molecular Physiology* **8**, 439-461

Storey, K.B., and Storey, J. (1983) 'Carbohydrate metabolism in cephalopod molluscs' *The Mollusca* (P.W. Hochachka, ed.). Vol. 1. pp. 91-130 (Academic Press, New York)

van den Thillart, G. and De Vries, I. (1985) 'Excretion of volatile fatty acids by anoxic *Mytilus edulis* and *Anodonta cygnea*' *Comparative Biochemistry and Physiology* Comp. Biochem. Phys. 80B, 299-301

Wieser, W. (1981) 'Responses of *Helix pomatia* to anoxia: Change of solute activity and other properties of the haemolymph' *Journal of Comparative Physiology* **141**, 503-509

Wijsman, T.C.M., Van der Lugt, M.C. and Hoogland, H.P. (1985) 'Anaerobic metabolism in the fresh water snail *Lymnaea stagnalis*: Haemolymph as a reservoir of D-Lactate and succinate' *Comparative Biochemistry and Physiology* **81B**, 889-895

Widdows, J., Bayne, B.L., Livingstone, D.R., Newell, R.I.E. and Donkin, P. (1979) 'Physiological and biochemical responses of bivalve molluscs to exposure to air' *Comparative Biochemistry and Physiology* **62A**, 301-308

Zaba, B.N. and Davies, I. (1984) 'Glycogen metabolism and glucose utilization in the mantle tissue of *Mytilus edulis*' *Molecular Physiology* **5**, 261-282

Zammit, V.A. and Newsholme, E.A. (1976) 'The maximum activities of hexokinase, phosphorylase, phosphofructokinase, glycerol phosphate dehydrogenase, lactate dehydrogenase, octopine dehydrogenase, phosphoenolpyruvate carboxykinase, nucleoside diphosphate kinase, glutamate-oxaloacetate transaminase and arginine kinase in relation to carbohydrate utilization in muscles from marine invertebrates' *Biochemical Journal* **160**, 447-462

Chapter 11

Arthropods

E. Zebe

Introduction

The major characteristic of arthropods is an external skeleton which dictates the organisation of the body and profoundly influences several important functions. With the exoskeleton, true appendages evolved, equipped with joints to act as levers against the substratum. Optimal functioning of these appendages required a special type of muscle which, unlike those of soft-bodied invertebrates, must be able to contract very rapidly and precisely but with only a modest amount of shortening. This requirement resulted in the co-evolution of cross-striated muscles, and a form of locomotion dependent on the efficient use of appendages.

Arthropods (and the vertebrates) are thus among the most mobile of all animals. Because of their high mobility, most arthropods avoid unfavourable environments or leave them when, for instance, the supply of oxygen becomes inadequate. The possession of a rigid and more or less impermeable integument also may have been helpful to some marine arthropods in establishing efficient ionic and osmotic regulatory systems that enabled them to invade fresh water. It is likely that this was an important step in gaining direct access to atmospheric oxygen. Since the oxygen content of air is at least 30 times higher than that of water and, in addition, virtually constant, aerial respiration, unlike aquatic respiration, guarantees an ample and reliable supply of oxygen.

Although generally arthropods encounter environmental hypoxia rather rarely, they may have to rely on anaerobic energy production in situations which require extreme muscular activity, such as flight. Then, an animal must undergo short-term functional anaerobiosis because its aerobic capacity is inadequate to supply the energy necessary.

Crustacea

Crustaceans in Hypoxic Environments

The crustaceans are an extremely diverse group of animals that have adapted to very different and, sometimes, to extreme environments. The latter include habitats where, temporarily or permanently, the pO_2 is very low.

The thalassinids *Callianassa californiensis* and *Upogebia pugettensis* inhabit muddy sediments in the tidal zone of the North American Pacific coast, where they build extensive systems of burrows and tubes. By continuously beating their paddle-like pleopods, they circulate water within their burrows and also accomplish an exchange with the water overlaying the sediment. This was demonstrated by monitoring pO_2 and pH in the burrows (Torres, Gluck and Childress, 1977). As a consequence of this activity, the previously anoxic layers close to the network of burrows appear to receive at least some oxygen. At low tide, the water in the burrows probably becomes stagnant. Then the shrimps seem to settle close to the interface between air and water within the burrows and, by action of their pleopods, cause more oxygen to dissolve in the water. They are able to maintain a low rate of respiration because of this activity and, as a consequence, only become more or less hypoxic instead of fully anoxic (Hill, 1981; Zebe, 1982).

Upogebia pugettensis, which has a relatively high metabolic rate, can survive for 24 h in oxygen-free water at 10°C, whereas *Callianassa californiensis*, with a substantially lower metabolic rate, tolerated at least 48 h of anoxia (Thompson and Pritchard, 1969; Zebe, 1982). *Callianassa jamaicense* is still more tolerant of anoxia (maximally 3 days at 25°C, Felder, 1979). The amphipods *Corophium volutator* and *C. arenarium* are typical inhabitants of the tidal zone. They live in burrows that extend to depths of 10 to 15 cm in anoxic mud or sand. They also obtain oxygen-rich water by the ventilatory activity of their pleopods and, very likely, do not remain long in environments that are truly anoxic. Even at low temperatures (5°C) they do not survive longer than 2 days of experimental anoxia. The semi-sessile tanaid *Tanais chevreuxi* may occasionally encounter hypoxia since it lives in tubes in very confined crevices. As it has a very low metabolic rate it is more tolerant of anoxic conditions than the very active species of the genus *Corophium* (Gamble, 1970).

Several bathypelagic copepods, mysids, decapods and amphipods permanently inhabit water containing less than 0.5 ml O_2/l. These crustaceans seem able, however, to maintain a fully aerobic metabolism even at such low oxygen concentrations. Their anaerobic capacity is only moderate, since at the normal temperatures of their habitats (7 to 15°C) they survive only 12 to 30 h in oxygen-free water (Childress, 1971, 1975).

Limnoria lignorum, an isopod species, is an interesting example because, like the shipworm *Teredo navalis*, a bivalve mollusc, it lives in submerged wood. The isopods are confined, however, to a peripheral zone and leave their burrows when hypoxia occurs. Shipworms, on the other hand, penetrate deeply into the wood and there they may sustain weeks of anaerobiosis. *Limnoria*, unlike *Teredo*, has only small stores of glycogen (George, 1966).

Barnacles are a special case. Several species are regularly exposed to air, sometimes for several days. Out of water, they do not seem to close their shells completely, so that a narrow gap remains. Since there is only a small amount of water within their shells, respiratory gas exchange can take place. Thus, even out of water, barnacles may obtain enough oxygen to maintain low rates of aerobic metabolism. In the absence of oxygen some species are reported to survive 3 to 4 days or even longer (Barnes, Finlayson and Piatigorsky, 1963).

The various examples described here clearly demonstrate that several crustaceans have found specific ways to cope with low oxygen concentrations in their environment. These are especially interesting from a comparative point of view. As a rule, species living in free water or on rocky shores, where there is always plenty of oxygen, are much more sensitive to hypoxia than those inhabiting less favourable places as, for instance, muddy sediments (Burke, 1979; Morris and Taylor, 1985). A correlation seems to exist between tolerance of hypoxia and metabolic rate of a species. It is not clear whether, in more tolerant crustaceans, hypoxia evokes a pronounced reduction of the normal metabolic rate as occurs in bivalves and annelids. Barnes and Barnes (1964) claim that barnacles also respond in this manner when anoxia arises but fail to mention the observations or measurement supporting their statement.

The anaerobic capacity of crustaceans is thus substantially lower than that of facultative anaerobic bivalves and annelids. Obviously, because of their high mobility, crustaceans do not have to endure hypoxia to the same degree as animals that are sluggish or even semi-sessile. This conclusion is confirmed by observations which show that at least some species respond to declining pO_2 by a marked increase in motility, apparently in an effort to move to normoxic conditions (Cook and Boyd, 1965; Anderson and Reisch, 1967; Costa, 1967).

Metabolic Effects of Hypoxia

The effect of experimental hypoxia or anoxia on various metabolic processes has been studied, mainly in decapods, by numerous investigators. In all species, lactate accumulates and is, quantitatively, the most important end-product of anaerobic metabolism. The concentrations measured in different species, in different tissues of the same animal and under different conditions may vary widely, however. As yet this has not been studied systematically.

The concentration of lactate in haemolymph may rise to 40 or even 60 mmoles/ml (Teal and Carey, 1967; Pritchard and Eddy, 1979; Gäde, 1984; Albert and Ellington, 1985). It is noteworthy that the level of lactate in the haemolymph sometimes exceeds that in the tissues, as in *Orconectes limosus* and *Menippe mercenaria* (Gäde, 1984; Albert and Ellington, 1985). This is presumed to be the result either of active excretion of lactate by the tissue cells or of the formation of complexes between lactate and Ca^{2+} or Mg^{2+} (Albert and Ellington, 1985). The phenomenon is not a general one, however, as in other species the rise of lactate in the haemolymph during anoxia is considerably lower than in the tissues.(Pritchard and Eddy, 1979; Zebe, 1982).

In some cases, crustaceans subjected to prolonged anoxia were observed to excrete substantial quantities of lactate into the ambient water (de Zwaan and Skjoldal, 1979; Zebe, 1982). It seems doubtful, however, that this occurs when the animals are living in their natural habitats. More likely, lactate excretion is due to the accumulation of extreme quantities under experimental conditions.

It is noteworthy that cirripedes, unlike all other groups of Crustacea studied so far, produce D-lactate instead of L-lactate. In *Balanus balanoides* it was found to rise from 40 to 85 mmol/g fresh weight after 2 to 4 days of experimental anoxia (Barnes et al., 1963).

Glycogen is the most important substrate of anaerobic metabolism in the crustacea but the amounts stored seem to be low in comparison with some bivalves (Hochachka, Freed, Somero and Prosser, 1971). At least in one case, the quantity of glycogen degraded during experimental anoxia corresponded well with the production of lactate (Zebe, 1982). Aspartate is utilised as an additional substrate by some species and results in the formation of alanine and succinate but, as rather small amounts are present, the process seems to be insignificant. On the other hand, muscles have large stores of phosphoarginine, so that degradation of this phosphagen may be an important additional source of energy (Gäde, 1984; Albert and Ellington, 1985).

So far, the formation of anaerobic end-products other than lactate has been reported in only one species of crustacean. *Cirrolana borealis* is a scavenging isopod that accumulates substantial quantities of succinate, propionate and acetate in addition to large amounts of lactate, and these end-products also appear in incubation water. It is noteworthy, however, that large quantities of succinate, propionate and acetate were found only in individuals that had been fed before the experiment. It is therefore likely that succinate, propionate and acetate were formed by microorganisms present in the food and/or the digestive tract of the isopods (de Zwaan and Skjoldal, 1979).

It seems safe to conclude that, in environmental hypoxia, energy production in Crustacea proceeds only by glycolytic degradation of glycogen; that is, the special modes of anaerobic metabolism characteristic of many bivalves and annelids are not present. Likewise, the formation of opines instead of lactate by a modification of glycolysis, as occurs in bivalves and annelids, has not yet been detected in crustaceans. Since glycolysis yields only 3ATP per glucose unit of glycogen as compared to six or seven in the formation of volatile fatty acids, the low efficiency of anaerobic energy production resulting in the accumulation of very large quantities of lactate may be responsible, at least in part, for the comparatively modest anaerobic capacity of crustaceans.

The results reported above were obtained by subjecting crustaceans to experimental hypoxia or anoxia. It is questionable whether such metabolic changes also occur in animals living in their natural habitats. Only a few direct observations have been published. In *Uca pugnax*, the level of lactate was shown to rise when, at low tide, the crabs remained in their burrows in stagnant water (Teal and Carey, 1967). In *Upogebia pugettensis* and in *Callianassa californiensis*, substantial accumulations of lactate occurred during emersion of their habitats. However, the quantities measured after 6 h were considerably lower than those produced in the same period of experimental anoxia (Zebe, 1982). This suggests that, normally, these shrimps encounter only hypoxia and not anoxia. But even if true anoxia should arise due to extreme weather conditions, it would not last longer than a few hours. In barnacles exposed to air no significant accumulation of lactate could be detected (Barnes and Barnes, 1964).

It seems evident that, in their normal environments, most crustaceans have to cope, at most, with only short periods of hypoxia.

Functional Anaerobiosis

In some situations, crustaceans may perform excessive muscular work that results in functional anaerobiosis. This can be demonstrated, for instance, by the marked rise of lactate in the haemolymph of crabs running in a tread-mill. The quantity of lactate produced, however, depends very much on the species. In *Gecarcinus* the level of lactate after running was only 2 mmole/ml (Smatresk, Preslar and Cameron, 1979), in *Cardisoma* it rose to 5 or 7 mmole (Wood and Randall, 1981) and in *Carcinus* concentrations as high as 13 mmolar were observed (Burke, 1979). No data are available on the accumulation of lactate in the muscles of these crabs and the extent of phosphagen utilisation is also unknown. In *Callinectes sapidus* the lactate concentration in the haemolymph was found to rise 14-fold during 30 to 60 minutes of swimming, while the pO_2 remained constant due to an increase of the respiratory rate. It appears that, in spite of functional anaerobiosis, the leg muscles of this crab can work for extended periods (Booth, McMahon and Pinder, 1982).

Escape swimming of macruran decapods is an example of burst-like muscular activity which, by necessity, results in functional anaerobiosis. Powerful contractions of the abdominal muscles flex the abdomen towards the cephalothorax and thus accelerate the animal backwards with great force.

The metabolic processes in the abdominal muscles accompanying escape swimming were analysed in detail in *Cherax destructor* (England and Baldwin, 1983). The first phase of escape consisted of about 18 vigorous tail flips and is characterised by a drop in phosphoarginine from 25 to 7 mmoles/g fresh weight. No lactate was detected. Thus, energy-rich compounds stored in the muscles are the sole source of energy in this phase. Tail-flips evoked by continuing stimulation were less powerful than those in the initial phase. While the remaining phosphagen was used up and the energy charge decreased somewhat, the lactate concentration started to rise. In the third phase, large-scale accumulation of lactate continued and resulted in rapid fatigue.

Similarly, in the shrimp *Crangon*, 50 per cent of the phosphagen was found to be utilised after performing about 10 tail-flips, whereas only small quantities of lactate accumulated (Onnen and Zebe, 1983; Kamp and Juretschke, 1987). The role of phosphagen in escape swimming was also confirmed in *Orconectes limosus* (Gäde, 1984).

The results of these investigations clearly demonstrate that phosphoarginine is the most important, if not the only energy source of the abdominal muscles performing normal work rather than the extremes that frequently occur in experiments.

Recovery from Anaerobiosis

When recovery sets in after anaerobiosis, various processes are initiated to restore the normal state of the tissues. It is most important to the survival of the organisms to replenish stores of phosphagen, and this is accomplished particularly quickly. Indeed, recharging of phosphagen may result in a second phase of anaerobic energy production by glycolysis, if the aerobic capacity of a muscle is low. Thus the concentration of lactate in the abdominal muscles of *Crangon* continued to increase in the initial phase of recovery (Onnen and Zebe, 1983). However, it is not clear whether this is a normal process or happens only after extreme muscular work as performed in experiments.

Usually, the respiratory rate rises to a maximum as soon as recovery begins. This indicates an increased rate of energy production which gradually returns to the normal resting level. Reoxidation of the lactate accumulated during anaerobiosis proceeds very slowly and frequently lactate concentrations in the haemolymph and in the tissues remain markedly elevated long after the oxygen consumption has become normal (Philipps, McKinney, Hird and MacMillan, 1977; McDonald, McMahon and Wood, 1979; Gäde, 1984; Albert and Ellington, 1974; Onnen, pers. comm.).

The very slow disappearance of lactate during recovery may suggest that its large-scale accumulation occurs only rarely in crustaceans living under natural conditions. However, striking differences between species seem to exist as Bridges and Brand (1980) have shown in a comparative study. In species burrowing in sediments, high levels of lactate after hypoxia declined to normal values within a few hours, whereas in non-burrowing species lactate was still elevated after 24 h. Obviously, these differences indicate an adaptation of the former to environments where hypoxia may occur.

In vitro *Investigations*

In addition to *in vivo* studies, several *in vitro* experiments have been performed to elucidate some details in the energy metabolism of crustaceans. The abdominal muscles of macruran decapods were found to be highly specialised for burst-like activity as they contain large stores of phosphagen and, in addition, the highest activities of arginine kinase measured in any muscle. Thus rephosphorylation of ADP can take place rapidly in spite of a rather low glycolytic capacity found at least in some species (Boulton and Huggins, 1970; Beis and Newsholme, 1975; Newsholme, Beis, Leech and Zammit, 1978; Onnen and Zebe, 1983; Morris and Baldwin, 1984).

In some cases enzymes important in anaerobic energy production have been isolated and their physical and catalytic properties analysed. Only a few of the investigations concerning lactate dehydrogenase (LDH) can be mentioned here. In the lobster, the presence of two LDH isoenzymes was demonstrated: one, found in abdominal muscles, has kinetic properties similar to heart-type LDH of vertebrates and a second one, typical of leg muscles, is similar to the muscle-type LDH (Eichner and Kaplan, 1977a, b). In the crayfish *Orconectes,* LDH isolated from the tail appears to be only slightly different from that in the heart (Scislowski, Biegniewska and Zydowo, 1982). England and Baldwin (1983) reported that the kinetic properties of LDH in the tail muscle of *Cherax* depend on the pH. At pH 7.0 it is similar to the M4 isoenzyme. These authors suggest that the change in catalytic properties may reflect the function of LDH in the different physiological states of the abdominal muscles. It thus would catalyse pyruvate reduction until the pH decreases to 6.5, when the enzyme would act as a lactate oxidase.

D-LDH of *Balanus nubilus* was shown to be a tetramer (like L-LDH) with kinetic properties characteristic of a pyruvate oxidase (Ellington and Long, 1978).

Chelicerata

Xiphosura

Limulus is the best-known representative of this small and ancient group. It lives in the subtidal region of the US Atlantic coast. In its search for prey it digs into muddy sediments and may thus encounter hypoxia. The eggs and young larvae which develop in sheltered beaches in the tidal zone also may have to cope with hypoxic conditions. As may be expected from its mode of life, *Limulus* is very tolerant of anoxia and survives 36 to 60 h without oxygen at 22°C (Falkowski, 1974). Likewise, Palumbi and Johnson (1982) reported that embryonic development continued and a high percentage of larvae moulted after up to 9 days of exposure to hypoxia (inadequately defined). It therefore seems surprising that adult *Limulus* appears to switch to anaerobic metabolism at a rather high residual pO_2 (50 mm Hg) as indicated by the rise of lactate concentration in the haemolymph, up to eight times the resting value (8.5 mmol/ml) within 12 h (Towle, Mangum, Johnson and Mauro, 1982).

In a more detailed study of *Limulus,* Carlsson and Gäde (1986) have shown that anaerobic metabolism is essentially like that described for crustaceans. During muscular activity as well as in anoxia, ATP is supplied by glycolysis and to some extent by the utilisation of phosphagen. The quantities of lactate found to accumulate (4 to 6 mmol/g fresh weight after 48 h of anoxia) are very small; however, in addition to lactate, the level of alanine rose after the onset of anoxia, whereas in functional anaerobiosis it remained unchanged. No significant accumulation of succinate was detected.

Carlsson and Gäde (1986) offer an explanation for the accumulation of alanine as well as lactate during anoxia. A large increase in the concentration of lactate seems impossible, because *Limulus* LDH has a relatively low K_m for lactate (10 mM). However, glycolytic flux can still continue if, by the action of glutamate-pyruvate transaminase (GPT), pyruvate is transformed to alanine. Glutamate can be regenerated from ketoglutarate, ammonia and NADH in the glutamate dehydrogenase reaction. The enzymes catalysing these reactions were shown to be present in relatively high activities. The authors propose that the rise in the alanine level, in turn, inhibits pyruvate kinase and thus causes a decline of the metabolic rate. However, in view of the high mobility of horse-shoe crabs it seems questionable whether, under natural conditions, they ever encounter periods of anoxia lasting many hours.

LDH of *Limulus* is a dimeric enzyme (m.wt. 70000-80000) catalysing the formation of D-lactate. Presumably, it is not homologous to L-LDHs of other animals (Siebenaller, Orr, Olwin and Taylor, 1983). It occurs in tissue-specific isoenzymes (Long and Kaplan, 1973; Carlsson and Gäde, 1985) which differ in their catalytic properties and may be functionally similar to the muscle-type and the heart-type enzymes, respectively, found in vertebrates. Carlsson and Gäde suggest that lactate produced by the working muscles may be transported to the hepatopancreas for reoxidation. Since the circulatory system of *Limulus* is highly developed such a 'Cori cycle' seems possible.

Arachnida

All arachnids are true terrestrial animals and, therefore, should never encounter environmental hypoxia except when inhabiting soil that is flooded. On the other hand, functional anaerobiosis probably occurs regularly as a consequence of the burst-like muscular activity necessary for catching and overwhelming prey or for escaping. Although the haemolymph of spiders has a comparatively high capacity for oxygen transport due to the presence of haemocyanin, circulation is impeded by the movements of legs which after being flexed by the contraction of the respective muscles must be extended hydraulically by haemolymph under high pressure. This may severely impair the supply of oxygen to the tissues (Anderson and Prestwich, 1975; Angersbach, 1978). Therefore, it seems likely that energy production in working muscle is mainly anaerobic. Leg muscles of spiders also contain very few mitochondria (Zebe and Rathmeyer, 1968) and very low activities of the enzymes involved in aerobic energy production, whereas their glycolytic capacity is high (Linzen and Gallowitz, 1975). Also, in the prosoma, energy production seems to be predominantly anaerobic, as suggested by the pattern of enzyme activities, which is similar to that in leg muscles (Prestwich and Ing, 1982).

Prestwich (1983a) demonstrated that, after running, the concentration of lactate was considerably elevated in the legs and in the prosoma of some spiders, but not in the opisthosoma. The extent of this accumulation depended on the species. Up to 15 mmol/g fresh weight were measured in the muscles and in the haemolymph of tarantulas after 10 min of running (Anderson and Prestwich, 1985). Active spiders also accumulated small quantities of glycerol phosphate besides lactate.

In a comparative study, Prestwich (1983b) tried to estimate the relative importance of anaerobic metabolism in the total energy production by calculations based on the data for oxygen consumption during short periods of maximal work and subsequent recovery. He found the proportion to vary between 55 and 93 per cent in different species and to depend on the surface of the book lungs.

More recently, Prestwich (1988a, b) succeeded in demonstrating that the spiders could maintain maximal running speeds for 20 seconds only. Thereafter, fatigue commenced until, after 2 minutes, they were unable to move at all. In the initial phase of maximal running, the phosphagen reserves were virtually depleted and also the level of ATP was clearly lowered. Prestwich calculated that the contribution of glycolysis to energy production was only 20%, due to the comparatively low glycolytic rate.

Recovery after complete fatigue proceeded only slowly. After 10 minutes, the concentrations of ATP and P-arginine were still far from normal and, after 20 minutes, the level of lactate in the prosoma had only declined to 50% of maximum, while in the opisthosoma it was not at all reduced. This observation indicates that lactate is transported to the opisthosoma because of the low oxidative capacity of the prosoma.

To summarise, a high anaerobic capacity and a low aerobic capacity appear to be characteristic features of energy metabolism in spiders. Thus, locomotion in spiders generally coincides with functional anaerobiosis.

Insecta

Environmental Hypoxia in Terrestrial Insects

Generally insects are perfectly adapted to aerial respiration and to terrestrial life and never experience environmental hypoxia. However, there are numerous exceptions, including species living in soil, endoparasites, saprophages and many larvae and adult insects that are specialised to aquatic life and sometimes also have 'returned' to aquatic respiration. Very little is known about the significance of hypoxia and anaerobic metabolism in most of these cases.

The carabid beetle *Pelochila borealis* is an example of a soil-living insect which regularly undergoes environmental anaerobiosis. As an imago, it hibernates in the Scandinavian mountains in places that are flooded periodically and subsequently freeze so that the beetles are enclosed in ice for several months (Conradi-Larsen and Somme, 1973). In the laboratory, they survive in nitrogen at 0°C for at least 100 days. During experimental anoxia the concentration of lactate in the haemolymph rises slowly until, after 40 to 60 days, it reaches a plateau at 25 to 30 mmol/ml. In addition, alanine also accumulates and may even exceed lactate concentrations, whereas glycerol phosphate and glycerol concentrations do not change. The authors also demonstrated that beetles dug out of their natural habitat in spring, while still embedded in ice, had accumulated lactate in quantities similar to those measured after several weeks of experimental anoxia.

The larvae of blowflies that feed on decaying meat are examples of saprophagous insects. However, no concrete information exists as to whether they do indeed encounter hypoxia while inhabiting carrion. They are supposed to survive 24 h of experimental anoxia (Meyer, 1977) accumulating alanine, glycerol phosphate, glycerol and sorbitol. The significance of these changes remains obscure, however, since at least 75 per cent of the larvae subsequently failed to go through metamorphosis (Meyer, 1980).

Gasterophilus larvae are parasites living in the stomach and in other parts of the digestive tract of horses, presumably under hypoxic conditions. They are reported to survive 17 to 25 days of experimental anoxia at 38°C (Dinulescu, 1932; Blanchard and Dinulescu, 1932). No recent investigations have been published.

Environmental Hypoxia in Aquatic Insects

Several aquatic insect larvae inhabiting the bottom of eutrophic lakes or ponds have adapted to aquatic respiration and must endure periods of hypoxia at least occasionally. As is well known, a thermocline may regularly arise in some lakes in summer which prevents the mixing of the water column. As a consequence the water beneath the thermocline (hypolimnion) becomes hypoxic or anoxic for periods of several weeks (Cole, 1921; Lindeman, 1942). Ice covering a lake in winter may also cause the oxygen content of the water to decline to zero and to remain anoxic for weeks. In such a situation, insect larvae living in the benthos - as, for instance, several chironomids - depend on anaerobic energy production. In experiments simulating natural conditions, an extraordinary anaerobic capacity of the larvae of *Chironomus plumosus* and *Ch. anthracinus* was demonstrated. These chironomids inhabit the profundal zone of eutrophic lakes. At 4°C many survived longer than 100 days in the absence of oxygen (Nagell and Landahl, 1978). Under anaerobic conditions, chironomids remain motionless and their glycogen reserves decrease with the duration of anaerobiosis (Augenfeld, 1967).

Anaerobic metabolism has been analysed in the larvae of *Chironomus thummi* which inhabit small ponds (Wilps and Zebe, 1976). At 12°C, they do not survive much longer than 2 days in the absence of oxygen; that is, they are not as tolerant of anoxia as some other chironomids are. They mobilised their large stores of glycogen (more than 1000 mmol glucose units/g dry weight) initially at a high rate which gradually decreased. Several metabolites accumulated. Within 48 h of anoxia, levels of lactate, ethanol and succinate rose to 20-40 mmol/g dry weight and that of alanine rose as high as 90 mmol. In addition, a high proportion of the ethanol was excreted into the ambient water. Altogether, the larvae produced more than 1000 mmol/g dry weight of ethanol within 48 h of anoxia, corresponding to about 80 per cent of the glycogen utilised. Clearly the anaerobic energy production of *Chironomus* larvae is characterised by alcoholic fermentation, whereas lactate and alanine are minor end-products of glycogen degradation. Succinate very likely originated from the oxidation of malate, which is present in considerable quantities in the larvae. It is not known, however, whether anaerobic metabolism proceeds uniformly in the whole body or whether differences exist between the tissues with respect to the end-products accumulated.

A more detailed investigation revealed that lactate and alanine were accumulated only in the initial phase of experimental anoxia. The rate gradually decreased until, after 6 h, the levels remained constant (Redecker and Zebe, 1988). Also, the utilisation of malate was maximal after the onset of anaerobiosis, and subsequently decreased. On the other hand, large-scale formation (and excretion) of ethanol, which started after a lag period, proceeded at a constant rate during the entire period of incubation. It is thus characteristic of long-term anaerobiosis.

In larvae incubated at constant, reduced oxygen concentrations, the accumulation of lactate and alanine was observed to commence at pO_2 of 8 Torr, indicating that, at this level, they become hypoxic. The sum of all anaerobic end-products produced at this pO_2 was still rather modest, as it only amounted to about 10% of that measured in a parallel experiment under full anoxia. These results demonstrate a remarkable ability of *Ch. thummi* larvae to utilise very low oxygen concentrations for maintaining a largely aerobic metabolism. In this respect, the presence in the haemolymph of a haemoglobin exhibiting a particularly high affinity for oxygen seems to be important. When the function of the haemoglobin was inhibited by carbon monoxide, the onset of hypoxia occurred sooner, at a pO_2 of 35 Torr (Zebe, unpublished data).

The anaerobic formation of ethanol by chironomid larvae was analysed in detail using homogenates and isolated mitochondria (Wilps and Schöttler, 1980). Homogenates were shown to degrade fructose-1,6-bisphosphate to ethanol and acetate. Arsenite inhibited the process, presumably by its effect on pyruvate dehydrogenase; the formation of acetate virtually stopped, and the accumulation of ethanol decreased substantially. A cytosolic fraction was found to degrade fructose-1,6-bisphosphate to pyruvate only. No ethanol was detected in this experiment, although the cytosol contains alcohol dehydrogenase at high activity. Evidently the presence of mitochondria is essential for the formation of ethanol by chironomid larvae. Isolated mitochondria were shown to transform pyruvate to ethanol and acetate at a high rate. The proportion of ethanol seemed to depend on the availability of NADH, since adding NADH together with alcohol dehydrogenase (ADH) to the medium resulted in a shift towards ethanol formation. These results show that the degradation of pyruvate to acetaldehyde proceeds only in the mitochondria because only they possess decarboxylating activity. The subsequent reduction of acetaldehyde to ethanol may also occur in the mitochondria, which contain ADH activity, as well as in cytosol. *In vitro,* the reducing equivalents are provided by oxidative decarboxylation of pyruvate by pyruvate dehydrogenase. It seems likely, however, that in the intact cell ethanol may result mainly from the reoxidation of cytosolic NADH.

It is evident that the mode of ethanol formation in chironomid larvae is different from that found in yeast, at least in its details. In addition, it also differs from the various mechanisms of alcoholic fermentation in bacteria. As Wilps and Schöttler (1980) point out, the conditions of anaerobic utilisation of endogenous glycogen by a highly differentiated metazoon must be quite unlike those found in single cells which utilise exogenous substrates. It seems likely that ethanol formation as a variety of anaerobic metabolism evolved rather recently in these insects, as the larvae have secondarily adapted to the conditions of life in the benthos of ponds or lakes.

Alcoholic fermentation, as opposed to normal glycolysis, may be advantageous for aquatic animals because ethanol easily permeates membranes and, therefore, can be excreted rapidly. Thus, no accumulation of large quantities of metabolic end-products occurs and such negative effects as, for instance, a disturbance of the osmotic balance are avoided. The price which has to be paid is the loss of energy-rich metabolites.

The larvae of *Chironomus thummi* and the oligochaete *Tubifex* live in similar habitats. Since the tubificids are much more tolerant of anoxia than the midge larvae it is interesting to compare the anaerobic metabolism of the two organisms. A striking difference in the rate of glycogen mobilisation is evident. In the larvae of *Chironomus thummi* 260 mmol of glucose units/g dry weight were utilised during 12 h of anoxia and 650 mmol in 48h, as compared with 75 and 260 mmol respectively in *Tubifex* incubated under identical conditions (Schöttler and Schroff, 1976; Wilps and Zebe, 1976). It is probable that this difference is due - at least to some extent - to the fact that *Tubifex* employs a mode of anaerobic metabolism which yields six to seven ATP molecules per glucose unit instead of the three of alcoholic fermentation. It seems that the larvae of *Ch. thummi* have specialised in living under conditions of very low oxygen concentrations rather than increasing their anaerobic capacity.

The aquatic larvae of the midge *Chaoborus*, which also have adapted to aquatic respiration, provides another example of a special mechanism of anaerobic metabolism. Living in lakes or ponds, these larvae undertake characteristic diurnal vertical migrations. During daylight they dig in the bottom-mud which frequently is anoxic. After dusk they appear in the free water and ascend to the layers of oxygen-rich water. There they stay to prey on zooplankton until they descend before dawn. The anaerobic capacity of *Chaoborus* larvae is low as they hardly survive longer than 24 h of experimental anoxia. This may be due to their exceptionally low glycogen reserves (about 100 mmol glucose units/g dry weight). On the other hand malate is present in extraordinary high concentrations and may function as a substrate for anaerobic energy production (Englisch, Opalka and Zebe, 1982).

After the onset of experimental anoxia malate is indeed found to decrease, at first rapidly, later more slowly. Concomitantly, succinate concentration rises in a reciprocal relationship with malate concentration. In addition, lactate and particularly alanine accumulate in large quantities (80-100 mmol and 160-180 mmol/g dry weight, respectively). However, there is no significant excretion of metabolites into the ambient water. During recovery, the concentrations of succinate, malate and phosphoarginine are restored to their normal values within 3h. In contrast alanine and lactate remain elevated even after 12h of recovery.

The nature of the metabolic processes resulting in the formation of the end-products described above is not yet fully understood. A remarkable discrepancy remains between the decrease of the glycogen content and the quantities of metabolites accumulated. Presumably the larvae utilise additional carbohydrates (trehalose?) which so far have evaded analytical procedures. Also, the origin of the amino groups (or ammonia) necessary for the production of the enormous quantities of alanine is not known. The reciprocal changes of malate and succinate suggest that malate is transformed into succinate via fumarate as the *Chaoborus* larvae switch from aerobic to anaerobic metabolism and *vice versa*. Presumably the reduction of fumarate to succinate is coupled to oxidative phosphorylation, as in facultative anaerobic annelids (Schroff and Schöttler, 1977). How the redox balance is maintained in this case is still obscure. It is important to note that, in *Chaoborus* larvae, succinate (unlike lactate and alanine) does not arise from the degradation of carbohydrates.

In their natural habitat these midge larvae certainly do not encounter periods of anoxia lasting 24 h. Therefore, the changes in the concentrations of the metabolites described are unphysiological, quantitatively at least. However, there is evidence that, in free-living larvae, the ratio of malate/succinate shifts characteristically, correlated with their diurnal migrations. In animals caught during the day, burrowing in the mud, this ratio was always close to 1. It steadily increased after dawn, as the larvae started swimming. Several hours later the ratio of concentrations became constant at values between 35 and 40. In larvae found in the mud shortly after dusk the ratio had already returned to 1. This rapid shift may have resulted from muscular work performed as the larvae had to burrow (Englisch and Zebe, 1983). It is not known whether the concentrations of other metabolites also change in the diurnal migrations as they do in experimental anoxia. The 'malate/succinate shift' probably is a metabolic specialty of the *Chaoborus* larvae, a consequence of their particular way of life.

Another type of aquatic insects is represented by the mosquito larvae. They respire air and, therefore, usually stay at the water surface. Only in the escaping response or while searching for food do they descend and thus interrupt the uptake of atmospheric oxygen. It seems unlikely that in their natural habitat they are ever exposed to environmental hypoxia. Accordingly, their anaerobic capacity is very low and they survive at most 4 h of experimental anoxia at $10°$. During this period they deplete more than 90% of their phosphagen reserves and reduce the level of ATP by 50% (Redecker and Zebe, 1988). Lactate is accumulated in large quantities, while alanine is a minor end-product. As well, a malate/succinate shift occurs, but it is quite insignificant. Evidently, mosquito larvae depend entirely on phosphagen utilisation and glycolysis for anaerobic energy production. Because of the ample and reliable supply of atmospheric oxygen, they did not have to adapt to environmental hypoxia and their short excursions into deeper layers of water result only in functional anaerobiosis.

The examples described here demonstrate that, rather recently in their evolution, insect larvae have developed remarkable modifications of normal energy metabolism as a means of invading hypoxic environments. It seems likely that still other interesting metabolic varieties will be discovered by future investigations of aquatic insect larvae.

Functional Hypoxia

The flight muscles of insects require extremely high rates of energy production for their functioning and, in addition, must work continuously for minutes or even hours. Their energy metabolism, therefore, has to be fully aerobic, as indicated by the high proportion of mitochondria present in the fibres (taking up to 40 or 50% of the total volume) and by numerous tracheoles in close contact with the mitochondria. On the other hand, their anaerobic capacity is virtually nil, since LDH is present only in trace amounts and the concentration of phosphagen is as low as that of ATP (Zebe and McShan, 1957; Delbruck, Zebe and Bücher, 1959; Kirsten, Kirsten and Arese, 1963) or, in some cases, such as bumble-bees, is entirely lacking (Surholt, personal communication). However, anaerobic energy production seems to be important for the development of tracheae and tracheoles when the flight muscle fibres differentiate prior to metamorphosis. This is suggested by the appearance of LDH in considerable activity during this phase (Brosemer, Vogel and Bücher, 1963; Beenakkers, van den Broek and de Ronde, 1975).

The functions of other muscular systems of the insect body are undoubtedly as diverse as those found in vertebrates and differences in energy metabolism may be expected. Unfortunately, very few concrete data exist to substantiate this presumption. In the leg muscles of cockroaches almost no activity of LDH was found (Smit, Becht and Beenakkers, 1967), whereas high activities of LDH were measured in the leg muscles of waterbugs and grasshoppers, as well as in the intersegmental muscles of several insect larvae (Zebe and McShan, 1957; Gäde, 1970). No direct observations of functional anaerobiosis occurring in these muscles have been reported.

Anaerobic metabolism has been investigated only in one case: in the femoral muscles in the hind legs of locusts. These muscles are responsible for developing the enormous power necessary for the rapid backward movement of the tibia which results in a leap. The fibres of these muscles contain relatively few mitochondria which means that the contractile elements take a maximal share of the volume. The concentration of phosphoarginine is high and so is the activity of LDH.

When locusts were continuously stimulated to jump, they were only able to perform 15 to 20 leaps in succession. Leaps decreased considerably in length until fatigue occurred (Kirsten, Kirsten and Arese, 1963; Hitzemann, 1979). Phosphoarginine concentrations fell from 20-25 mmol/g fresh weight to 7-9 mmol after only five leaps. Concomitantly, lactate rose by about 2 mmol. In the state of fatigue, virtually all phosphagen had disappeared except when the locusts were allowed to pause for several seconds between leaps. Even then lactate accumulated to only 3 to 5 mmol/g fresh weight. Also, in leaping *Collembola,* phosphoarginine is utilised as sole energy source (Zinkler and Schroff, pers. comm.). These observations are striking demonstrations of the particular importance of phosphagen in burst-like muscle activity. It can immediately supply the energy that is crucial for escape. In contrast, glycolysis is initiated comparatively slowly and, therefore, has only a minor support role.

Conclusions

Generally arthropods are very mobile animals depending largely on aerobic metabolism. Anaerobic energy production by utilising stored phosphagen or glycolysis is used in particular for short-term muscular work. The rather low anaerobic capacity is, in many species, compensated for by different, specialised modes of behaviour which still enable them to exploit hypoxic environments.

A large number of publications concerned with 'anaerobic metabolism in insects' have appeared that were not mentioned in this section. Most of these either demonstrate that some insect species, although they appear to succumb within one or two minutes of oxygen deprivation, seem to recover from several hours of experimental anoxia, or they deal with the metabolic changes which occur in this situation - for instance, the accumulation of pyruvate and glycerolphosphate. These papers were not taken into consideration, because, in the opinion of the author, they are not related to normal physiological processes.

References

Albert, J.L. and Ellington, W.R. (1985) 'Patterns of Energy Metabolism in the Stone Crab, Menippe mercenaria, During Severe Hypoxia and Subsequent Recovery' Journal of Experimental Zoology 234, 175-183

Anderson, J.F. and Prestwich, K.N. (1985) 'The Fluid Pressure Pumps of Spiders (Chelicerata, Aranea)' Zeitschrift für Morphologie der Tiere 81, 257-277

Anderson, J.F. and Prestwich, K.N. (1985) 'The Physiology of Exercise At and Above Maximal Aerobic Capacity in a Teraphosid (Tarantula) Spider Brachypelma smithi' Journal of Physiology 155, 529-539

Anderson, J.W. and Reisch, D.J. (1967) 'The Effects of Varied Dissolved Oxygen Concentrations and Temperature on the Wood-boring Isopod Genus Limnoria' Marine Biology 1, 56-59

Angerbach, D. (1978) 'Oxygen Transport in the Blood of the Tarantula Eurypelma californicum: pO_2 and pH During Rest, Activity and Recovery' Journal of Comparative Physiology 123, 113-125

Augenfeld, J.M. (1967) 'Effects of Oxygen Deprivation on Aquatic Midge Larvae Under Natural and Laboratory Conditions' Physiological Zoology 40, 149-158

Barnes, H. and Barnes, M. (1964) 'Some Relations Between the Habitat, Behaviour and Metabolism on Exposure to Air of the High-level Intertidal Cirripede Chthamalus depressus' Helgoländer Wissenschaftliche Meeresuntersuchungen 10, 19-27

Barnes, H., Finlayson, D.M. and Piatigorsky, J. (1963) 'The Effect of Desiccation and Anaerobic Conditions on the Behaviour, Survival and General Metabolism of Three Common Cirripedes' Journal of Animal Ecology 32, 233-252

Beenakkers, A.M.T., van den Broek, T. and de Ronde, T.J.A. (1975) 'Development of Catabolic Pathways in Insect Flight Muscles. A Comparative Study' Journal of Insect Physiology 21, 849-859

Beis, I. and Newsholme, E.A. (1975) 'The Contents of Adenine Nucleotides, Phosphagens and Some Glycolytic Intermediates in Resting Muscles from Vertebrates and Invertebrates' Biochemical Journal 152, 23-32

Blanchard, L. and Dinulescu, G. (1932) 'Le Metabolisme Glucidique Chez les Larves de Gasterophiles au Cours de l'Inanition et de l'Anaerobiose' Comptes Rendue de l'Societe Biologie, Paris 110, 340-343

Booth, C.E., McMahon, B.R. and Pinder, A.W. (1982) 'Oxygen Uptake and the Potentiating Effects of Increased Hemolymph Lactate on Oxygen Transport During Exercise in the Blue Crab Callinectes sapidus' Journal of Comparative Physiology 148, 11-121

Boulton, A.P. and Huggins, A.K. (1970) 'Glycolytic Activity in Crustaceans' Comparative and Biochemical Physiology 33, 491-498

Bridges, C.R. and Brand, A.R. (1980) 'The Effect of Hypoxia on Oxygen Consumption and Blood Lactate Levels in Some Marine Crustacea' Comparative and Biochemical Physiology 65A, 399-409

Brosemer, R.W., Vogel, W. and Bücher, T. (1963) 'Morphologische und Enzymatische Muster bei der Entwicklung indirekter Flugmuskeln von Locusta migratoria' Biochemica Zeitschrift 338, 854-910.

Burke, E.M. (1979) 'Aerobic and Anaerobic Metabolism During Activity and Hypoxia in Two Species of Intertidal Crabs' *Biological Bulletin* **156**, 157-168

Carlsson, K.H. and Gäde, G. (1985) 'Isolation and Characterization of Tissue-specific Isoenzymes of D-Lactate Dehydrogenase from Muscle and Hepatopancreas of *Limulus polyphemus*' *Journal of Comparative Physiology* **155B**, 723-732

Carlsson, K.H. and Gäde, G. (1986) 'Metabolic Adaptation of the Horse-shoe Crab, *Limulus polyphemus*, During Exercise and Environmental Hypoxia and Subsequent Recovery' *Biological Bulletin* **171**, 217-235

Childress, J.J. (1971) 'Respiratory Adaptations to the Oxygen Minimum Layer in the Bathypelagic Mysid *Gnathophausia ingens*' *Biological Bulletin* **141**, 109-121

Childress, J.J. (1975) 'The Respiratory Rates of Midwater Crustaceans as a Function of Depth of Occurrence and Relation to the Oxygen Minimum Layer off Southern California' *Comparative and Biochemical Physiology* **50A**, 787-799

Cole, E.A. (1921) 'Oxygen Supply of Certain Animals Living in Water Containing No Dissolved Oxygen' *Journal of Experimental Zoology* **33**, 293-320

Conradi-Larsen, E.M. and Somme, L. (1973) 'Anaerobiosis in an Overwintering Beetle, *Pelophila borealis*' *Nature (London)* **245**, 388-390.

Cook, R.H. and Boyd, C.M. (1965) 'The Avoidance by *Gammarus oceanicus* (Amphipoda, Crustacea) of Anoxic Regions' *Canadian Journal of Zoology* **43**, 971-975

Costa, H.A. (1967) 'Responses of *Gammarus pulex* to Modifed Environment III. Reactions to Low Oxygen Tensions' *Crustaceana* **13**, 175-189

Delbruck, A., Zebe, E. and Bücher, T. (1959) 'Über Verteilungsmuster von Enzymen des Energie liefernden Stoffwechsels im Flugmuskel, Sprungmuskel und Fettkörper von *Locusta migratoria* und ihre Cytologische Zuordnung' *Biochemica Zeitschrift* **331**, 273-296

Dinulescu, G. (1932) 'Recherches sur la biologie des Gasterophiles' *Annales Scientifique de Nature (10eme Ser. Zoologie)* **15**, 1-180

Eichner, R.D. and Kaplan, N.O. (1977a) 'Physical and Chemical Properties of Lactate Dehydrogenase in *Homarus americanus*' *Archives of Biochemistry and Biophysics* **181**, 490-500

Eichner, R.D. and Kaplan, N.O. (1977b) 'Catalytic Properties of Lactate Dehydrogenase in *Homarus americanus*' *Archives of Biochemistry and Biophysics* **181**, 501-507

Ellington, W.R. and Long, G.L. (1978) 'Purification and Characterization of a Highly Unusual Tetrameric D-Lactate Dehydrogenase from the Muscle of the Giant Barnacle *Balanus nubilus*' *Archives of Biochemistry and Biophysics* **186**, 265-274

England, W.R. and Baldwin, J. (1983) 'Anaerobic Energy Metabolism in the Tail Musculature of the Australian Yabby *Cherax destructor* (Crustacea, Decapoda, Parastacidae): Role of Phosphagen and Anaerobic Glycolysis During Escape Behaviour' *Physiological Zoology* **56**, 614-622

Englisch, H., Opalka, B. and Zebe, E. (1982) 'The Anaerobic Metabolism of the Larvae of the Midge *Chaoborus crystallinus*' *Insect Biochemistry* **12**, 149-155

Englisch, H. and Zebe, E. (1983) 'Diurnal Migrations and Metabolic Changes in *Chaoborus* Larvae' *Verhandlung der Deutschen Zoologischen Gesellschaft* 283

Falkowski, P.G. (1974) 'Facultative Anaerobiosis in *Limulus polyphemus*: Phosphoenolpyruvate Carboxykinase and Heart Activities' *Comparative and Biochemical Physiology* **49B**, 749-759

Felder, D.L. (1979) 'Respiratory Adaptations of the Estuarine Mud Shrimp *Callianassa jamaicense* (Crustacea, Decapoda, Thalassinidae)' *Biological Bulletin* **157**, 125-137

Gäde, G. (1970) 'Zur Aktivität der "Schlüsselenzyme" des Energiestoffwechsels in Muskeln' Thesis (University of Münster)

Gäde, G. (1984) 'Effects of Oxygen Deprivation During Anoxia and Muscular Work on the Energy Metabolism of the Crayfish *Orconectes limosus*' *Comparative and Biochemical Physiology* **77A**, 495-502

Gamble, J.C. (1970) 'Anaerobic Survival of the Crustaceans *Corophium volutator, C. arenarium* and *Tanais chevreuxi*' *Journal of the Marine Biological Association of the U.K.* **50**, 657-671

George, R.Y. (1966) 'Glycogen Content in the Wood-boring Isopod, *Limnoria lignorum*' *Science* **153**, 1262-1264

Hill, B. (1981) 'Respiratory Adaptations of Three Species of *Upogebia* (Thalassinidae, Crustacea) with Special Reference to Low Tide Periods' *Biological Bulletin* **160**, 272-279

Hitzemann, K. (1979) 'Untersuchungen über den Energiestoffwechsel in der Sprungmuskulatur von *Locusta migratoria*' Thesis (University of Munster)

Hochachka, P.W., Free, J.M., Somero, G.N. and Prosser, C.L. (1971) 'Control Sites in Glycolysis of Crustacean Muscle' *International Journal of Biochemistry* **2**, 125-130

Kamp, .G and Juretschke, H.P. (1987) 'An in vivo ^{31}P-NMR Study of the Possible Regulation of Glycogen Phosphorylase a by Phosphagen via Phosphate in the Abdominal Muscle of the Shrimp *Crangon crangon*'. *Biochemica et Biophysica Acta* **929**, 121-127

Kirsten, E., Kirsten, R. and Arese, P. (1963) 'Das Verhalten von freien Aminosäuren, energiereichen Phosphorsaure-Verbindungen und einigen Glykolyse- und Tricarbonsaure-Cyclus-Substraten in den Muskeln von *Locusta migratoria* bei der Arbeit' *Biochemica Zeitschrift* **337**, 167-178

Lindeman, R.L. (1942) 'Experimental Simulation of Winter Anaerobiosis in a Senescent Lake' *Ecology* **23**, 1-13

Linzen, B. and Gallowitz, P. (1975) 'Enzyme Activity Patterns in Muscles of the Lycosid Spider, *Cupiennius salei*' *Journal of Comparative Physiology* **96**, 101-109

Long, G.L. and Kaplan, N.O. (1973) 'Diphosphopyridine Nucleotide-linked D-Lactate Dehydrogenases from the Horse-shoe Crab, *Limulus polyphemus* and the Seaworm, *Nereis virens*. I. Physical and Chemical Properties. II. Catalytic Properties' *Archives of Biochemistry and Biophysics* **154**, 696-710, 711-725

McDonald, D.G., McMahon, B.R. and Wood, C.M. (1979) 'An Analysis of Acid-base Disturbances in the Haemolymph Following Strenuous Activity in the Dungeness Crab' *Journal of Experimental Biology* **79**, 47-58

Meyer, S.G.E. (1977) 'Concentrations of Some Glycolytic and Other Intermediates in Larvae of *Callitroga macellaria* (Diptera, Calliphoridae) During Anaerobiosis' *Comparative and Biochemical Physiology* **58B**, 49-55

Meyer, S.G.E. (1980) 'Studies on the Anaerobic Glucose and Glutamate Metabolism in Larvae of *Callitroga macellaria*' *Insect Biochemistry* **10**, 449-455

Morris, G.M. and Baldwin, J. (1984) 'pH Buffering Capacity of Invertebrate Muscle: Correlations with Anaerobic Work' *Molecular Physiology* **5**, 61-70

Morris, S. and Taylor, A.C. (1985) 'The Respiratory Response of the Intertidal Prawn *Palaemon elegans* to Hypoxia and Hyperoxia' *Comparative and Biochemical Physiology* **81A**, 633-640

Nagell, B. and Landahl, C.C. (1978) 'Resistance to Anoxia of *Chironomus plumosus* and *Ch. anthracinus* (Diptera) larvae' *Holoarctic Ecology* **1**, 333-336

Newsholme, E.A., Beis, I., Leech, A.R. and Zammit, V.A. (1978) 'The Role of Creatine Kinase and Arginine Kinase in Muscle' *Biochemical Journal* **172**, 633-637

Onnen, T. and Zebe, E. (1983) 'Energy Metabolism in the Tail Muscle of the Shrimp *Crangon crangon* During Work and Subsequent Recovery' *Comparative and Biochemical Physiology* **74A**, 833-838

Palumbi, S.R. and Johnson, B.A. (1982) 'A Note on the Influence of Life-history Stage on Metabolic Adaptation: The Responses of Limulus Eggs and Larvae to Hypoxia' in Bonaventura et al. (eds.), *Physiology and Biology of Horseshoe Crabs* (Alan Liss, New York)

Phillips, J.W., McKinney, R.J.W., Hird, F.J.R. and MacMillan, D.L. (1977) 'Lactic Acid Formation in Crustaceans and the Liver Function of the Midgut Gland Questioned' *Comparative and Biochemical Physiology* **56B**, 427-433

Prestwich, K.N. (1983a) 'Anaerobic Metabolism in Spiders' *Physiological Zoology* **56**, 112-121

Prestwich, K.N. (1983b) 'The Role of Anaerobic and Aerobic Metabolism in Active Spiders' *Physiological Zoology* **56**, 122-132

Prestwich, K.N. (1988a) 'The Constraints on Maximal Activity in Spiders I. Evidence against the Fluid Insufficiency Hypothesis' *Journal of Comparative Physiology* **B158**, 437-448

Prestwich, K.N. (1988b) 'The Constraints on Maximal Activity in Spiders II. Limitations Imposed by Phosphagen Depletion and Anaerobic Metabolism' *Journal of Comparative Physiology* **B158**, 449-456

Prestwich, K.N. and Ing, N.H. (1982) 'The Activities of Enzymes Associated with Anaerobic Pathways, Glycolysis and the Krebs Cycle in Spiders' *Comparative and Biochemical Physiology* **72B**, 295-302

Pritchard, A.W. and Eddy, S. (1979) 'Lactate Formation in *Callianassa californiensis* and *Upogebia pugettensis*' *Marine Biology* **50**, 249-253

Redecker, B. and Zebe, E. (1988) 'Anaerobic Metabolism in Aquatic Insect Larvae: Studies on *Chironomus thummi* and *Culex pipiens*' *Journal of Comparative Physiology* **B158**, 307-315

Schöttler, U. and Schroff, G. (1976) 'Untersuchungen zum anaeroben Glykogenabbau bei *Tubifex tubifex*' *Journal of Comparative Physiology* **108**, 243-254

Schroff, G. and Schöttler, U. (1977) 'Anaerobic Reduction of Fumarate by Mitochondria from Body Muscle of *Arenicola marina*' *Journal of Comparative Physiology* **116**, 325-336

Scislowski, P.W.D., Biegniewska, A. and Zydowo, M. (1982) 'Lactate Dehydrogenase from the Abdomen and Heart Muscle of the Crayfish *Orconectes limosus*' *Comparative and Biochemical Physiology* **73B**, 697-699

Siebenaller, J.F., Orr, T.L., Olwin, B.B. and Taylor, S.S. (1983) 'Comparison of the D-Lactate Stereospecific Dehydrogenase of *Limulus polyphemus* with Active-site Regions of L-Lactate Dehydrogenases' *Biochimica et Biophysia Acta* **749**, 153-162

Smatresk, N.J., Preslar, A.J. and Cameron, J.N. (1979) 'Post Exercise Acid-base Disturbance in *Gecarcinus lateralis*, A Terrestrial Crab' *Journal of Experimental Biology* **210**, 205-210

Smit, W.H., Becht, G. and Beenakkers, A.M.T. (1967) 'Structure, Fatigue and Enzyme Activities in "Fast" Insect Muscles' *Journal of Insect Physiology* **13**, 1857-1868

Teal, J.M. and Carey, F.G. (1967) 'The Metabolism of Marsh Crabs Under Conditions of Reduced Oxygen Pressure' *Physiological Zoology* **40**, 83-91

Thompson, P.K. and Pritchard, A.W. (1969) 'Respiratory Adaptations of the Two Burrowing Crustaceans, *Callianassa californiensis* and *Upogebia pugettensis*' *Biological Bulletin* **136**, 274-287

Torres, J.J., Gluck, D.L. and Childress, J.J. (1977) 'Activity and Physiological Significance of the Pleopods in the Respiration of *Callianassa californiensis* (Crustacea, Thalassinidea)' *Biological Bulletin* **152**, 134-146

Towle, D.W., Mangum, C.P., Johnson, B.A. and Mauro, N.A. (1982) 'The Role of the Coxal Gland in Ionic, Osmotic and Regulation in the Horseshoe Crab, *Limulus polyphemus*' in Bonaventura et al (eds.), *Physiology and Biology of Horseshoe Crabs* (Alan R. Liss, New York)

Wilps, H. and Schöttler, U. (1980) '*In vitro* Studies on the Anaerobic Formation of Ethanol by the Larvae of *Chironomus thummi* (Diptera)' *Comparative and Biochemical Physiology* **67B**, 239-242

Wilps, H. and Zebe, E. (1976) 'The End-Products of Anaerobic Carbohydrate Metabolism in the Larvae of *Chironomus thummi thummi*' *Journal of Comparative Physiology* **112**, 263-272

Wood, C.M. and Randall, D.J. (1981) 'Haemolymph Gas Transport, Acid-base Regulation and Anaerobic Metabolism During Exercise in the Land Crab *(Cardisoma carnifex)*' *Journal of Experimental Zoology* **218**, 23-25

Zebe, E. (1982) 'Anaerobic Metabolism in *Upogebia pugettensis* and *Callianassa californiensis* (Crustacea, thalassinidea)' *Comparative and Biochemical Physiology* **72B**, 613-617

Zebe, E. and McShan, W.H. (1957) 'Lactic and α-Glycerophosphate Dehydrogenases in Insects' *Journal of General Physiology* **40**, 779-790

Zebe, E. and Rathmeyer, W. (1968) 'Elektronenmikroskopische Untersuchungen an Spinnenmuskeln' *Zeitschrift für Zellforschung und Mikroscopische Anatomie* **92**, 377-387

Zwaan, A. de and Skjoldal, H.R. (1979) 'Anaerobic Metabolism of the Scavenging Isopod *Cirrolana borealis*' *Journal of Comparative Physiology* **129**, 327-331

Chapter 12

The Metabolic Arrest and Channel Arrest Concepts of Defence against Hypoxia in Vertebrates

P.W. Hochachka

Origins of the Metabolic Arrest Concept

Vertebrate animals profoundly resistant to hypoxia can be broadly categorised into two groups on the criteria of their adaptive strategies. In one category, illustrated by high altitude adapted animals and, perhaps unexpectedly, by patients suffering from chronic hypoxia, metabolic mechanisms are directed towards sustained oxidative function despite potentially chronic oxygen limitations, with little or no extension in tissue anaerobic capacities (Hochachka, 1985). On the other hand, numerous vertebrates either at risk of total anoxia (for example, some fishes, diving aquatic turtles) or at the disadvantage of a limited and fixed amount of oxygen available over specified time periods (as in diving marine mammals), direct tissue metabolic strategies towards sustained anaerobic function. Many such ectothermic species display mechanisms of protection against hypoxia so effective that they are often referred to as 'good' animal anaerobes or as facultative anaerobes (Hochachka and Somero, 1984). In this adaptive response, the two most fundamental metabolic problems are (i) conservation of fermentable substrate, and (ii) avoidance of self-pollution by production of undesirable end-products. The first problem arises from the energetic inefficiency of anaerobic metabolism, for the yield of adenosine triphosphate (moles of ATP generated per mole of substrate fermented) is always modest compared with oxidative metabolism. For this reason in most animal tissues, glycogen or glucose utilisation rates vary inversely with oxygen availability As a result, if demands for ATP remain unchanged during oxygen lack, carbohydrate consumption rates necessarily have to rise drastically. (This is the Pasteur Effect, which is defined as the inhibition of carbohydrate consumption when O_2 concentrations are high and includes the opposite situation: increased anaerobic glycolysis when O_2 is limiting. A reversed Pasteur Effect is defined as decreased or unchanging glycolytic flux when O_2 is limiting. The Crabtree Effect is defined as the inhibition of O_2 consumption by activated carbohdyrate fermentation. Current information indicates that several kinds of controlling mechanisms may cause the Pasteur Effect and that these may be cell-line or species-specific. However, neither

O_2 nor carbohydrate *per se* play any direct regulatory roles in these effects. Metabolite and enzyme control mechanisms accounting for a reversed Pasteur Effect are not well understood (see Hochachka and Guppy, 1987; Hochachka, 1985).

In this situation, potentially large carbohydrate depletions are minimised in facultative anaerobes (i) by storing more glycogen, (ii) by utilising more efficient fermentation pathways, and/or (iii) by reversing the classical Pasteur Effect so as to allow precipitous drops in ATP turnover rates during periods when oxygen is limiting. All three mechanisms are demonstrably useful by the criterion of anoxic survival time; however, the first two in principle could not extend hypoxia tolerance by more than a 3-4 fold factor. In contrast, metabolic arrest processes in goldfish or lungfish can increase anoxia tolerance by some five fold and, in diving turtles, by some 60 fold. This strategy is expressed at its limit in anhydrobiotes whose maximum anoxia tolerance coincides with entrance into a fully arrested or ametabolic state (Hochachka and Guppy, 1987).

The 'end-products' problem is more difficult to unravel because it requires assessing the relative effects of organic (usually anionic) end-products against those of H^+ *per se,* as well as the metabolic sites of H^+ production, the pathways for later proton disposition, and the H^+ stoichiometry of fermentation pathways. In addition, any metabolic effects of concentration changes in other 'strong ions' such as Na^+, K^+, Ca^{2+} and Cl^- must be evaluated. From careful analysis of good animal anaerobes, however, it is now evident that the potentially perturbing direct or indirect effects of metabolic end-products are minimised by only a handful of mechanisms, which include (i) utilising fermentation pathways which allow more ATP to be turned over per mole of H^+ generated than in classical glycolysis, (ii) tolerating proton production by improved tissue buffering capacity, or by adjusting strong ion concentrations in the intracellular fluid (ICF) relative to the extracellular fluid (ECF), (iii) minimising end-product accumulation by recycling it for further metabolism or excretion, (iv) utilising H^+ consuming reaction pathways, and (v) depressing metabolic rates during anoxia (Hochachka and Somero, 1984; Hochachka, 1985).

Whereas any of the first four mechanisms listed above are obviously advantageous, potentially yielding up to a several fold improvement in anoxia tolerance, it is evident that, by harnessing metabolic arrest mechanisms, (v) above, an organism not only reduces rates of carbohydrate depletion in fruitless fermentations, it also automatically reduces rates of formation of anaerobic end-products, including H^+. In anoxic goldfish, for example, the rate of proton production is reduced by five fold due to metabolic depression, while in the turtle it is reduced by nearly two orders of magnitude.

In principle metabolic arrest could be achieved by blocking either the ATP generating or the ATP utilising arm of the ATP cycle (Figure 12.1). In practice, hypoxia-tolerant vertebrate tissues seem to rely on the former and therefore characteristically display a reversed Pasteur Effect (Hochachka, 1985). Why this should be so is unclear, nor is it known if this choice is universally made by animal anaerobes. Nevertheless, a fairly clear picture emerges from our analysis of

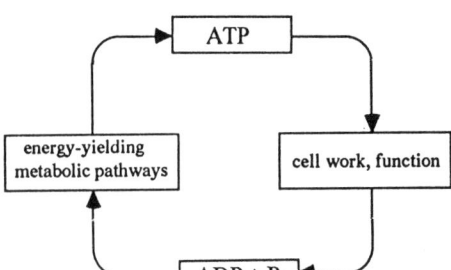

Figure 12.1 The ATP cycle.

phylogenetically quite diverse organisms. Of several processes contributing to the hypoxia tolerance of animal anaerobes, metabolic arrest provides, by far, the most effective protection against O_2 lack. Alone among known protective measures, metabolic arrest supplies resolution to both key problems faced during hypoxia - substrate conservation and end-product accumulation. That is why we concluded some time ago (Hochachka, 1982) that for extended survival without oxygen, hypoxia-tolerant systems (whether considered at the organismal, organ, or cellular level) must be able to switch-down metabolically or even switch-off; for convenience, we term this the metabolic arrest hypothesis of defence against hypoxia.

Is Metabolic Arrest a Realisable Intervention Strategy?

Interestingly, similar positions have been independently deduced by other investigators using entirely different systems. Thus, we have found 'arrest'-type concepts of protection against hypoxia expressed in the scientific literature on torpor and hibernation and in the clinical literature on cardiac arrest, stroke, acute renal failure (ARF) and liver ischaemia. It is commonly assumed, for example, that ischaemic myocardial damage is a function of work load or metabolic rate, which is why several intervention procedures all aim to minimise tissue damage by reducing myocardial energy requirements during the O_2 limiting period.

To some degree all such interventions are helpful, and so consistent with expectations based on an analysis of animal anaerobes (Hochachka, 1985). One of the most convincing tests of the metabolic arrest hypothesis is provided by studies using ischaemic rat kidney to model ARF (Brezis, Rosen, Spokes, Silva and Epstein, 1984). During ischaemic ARF, reductions in renal blood flow lead directly to reductions in delivery of O_2 and substrates to the tissue. Since the main ATP utilising processes of the kidney are membrane-coupled ion pumps, the metabolic arrest hypothesis leads to an explicit, testable prediction; that is, a fractional decrease in demands for ATP by ion pumps should yield an enhanced hypoxia tolerance of similar magnitude. In the mammalian kidney, the medullary thick ascending limb (mTAL) of Henle's loop is the most O_2 dependent segment of the nephron and during perfusion of the isolated organ, easily demonstrable, hypoxia-induced damage to mTAL cells occurs within about 90 minutes. However, this hypoxia sensitivity can be minimised by perfusion with ouabain (a specific inhibitor of $Na^+ K^+$ ATPase) or by reducing ion pumping work by preventing glomerular filtration. Conversely, experimentally increasing membrane permeability increases the ATP costs of ion transport and consequently increases hypoxia sensitivity; nevertheless, the hypoxia-induced damage is again avoidable by ouabain inhibition of the Na^+ pump. An impressive protection can even be effected against KCN-induced lesions by simultaneous ouabain inhibition of K^+ and Na^+ pumping (Brezis et al., 1984).

These exciting and provocative data can be taken as empirical attempts to establish in a mammalian organ a hypoxia tolerance rather more characteristic of hypoxia adapted ectothermic animals. While the results support the general predictions of the metabolic arrest hypothesis, they also emphasise a most significant difference: thus far, no metabolically-arrested mammalian system can sustain severe hypoxia for more than minutes to hours, while hypoxia tolerant ectotherms are able to sustain comparable or greater degrees of O_2 lack for days, weeks and, in some cases, even months (Hochachka and Somero, 1984). Such large differences in hypoxia tolerance between standard mammalian preparations and those from facultatively anaerobic animals means that something is still missing in our analyses of strategies for protecting tissues against O_2 lack. Further comparisons of hypoxia tolerant and hypoxia sensitive animals indicate that the missing element is to be found at the interface between cell membrane functions and cell metabolism; in effect, hypoxia sensitive systems behave as if these two critical cell functions become uncoupled (Hochachka, 1985, 1986).

Balancing Metabolism with Membrane Functions

To illustrate the critical role of close coupling between cell metabolism and cell membrane functions in hypoxia sensitive systems, we can find no more convenient a model that the mammalian brain, for which knowledge of events during various kinds of energy perturbations is fairly extensive (Siesjo, 1981). In the complete cerebral ischaemia of cardiac arrest (when, for example, the electroencephalograph or EEG becomes isoelectric within 15-25 seconds) an electrically silent period preceeds a massive outflux of K^+ from the neurons, and a flux of Na^+ into the ICF.

The breakdown in normal ion gradients is caused by ATP insufficiency (due to the failure of membrane ion pumps) and occurs when regional cerebral blood flows are reduced to 10% of normal or less. When the $[K^+]$ rises in the ECF to about 12-13 mM, membrane potential changes appear to be large enough (i) to activate (or open) voltage-dependent Ca^{2+} pores (or channels) and consequently (ii) to develop a large uncontrollable influx of Ca^{2+}, a cation which at abnormally high cytosolic concentration acts as a cellular toxin (Hochachka, 1986). Although high cytosolic $[Ca^{2+}]$ may disrupt various intracellular functions, its activation of phospholipases A1 and A2 may lead to the most far-reaching damage to cell structure and controlled cell metabolism (Siesjo, 1981).

Uncontrolled, Ca^{2+} activated phospholipase catalysis gradually hydrolyses membrane phospholipids, and so disrupts cell and mitochondrial membranes, releases telling free fatty acids (such as arachidonic acid) and further potentiates ion redistributions in the O_2-limited brain (Figure 12.2).

The debilitating impact of O_2 limitation to the brain is also evident in similar uncontrolled cascades in other hypoxia-sensitive mammalian tissues and organs. In myocardial ischaemia, for example energy-deficiencies and membrane failures are indicated by ICF and ECF changes in $[Na^+]$ and $[K^+]$ as well as by a large influx of Ca^{2+}, a loss of sarcolemmal Ca^{2+}, and a disruption of mitochondrial Ca^{2+}

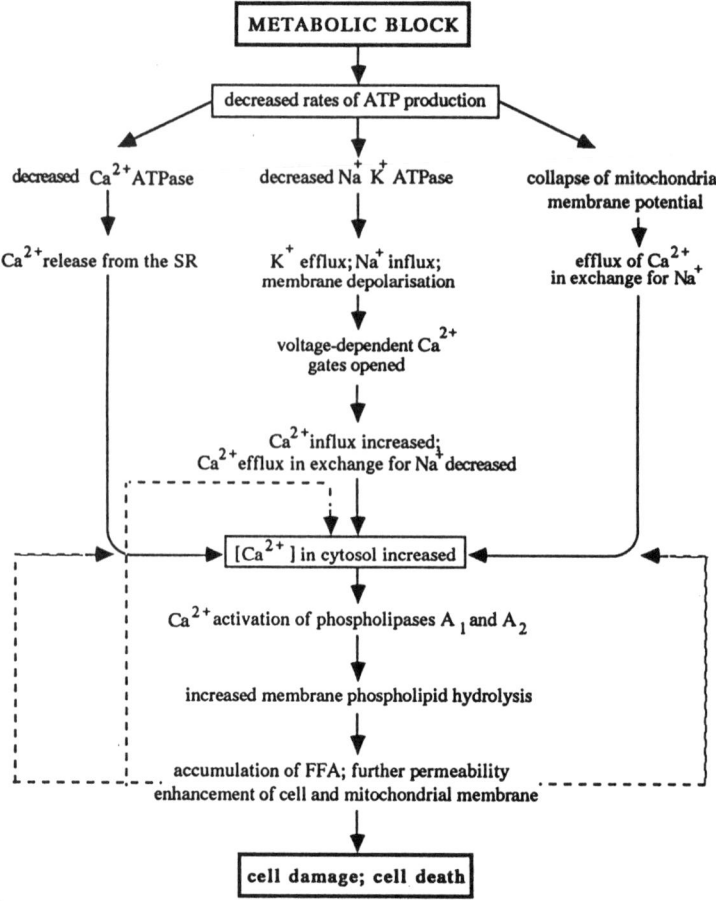

Figure 12.2 A summary of probable metabolic events progressing in hypoxia-sensitive cells from the initial energetic consequences of O_2 limitation to cell damage and cell death. (The summary is based on analysis in the text and is constructed from various studies of hypoxia-sensitive mammalian tissues. From Hochachka, 1986.)

homeostasis. Analogous disruptions of membrane function with consequent translocation of Ca^{2+} and other ions between intra- and extracellular pools occur in hypoxic liver, kidney, smooth and skeletal muscle and, presumably, generally in mammalian organs and tissues during O_2 lack (Hochachka, 1985, 1986).

Coupled Membrane and Metabolic Functions in Hypoxia Tolerant Systems

It is an instructive observation that, in the brain and other organs of good ectothermic anaerobes during hypoxia, the above kinds of failures of membrane functions do not occur at all, or develop very much more slowly. Something about cell membranes in facultatively anaerobic tissues is obviously different. When O_2 is lacking, these membranes either (i) are more impermeable to ions or (ii) their ionic pumping capacity can pace the drift of ions towards electro-chemical equilibrium of the ICF and ECF. Stabilised ion gradients almost certainly cannot be due to accelerated ion pumping because ATP turnover rates are lowered in the metabolically arrested states typical of animal anaerobes in anoxia. That is why we are led to the conclusion that the above membrane-based differences in the effects of hypoxia are primarily due to different permeability barriers, and are in fact an expression of a basic difference between cell membranes of hypoxia-tolerant ectotherms and hypoxia-sensitive endotherms. For convenience, and because membrane permeability to ions depends upon densities of functional, ion-specific pores or channels, we term this the channel arrest concept of defence against hypoxia. This channel arrest concept well explains studies showing that even if ion gradients do decline in hypoxia-tolerant ectotherms in hypoxia, the process is much slower than in mammals and that, in the extreme, ion gradients obviously remain stable, even after days of anoxia, despite metabolic arrest keeping ATP turnover rates too low to account for the stability with ion pumping (Hochachka, 1985; Hochachka and Guppy, 1987). This interpretation is also closely consistent with recent studies of the origins of endothermy (Else, 1984) which show (i) that ATP turnover rates and ouabain sensitivities of homologous tissues in mammals are about five times higher than in ectothermic reptiles, but (ii) that the permeabilities of mammalian cell membranes are also some several fold greater than in ectotherms. Along with this author, we argue that the latter explains the former; that is, that one cost of endothermy is a higher rate of thermogenesis arising in part from 'leaky' membranes, and from the consequent necessity for higher ion pumping rates and higher ATP turnover rates. In this view, 'leaky' membranes may be adaptive in most endotherms since they form part of an O_2-fuelled metabolic furnace while non-leaky, channel arrested membranes are adaptive in animal anaerobes because they allow metabolic arrest without the risk of a breakdown in ion regulation. In fact, the implication that leaky membranes are inherent to endothermy and in effect are an important cause of increased hypoxia sensitivity, is exactly what is predicted by the channel arrest concept and appears to supply the element -

stabilised membrane functions - missing in earlier attempts to extend the hypoxia tolerance of mammalian tissues and cells (Brezis et al., 1984).

Now we have arrived at the kind of generalisation that perhaps could only come from comparative studies; namely, that the coupling of metabolic arrest with channel arrest is a minimal requirement for establishing in mammalian tissues the hypoxia tolerance of ectothermic anaerobes. In effect, the idea is that intervention strategies should target the same two processes that are adjusted by ectothermic anaerobes to extend greatly their own hypoxia tolerance. It seems a simple enough strategy, one eminently successful in animal anaerobes; but is it realisable with endothermic systems?

Interestingly, interventions aimed precisely at such a goal have been tried, with various Ca^{2+} channels being targeted (using channel-specific pharmacological agents) in order to block uncontrolled Ca^{2+} fluxes and uncontrolled elevation of $[Ca^{2+}]$ in the ICF (Figure 12.1). Ca^{2+} channel arrest in such manoeuvres is thus one line of defence designed to protect tissues against hypoxia, and it is often coupled with cold-induced metabolic arrest. Although such interventions applied to any mammalian tissue or organ so far investigated usually are 'protective' against hypoxia for short time periods, for prolonged protection they are of limited success, presumably because hypothermia in itself strongly perturbs membrane and metabolic functions (Hochachka, 1986). Since hypothermia *per se* is damaging and its disrupting effects in combination with hypoxia are probably enhanced, it is not the metabolic arrest mechanism of choice.

An alternative mechanism of blocking ATP turnover by blocking ATP utilisation, as in the ischaemic ARF model, may be a more effective strategy but thus far has not been extensively assessed, and certainly places focus on the wrong side of the ATP = ADP + P_i cycle; that is, opposite to that favoured by animal anaerobes. So, as of this writing, no clinical interventions have been able successfully to extend the hypoxia tolerance of tissues of terrestrial mammals anywhere near the limit achievable by ectothermic anaerobes. In fact, because of the fundamental metabolic and membrane adjustments seemingly required, it is tempting to conclude that this goal is not realisable with mammalian tissues. We do not succumb to this temptation, however, for a very good reason: because one group of endotherms apparently are able to resolve the problems of coupling a reversed Pasteur Effect with stabilised membrane functions. These are diving marine mammals, to which we will now briefly turn.

Diving Response as a Physiological Defence Against Hypoxia

In the case of breath-hold diving in marine mammals, the two key requirements for any O_2-limited cell, tissue, or organism (conserving fermentable substrate and minimising end-product accumulations) in part are met by a set of physiological reflexes which taken together are termed the diving response and which can be viewed as the animal's first line of defence against potential hypoxia.

In laboratory studies of simulated diving in aquatic animals, the diving response involves apnoea, bradycardia, and peripheral vasoconstrictions, all of which are always observed. Metabolic effects of the diving response include (i) preferential redistribution of O_2 and blood-borne substrates to specific organs and tissues, (ii) increased accumulation of anaerobic end-products such as lactate in hypoperfused regions of the body, concomitant with declining plasma glucose levels, and (iii) a distinct lactate washout profile during recovery with a concomitant hyperglycaemia evident in the plasma (Hochachka and Somero, 1984).

Although some of these patterns during simulated diving are well described in studies over 40 years old, controversy has now arisen concerning when, and even if, this reflex diving response is used during voluntary diving of aquatic mammals and birds. Recent work on grey seals suggests that a large part of this controversy may arise from comparing a similar process (breath-hold diving) in the presence and absence of exercise. This may explain why studies using only diving as a variable tend to favour the validity of the classical diving response, while studies involving both variables tend to favour an 'exercise' model of diving, viewing diving as rather analogous to any other exercise, such as surface swimming, for example (Castellini, Murphy, Fedak, Ronald, Gofton and Hochachka, 1985).

At least for the Weddell seal, this controversy is now resolved, thanks to the development of a compact microcomputer backpack which allows monitoring of heart rate, swimming speed, diving depth and diving duration, and which can operate a peristaltic withdrawal pump for obtaining sequential blood arterial samples (before and after injections of isotopically labelled metabolites) at any time during voluntary diving under the sea ice (Guppy, Hill, Schneider, Qvist, Liggins, Zapol and Hochachka, 1986; Qvist, Hill, Schneider, Falke, Liggins, Guppy, Elliot, Hochachka and Zapol, 1986). All of the physiological and biochemical data arising from these studies are fully consistent with the hypothesis (i) that the classical Irving-Scholander diving response is employed to reduce cardiac output and redistribute flow during both short (feeding) and prolonged (exploratory) diving, and (ii) that perfusion rates are graded according to diving modes (short *versus* long; fast *versus* slow).

The Internal SCUBA Tank

An interesting extension of this diving response involves the control of haemoglobin (Hb) content of the blood (Qvist et al., 1986). During the first 10-15 min of both long and short dives, blood Hb content increases by nearly 60% (haematocrit increases from about 40% to as high as 65%). In terrestrial species, such as the sheep and horse, the spleen is known to function as a dynamic red blood cell (RBC) reservoir, at rest containing 26% and 52% of the total RBC mass, respectively. Excitement, exercise or catecholamines cause the spleen to contract and increase the haematocrit by 25% or even more. It appears that a similar mechanism operates during diving, when the Weddell seal capitalises upon sympathetic vasoconstriction of the peripheral vaculature to induce constriction of its very large spleen. In this way, over the first 10-15 min of diving the seal gradually injects more and more of the 50-60% fraction of RBC mass which is stored (oxygenated) in the spleen. The process is so finely tuned (O_2 uptake balancing O_2 delivery) that for the first 10-15 min, the O_2 content of the blood, which would otherwise drop due to O_2 consumption, is maintained at a constant level by precise pacing of injection rates of oxygenated RBC. Once the full RBC mass is circulating, O_2 content declines and, because the lungs are collapsed and non-functional in gas exchange, the decline allows approximate calculations of metabolic rate.

Metabolic Rates of Voluntary Diving Seals

To illustrate metabolic conditions during the dive state, it is useful to analyse quantitatively an hypothetical 30 minute dive for a Weddel seal, during the first 15 minutes of which the vertical velocity averages 0.23 m sec^{-1}, the heart rate averages 15 beats min^{-1} (BPM), and an initial blood O_2 content of 345 ml O_2 l^{-1} is reduced to 230 ml l^{-1}. These are by no means unusual conditions (Guppy et al., 1986). Assuming a blood volume of 60 l, the O_2 uptake from the blood can be shown to be 6900 ml, of which about 1815 ml O_2 is estimated to be due to heart, lung and brain metabolism. If we assume that all the rest of the available O_2 is utilised by working muscle, its maximum aerobic metabolic rate would be 5085 ml O_2 over the first 15 min of diving. If all 135 kg of skeletal muscle are utilised in swimming, the mass specific rate of ATP turnover would be 0.75 mmol ATP kg^{-1} min^{-1}; under these conditions, whole-organism O_2 uptake rate ($\dot{V}O_2$) equals 460 ml O_2 min^{-1}, or about 24% of a resting VO_2 of 1900 ml O_2 min^{-1}. In the second half of such a dive, heart rate typically increases as does swim speed. At a heart rate of 25 BPM, the heart, lung and brain utilise about 3000 ml O_2 in 15 min. If the remaining blood O_2 supplies, plus all myoglobin bound O_2 stores (which in 450 kg Weddell seals are estimated at about 10,000 ml O_2) were used to support the work of 135 kg of muscle, the muscle metabolic rate would be 4.9 ml O_2 kg^{-1}, equivalent to an ATP turnover rate of about 1.5 mmol ATP kg muscle^{-1} min^{-1}, at a $\dot{V} O_2$ for a 450 kg seal of 863 ml O_2 min^{-1}. These metabolic rates could vary due to changing heart rates and swim speeds and admittedly (Guppy et al., 1986) are approximations (probably too high because all myoglobin O_2 could not be fully depleted under these conditions and the assumed O_2 content at the end of diving is lower than average (Qvist et al., 1986). Yet the instructive insight arising from the calculations is that in terms of mammalian working muscle and in terms of the whole-organism resting metabolism, the energy requirements of the diving Weddell seal obviously are low, surprisingly low.

Metabolic Arrest and Channel Arrest Strategies During Diving

How does a 450 kg Weddell seal manage on only fractions (24% and 45%, for example) of resting metabolic rate during descent and ascent, respectively? One possibility is that it makes up the energetic shortfall with anaerobic glycolysis in hypoperfused organs and tissues; that is, it activates the classical Pasteur Effect in these tissues. Since resting $\dot{V}O_2$ is known, it can be shown that making up the energetic shortfall by anaerobic glycolysis in a 450 kg seal would, in 30 minutes, lead to an unusual accumulation of lactate, in excess of about 20 mmol kg^{-1} when completely equilibrated throughout the 450 kg body. Not surprisingly, this is much higher than ever observed in dives of such duration. As in ectothermic anaerobes, the mystery of the missing lactate appears to be resolved by a reversed Pasteur Effect which is sustained in many of the Weddell seal's organs and tissues whenever the diving reflex is employed, in effect allowing energy metabolism to be switched down well below resting rates. This is the reason why the 'working' metabolic rate of Weddell seals, averaging diving and recovery periods over many hours per day, does not exceed the resting metabolic rate of seals, a paradox that has not been previously properly explained (Guppy et al., 1986; Hochachka and Guppy, 1987).

Metabolic arrests in diving animals, profound enough to reduce energy metabolism to levels ranging from 10% to about 60% of normoxic resting rates, have been noted before in several species, so these estimates for the Weddell seal are unique only in that they derive from animals diving voluntarily at sea. For this reason, we felt that two conclusions can be made with some assurance. First, marine mammals use the diving response to control precisely the rates and sites of utilisation of fixed amounts of O_2 over fixed periods of time. Secondly, metabolic arrest mechanisms (apparently involving a reversed Pasteur Effect) are activated in hypoperfused tissues and organs to minimise over-depletion of carbohydrate in inefficient fermentations and to minimise end-product accumulation. Of two best defence strategies used by ectothermic anaerobes, at least one (metabolic arrest) can, therefore, also be harnessed by endothermic marine mammals. What about the second strategy, the coupling of low-permeability membrane functions to metabolic arrest capacities?

In that hypoperfused organs and tissues can sustain hundreds of bouts per day of partial metabolic arrest, it may be possible to suggest that the hypoxia-induced membrane failure typical of terrestrial mammals is somehow avoided at these sites in diving species. Unfortunately, not enough data are available to assess this situation quantitatively for most hypoperfused tissues. An exception, however, is the seal kidney where metabolic arrest capacities are clearly coupled to low-permeability membrane functions (Halasz, Elsner, Garvie and Grotke, 1974), and the anoxia tolerance of the organ is predictably greatly expanded towards the kinds of limits observed in ectothermic anaerobes. In at least one mammalian organ in one group of endotherms, then, we can observe the same defence mechanisms against hypoxia as are widely used in ectothermic anaerobes. If the underlying mechanisms can be more fully described, they may serve as a point of departure for further refining a potentially exciting intervention strategy.

Summary

In many animals, the rates of glucose and O_2 consumption are inversely related (Pasteur Effect). Under O_2 limiting conditions the demands for glucose (glycogen) in such cells may drastically rise as a means of maintaining ATP turnover close to normoxic rates. Yet ion and electrical potentials typically dissipate because of an energy deficit and high membrane permeabilities; metabolic and membrane functions in effect are decoupled. 'Good' animal anaerobes resolve these problems with a number of biochemical and physiological mechanisms; of these (i) metabolic arrest and (ii) stabilised membrane functions are the most effective strategies for extending hypoxia tolerance. Metabolic arrest appears to be achieved by means of a reversed or negative Pasteur Effect (reduced or unchanging glycolytic flux at reduced O_2 availability) while coupling of metabolic and membrane function is made possible despite the lower energy turnover rates by maintaining membranes of low permeability (possibly by means of reduced densities of ion-specific membrane channels). Although the dual harnessing of metabolic arrest with channel arrest has been attempted as an intervention strategy, thus far success has been minimal, probably because hypothermia is the usual metabolic arrest mechanism employed but this in itself perturbs controlled cell function in most endotherms. The only endothermic systems currently known which seem successfully to employ the above dual strategy for surviving hypoxic episodes are hypoperfused hypometabolic tissues and organs of diving marine mammals.

Acknowledgements

This work was supported by NSERC (Canada) through operating grants. Special thanks are due to my colleagues Drs W.M. Zapol, J. Qvist and G.C. Liggins for numerous ice-discussions of diving in Weddell seals.

References

Brezis, M., Rosen, S., Spokes, K., Silva, P., and Epstein, F.H. (1984) 'Transport-dependent anoxic cell injury in the isolated perfused rat kidney'. *American Journal of Pathology* **116**, 327-341

Castellini, M.A., Murphy, B.J., Fedak, Ronald, K., Gofton, N., and Hochachka, P.W. (1985) 'Potentially conflicting metabolic demands of diving and exercise in seals'. *Journal of Applied Physiology* **58**, 392-399

Else, P.L. (1984) 'Studies in the Evolution of Endothermy: Mammals From Reptiles'. PhD thesis (University of Wollongong, Sydney, N.S.W., Australia)

Guppy, M., Hill, R.D., Schneider, R.C., Qvist, J., Liggins, G.C., Zapol, W.M., and Hochachka, P.W. (1986) 'Microcomputer-assisted metabolic studies of voluntary diving of Weddell seals'. *American Journal of Pathology* **250**, R175-R187.

Halasz, M.A., Elsner, R., Garvie, R.S., and Grotke, G.T. (1974) 'Renal recovery from ischemia: a comparative study of harbor seal and dog kidneys'. *American Journal of Pathology* **227**, 1331-1335

Hochachka, P.W. (1982) 'Metabolic arrest as a mechanism of protection against hypoxia'. In A. Wauquier, M. Borgers, and W.K. Amery (eds.) *Protection of Tissues Against Hypoxia* (Elsevier Press, Amsterdam)

Hochachka, P.W. (1985) 'Assessing metabolic strategies for surviving O_2 lack: role of metabolic arrest coupled with channel arrest'. *Molecular Physiology* **8**, 331-350

Hochachka, P.W. (1986) 'Defence strategies against hypoxia and hypothermia'. *Science* **231**, 234-241

Hochachka, P.W. and Guppy, M. (1987) *Metabolic Arrest and the Control of Biological Time* (Harvard Univ. Press, Cambridge, Mass), pp 1-237.

Hochachka, P.W. and Somero, G.N. (1984) *Biochemical Adaptation* (Princeton University, Princeton, N.J.), pp 1-537

Qvist, J., Hill, R.D., Schneider, R.C., Falke, K.J., Liggins, G.C., Guppy, M., Elliot, R.L., Hochachka, P.W., and Zapol, W.M. (1986) 'Hemoglobin concentrations and blood gas tensions of free-living Weddell seals'. *Journal of Applied Physiology* **61**, 1560-1569

Siesjo, B. (1981) 'Cell damage in the brain: a speculative synthesis'. *Journal of Cerebral Blood Flow and Metabolism* **1**, 155-185

Chapter 13

Ruminant Animals and the Exploitation of Anaerobiosis

R.A.F. Chevis

Introduction

There is an ancient adage that 'all flesh is grass' and, for domesticated ruminant animals, as well as those that consume the meat, milk, cheese or butter that they produce, the saying contains more than a grain of truth. Indeed, the importance of ruminant animals extends beyond their role in food production since the fibre (wool and hair) and leather which they provide have, for many centuries, helped to protect mankind from the elements.

During the Eocene, forests covered most of the available land and the dominant herbivores were perissodactyls, now represented by the horses, tapirs and rhinoceroses a mere six genera of the 158 known to have existed. By the end of the Eocene, however, the artiodactyls (the group that includes the ruminants) had increased and, with the appearance of grasslands as the climate became drier and cooler during the Miocene (Stirton, 1959), the increase continued. Since the end of the Oligocene the ruminants have come to dominate the artiodactyls with over 70 genera still extant. The ascending dominance of ruminant over non-ruminant herbivory has been cited as evidence of the superiority of fermentation in the fore-stomach over that in the caecum (Clarke, 1968).

Vertebrates are incapable of digesting cellulose and related plant polymers but many species of microorganisms have this capability (Clarke, 1968). Microorganisms have been adopted by herbivores as cellulose digesting symbionts. While ruminants provide accommodation for their microbial partners in a modified stomach, where fermentation occurs before gastric digestion, non-ruminant herbivores have evolved a large caecum, as an appendage to the hind-gut, wherein fermentation occurs. Hind-gut fermentation provides the host animal with volatile fatty acids as a source of energy, but the lack of a subsequent digestive process and the limited opportunity for the absorption of other products (especially vitamins) puts the non-ruminants at a relative disadvantage.

Introduction 251

Cattle, sheep and goats, in particular, share three characteristic adaptations; they lack upper incisor teeth, having instead a dental pad; they have a complex four chambered stomach; and they regurgitate ingesta, from the first chamber of the stomach and subject it to a lengthy period of mastication (i.e. they ruminate). It is the complex stomach which is of most interest since the first and largest chamber (the rumen) receives newly ingested fodder which is then attacked by a mixed population of anaerobic bacteria, protozoa and fungi. This is pre-gastric digestion, in which fodder carbohydrates and proteins are metabolised to a mixture of volatile fatty acids, ammonia and, of course, microbial cellular constituents, about one half of which is protein, the remainder being fat and carbohydrate (Oxford, 1958). As the ingesta is reduced to small particles it passes with ruminal fluid to the reticulum (second chamber) and omasum (third chamber) where further physical maceration occurs together with the absorption of much fluid. True gastric digestion occurs in the abomasum (fourth chamber) which is functionally and biochemically the equivalent of the stomach of other animals. Since large numbers of microorganisms accompany the ingesta and are killed and digested in the abomasum they represent an appreciable proportion of the solid part of the host animal's real ration.

Some fibre remains intact throughout both ruminant and gastric digestion but this is subjected to further microbial action in the caecum to produce more volatile fatty acids (Figure 13.1).

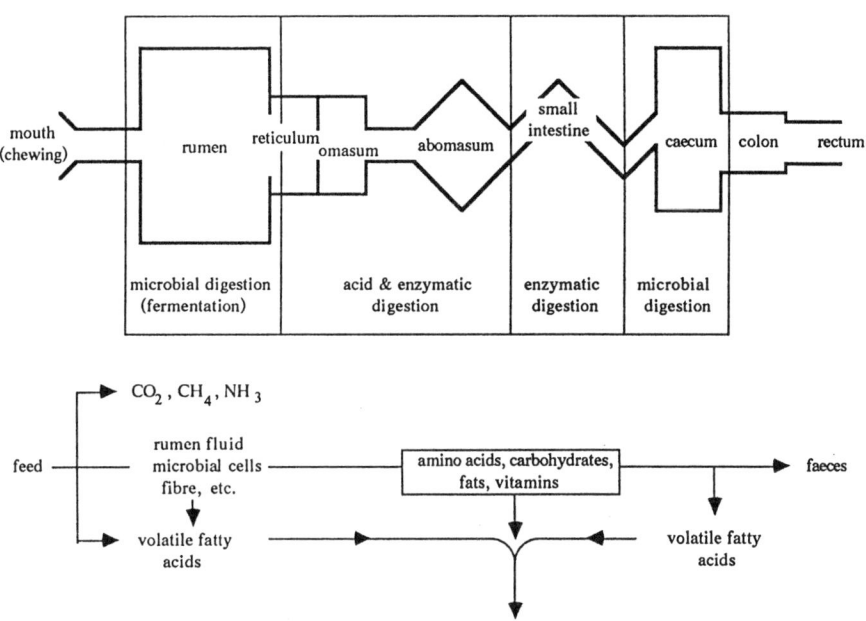

Figure 13.1 Diagrammatic representation of ruminant digestion.

Rumen Activity

As a rule, it is accepted that the rumen accounts for about 60% of digestion in cattle and sheep and this determines the nature of the nutrients which may be used. The particular advantages of pre-gastric or ruminal digestion are: cellulose and other structural plant polymers are brought into solution; the microorganisms carrying out the process can use non-protein nitrogen for growth, so that inorganic nitrogen or urea from the host's protein metabolism may be converted to microbial protein, which eventually becomes available to the host animal; and the vitamin content of ingesta is increased by microbial synthesis. Ruminants are independent of all dietary vitamins, other than A and D (Clarke, 1968).

Metabolism of forage produces acetic, propionic and butyric acids roughly in the proportions 70:20:10, together with some lactate, formate, valerate, isovalerate, ammonia, methane, carbon dioxide and hydrogen (Esdale and Satter, 1972; Wolin 1975). The temperature of the rumen is about $39^o,$ maintained by heat produced during fermentation; the pH varies between 7.4 and 5.5 with a buffer system provided by the bicarbonate introduced in saliva and the volatile fatty acids; the gas phase consists of about 27% CH_4, 65% CO_2, 7% N_2 and a trace of H_2; and the environment is anaerobic, the redox potential reaching as low as -350 to -400 mV at pH7 (Clarke, 1968) (See Figure 13.2).

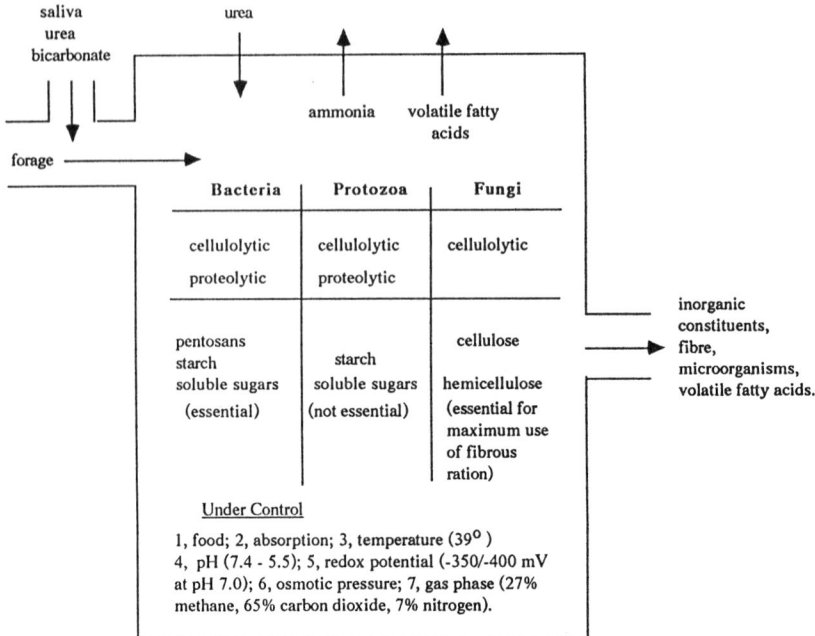

Figure 13.2 A summary of physiological conditions in the rumen.

Methane present in the rumen represents an electron sink for the considerable amount of hydrogen produced in fermentation reactions; methane is produced rather than ethanol and the partial pressure of hydrogen is maintained at a low level (approximately 3×10^{-4} atmospheres, Wolin, 1975). Wolin (1975) derived a chemical equation to describe rumen fermentation which produced acetate, propionate and butyrate in mole ratios as 65:20:15 :

$$57C_6H_{12}O_6 \rightarrow 65CH_3COOH + 20CH_3CH_2COOH + 15CH_3CH_2CH_2COOH + 60CO_2 + 35CH_4 + 25H_2O$$

Ammonia in the rumen is derived from the deamination of amino-acids and, on average, some 30% of dietary nitrogen intake is degraded to ammonia (Nolan, 1975). Much of the ammonia is used for the synthesis of microbial proteins and most of the rest is absorbed and used for the synthesis of urea (via the ornithine cycle), predominantly in the liver. Some urea is then returned to the rumen in saliva and absorbed by direct diffusion across the rumen wall, while the rest is, of course, excreted in urine.

Rumen Microorganisms

Bacteria

Ruminal digestion is carried out by a mixed population of anaerobic bacteria, protozoa and fungi and while the former two groups have been investigated over many years (Oxford, 1958; Clarke, 1968; Coleman, 1980) the last has attracted attention only in very recent times (Bauchop, 1979).

Oxford (1958) provided a list of some 12 genera of bacteria which, according to several criteria, were regarded as true rumen organisms. Bryant (1963) extended the list by two unnamed groups; 22 species of true rumen dwellers were identified. While it has been noted that most functional ruminal bacteria stain gram negative (Clarke, 1968) the accuracy and importance of this observation has been questioned by Bryant (1963) who recommended that more detailed cytological and biochemical investigations be carried out in order to identify the various strains of organisms.

Oxford (1958) assembled ruminal bacteria into five functional groups according to their ability to ferment various constituents of fodder. The groups are indicated below.

As Oxford (1958) pointed out, assignment to a relevant group was dependent upon activity displayed by a strain of organism *in vitro* under conditions bearing only a remote resemblance to those prevailing in the functioning rumen. It was held, however, that if, *in vitro,* an organism attacked a carbohydrate to yield a product, not being a rumen end-product, but which, in turn, was attacked by a different organism with the production of a recognised rumen end-product, then the same symbiotic relationship might actually occur in the rumen (Oxford, 1958).

The products which emerged as the result of fermentation by the members of each of the five groups (Oxford, 1958) were:

    Group 1.   **cellulose fermenters**
        Products: formate, acetate, butyrate, lactate, succinate, ethanol, CO_2, H_2.

    Group 2.   **pentosan fermenters**
        Products: acetate, butyrate, formate, lactate.

    Group 3.   **starch/sugars fermenters**
        Products: acetate, butyrate, formate, lactate, propionate, succinate, CO_2, H_2.

    Group 4.   **lactate fermenters**
        Products: acetate, butyrate, propionate, higher volatile fatty acids, CO_2

    Group 5.   **proteolytic bacteria**
        Products: unknown

All of the products named above have been identified as constituents of normal rumen liquor and while acetate, butyrate and propionate are absorbed by the host animal, and possibly a small amount of lactate, other products such as formate, succinate, higher volatile fatty acids (e.g. valerate and isovalerate), ethanol and hydrogen, are not. Yet members of the latter group of products do not accumulate in the rumen, so that effective processes for their removal must exist.

Wolin (1975) stated that 'the nutrition and physiology of ruminants is intimately dependent on the result of interacting bacterial activities' and illustrated the point with a series of experiments in which various bacterial species were cultured alone or in combination. He showed, by using defined substrates, that the end-products of fermentation differed according to whether particular species were in single or combined cultures. Thus, when *Bacteroides succinogenes* and *Selenomonus ruminantium* were placed in mixed culture with cellulose as an energy source, fermentation resulted in the production of acetate, propionate and CO_2. When *B. succinogenes* was cultured alone the products were succinate and acetate. Since *S. ruminantium* does not ferment cellulose it relied on the cellulolytic *B. succinogenes* to produce an energy source and decarboxylated the succinate, produced by the latter organism, to propionate. Combined cultures of *Ruminococcus flavifaciens* and *S. ruminantium* produced essentially the same results (Wolin, 1975). Succinate did not support the growth of *S. ruminantium* in pure culture, nor increase the growth yield when glucose was the limiting energy source, so the energy source in the combined cultures was probably a small molecular weight carbohydrate, either glucose or glucose saccharides (Wolin, 1975). *B. succinogenes* and *R. flavifaciens* are two of the most important cellulolytic species found in the rumen and both produce large amounts of succinate from cellulose. Similarly, *Bacteroides rumincola* and several other important rumen species produce succinate as a major end-product of carbohydrate fermentation. (Wolin, 1975). All must, therefore, interact with other organisms which have the capacity to decarboxylate succinate; indeed it would appear that the production of propionate depends to a large degree on the ability of some bacterial species to decarboxylate exogenous succinate (Figure 13.3)

The method by which exogenous succinate is decarboxylated by *S. ruminantium* has not been explained but is considered to be similar to that employed by *Veillonella alkalescens,* that is through oxaloacetate, methylmalonyl-CoA decarboxylate to form propionyl-CoA, followed by loss of the CoA.

Lactate does not accumulate in the rumen because there are bacteria such as certain *S. ruminantium* strains, capable of fermenting lactate with the production of propionate (Figure 13.4). As discussed later, the presence of such strains of bacteria is of value when ruminants are fed rations rich in soluble carbohydrates. At least two species of strictly anaerobic methanogenic bacteria occur in ruminants, their preferred substrates being $H_2 + CO_2$ or formate (Demeyer and Vam Nevel, 1975) (see Figure 13.3). Methane production may be regarded as an energy sink into which the hydrogen from all rumen organisms drains, allowing a higher yield of ATP (Hungate, 1962 cited by Demeyer and Van Nevel, 1975). Maintenance of low partial pressure of H_2 by methanogenesis, also best explains the discrepancy between production of ethanol by pure cultures of some bacteria and its insignificance as an intermediate in the rumen. (Demeyer and Van Nevel, 1975). Formate is actively converted into methane by rumen contents following the reaction:

$$4HCOOH \rightarrow 3CO_2 + CH_4$$

and, probably after initial conversion of the formate to intercellular CO_2 and H_2 by some non-methanogenic organisms (Demeyer and Van Nevel, 1975); methane, can, however, be formed from formate without prior oxidation to CO_2 (Demeyer and Van Nevel, 1975) especially where it arises as a by-product of pyruvate metabolism to acetate and ATP where electrons are transferred to protons (H_2) and protons and CO_2 (HCOOH). Only the carboxyl carbon, of pyruvate, is significantly converted to methane (Demeyer and Van Nevel, 1975).

About 30% of dietary nitrogen intake is degraded to ammonia. The remainder is either assimilated by microorganisms as peptides, amino-acids or nucleic acid bases or passes intact from the rumen (Nolan, 1975). Some ammonia present in the ruminal fluid becomes incorporated in microorganisms with some 30-80% of bacterial nitrogen derived from this source (Nolan, 1975). Indeed, it is generally agreed that ammonia is, quantitatively, the most important nitrogenous nutrient for

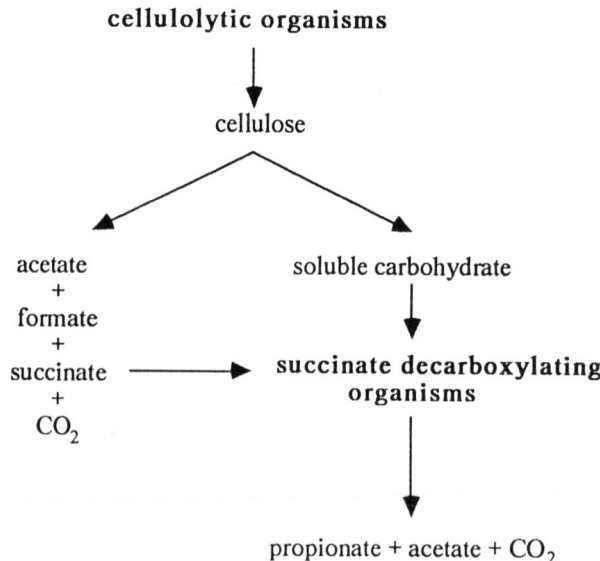

Figure 13.3 The interaction of cellulolytic and succinate decarboxylating organisms.

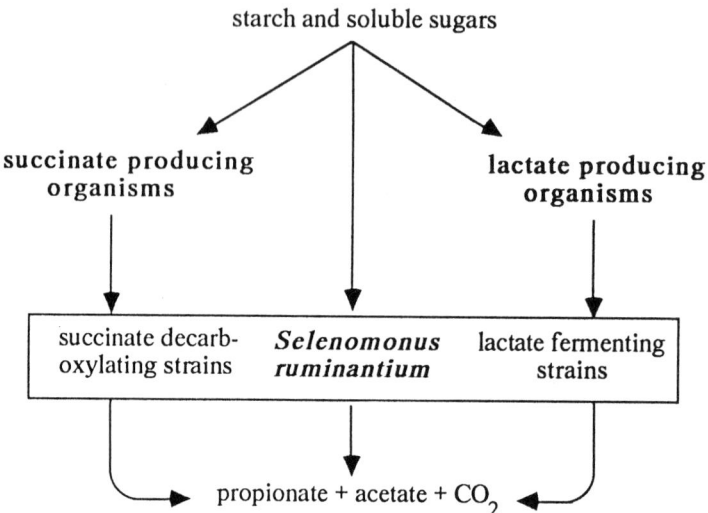

Figure 13.4 Further interactions between rumen organisms.

rumen bacteria (Smith, 1975) and the rate of microbial growth increases with ammonia concentration up to 5 nM (Smith, 1975). It is, then, not surprising that the amino-acid composition of bacterial protein is not, to any great degree, affected by the amino-acid composition of ruminant diets (Purser and Beuchler, 1966; Ibrahim and Ingalls, 1972). Urea secreted through the rumen wall, or contained in saliva represents an endogenous supply of ammonia since ureolysis occurs quite rapidly in rumen contents (Smith, 1975) and there are strains of bacteria which possess a very active urease (Wolin, 1975).

Protozoa

Two groups of Protozoa are found in the rumen, the entodiniomorphs and the holotrichs. The former belong to the Family Ophryoscolecidae, Order Entodiniomorpha and the latter in the Family Isotrichidae, Order Trichostomatida. (Clarke, 1968). Both groups are ciliated, with the cilia of the entodiniomorphs confined to the peristome and some other specialised areas, while the holotrichs are completely covered with cilia (Coleman, 1980)

All rumen ciliate protozoa are anaerobic, the holotrichs resemble free-living paramecia, while the entodiniomorphs are amongst the most complex of ciliates and are paralleled only by species occuring in the horse caecum and some marine protozoa (Clarke, 1968) Most rumen protozoa can ferment carbohydrates with the production of volatile fatty acids, CO_2 and H_2 (Clarke, 1968; Coleman, 1975; Coleman, 1980). More importantly, however, they store reserve polysaccharides as an amylopectin; the holotrichs throughout the body and the entodiniomprphs in rigid, so-called skeletal plates (Clarke, 1968; Coleman, 1975; Coleman, 1980).

Oxford (1958) noted that many rumen protozoa engulfed starch granules, plant chloroplasts, bacteria and even other protozoa and pointed out that the engulfed bacteria supplied amino acids, purines and B vitamins which the ruminal protozoa were not capable of synthesising, Coleman (1975) reviewed the data pertaining to the engulfment of ruminal bacteria by protozoa and showed that all engulfed large numbers of a wide range of bacterial species. In sheep fed on a grain diet (and thus rich in readily available carbohydrate) there were about 2.5×10^6 protozoa and 10^{10} bacteria/ml of rumen fluid and bacteria were engulfed at the rate of 3.8×10^{14} organisms/day. When supplied with a roughage (hay and grass) diet with less available carbohydrate sheep rumen fluid harboured 2.2×10^5 protozoa and 2×10^{10} bacteria which were engulfed at the rate of 3.6×10^{13} organisms/day. It was estimated that there were about 4×10^{12} bacteria/g (dry weight) so that the protozoa in grain fed animals engulfed 95 g dry weight of bacteria/day while those in roughage fed animals engulfed 9g/day (Coleman, 1975).

Coleman (1975) quoted work which showed that sheep maintained on pasture had about 5.77×10^5 protozoa and 4.1×10^{10} bacteria/ml of rumen fluid, but no estimate of engulfment rate was made. Since, however, the protozoal population was dominated by a species known to have a high degree of bacterial engulfment the actual rate can be calculated to be about 25% of that for grain fed animals, that is, about 23 g/day.

Engulfed bacteria are digested, some more rapidly than others, and their amino acids incorporated in protozoal protein unchanged; bacterial purines and pyrimidines are similarly incorporated into protozoal nucleic acids (Coleman, 1975). Energy for growth and development is derived from a range of carbohydrates (starch, solubilised cellulose derivatives and bacterial carbohydrate) by glycolysis (Coleman, 1980).

It has been suggested that rumen protozoa control fermentation, first by the removal of starch grains from rumen fluid so that bacterial attack and the rapid production of lactic acid is obviated (Oxford, 1958; Krogh, 1959) and second, by engulfing bacteria and thus regulating fermentation rate (Mackie, Gilchrist, Roberts, Hannah and Schwartz, 1978). The presence of protozoa certainly maintains bacterial numbers at a level below that which obtains when protozoa are absent (Eadie and Hobson, 1962) but the relevance of these factors is unclear.

Fungi

Clarke (1968) noted that there was a small population of flagellates present in rumen fluid, but regarded them as being of little importance. Orpin (1975, 1977) first recognised obligate anaerobic fungi in the rumen but failed to appreciate that they might be an important element of the rumen microbiota. This despite the work of Halliwell (1957) which demonstrated that the rate of cellulose degradation, revealed by pure cultures of rumen bacteria, fell far short of that which occurred in the rumen.

Bauchop (1979) showed that the rumen flagellates referred to by Clarke (1968) were, in fact, fungal zoospores and four distinct ruminal fungi have now been described (Orpin, 1975, 1976, 1977; Orpin and Munn, 1986). Being anaerobic, these fungi differ from most known species, other than those identified from the large intestine of some large herbivores (Bauchop, 1979). Furthermore, their vegetative structures were attached exclusively to plant material in the rumen, which militated against their recognition because research on rumen microorganisms had concentrated on those located in rumen liquor.

Rumen fungi are found in their greatest profusion in animals subsisting on fibrous fodder and are either absent, or present in very low numbers, in animals receiving a high grain diet (Bauchop, 1985). Additionally, they are predominantly associated with the vascular tissue of stems and leaves, or those parts of the plant which remain longest in the rumen; some softer tissue is, however, also colonised. Initial entry to plant tissue is made through areas in which the epidermis has been broken (Bauchop, 1979) so that rumination, with its continual maceration of fodder material, facilitates colonisation.

Following fungal invasion of fodder material, rhizoids soon extend to vascular tissue which is penetrated and it has been postulated that continued growth gives rise to mechanical disruption which assists in the reduction in size of fodder fragments. Akin and Rigsby (1987) record that the rapid degradative action of ruminal fungi on lignocellulose appears to arise from the action of secreted cell-wall-degrading enzymes together with penetration and mechanical disruption by rhizoids. Indeed the relative lack of degradation of plant cell walls by pure bacterial cultures is interpreted as indicating that ruminal fungi are essential to the process *in vivo* where bacteria play a secondary role (Akin and Rigsby, 1987).

It has been established that rumen fungi actively ferment cellulose (Bauchop and Mountfort, 1981). In monocultures fungi metabolised cellulose to acetate, formate, lactate, ethanol, CO_2 and H_2, while in combined cultures with rumen methanogenic bacteria, acetate production increased considerably and that of lactate, ethanol and formate was much reduced; the CO_2 and H_2 were converted to methane (Bauchop and Mountfort, 1981). Hydrogen production was, thus, added to the list of attributes which distinguished rumen fungi from all others.

The glycosidase activity of rumen fungi has been investigated using such substrates as avicel, carboxymethylcellulose, xylan, starch, polygalacturonic acid and p-nitrophenyl-derivatives of galactose, glucose and xylose. Pectinase activity appears to be absent from anaerobic fungi but the ability to produce a range of enzymes necessary to degrade cellulose and hemicellulose to simple sugars, most certainly exists (Bauchop, 1985).

The Large Intestine

Ruminants possess a voluminous caecum and proximal colon though their capacity, in relation to body size, is less than that of the horse. Digestion in the large intestine deals with residues (mostly cellulose and hemicelluloses) not degraded further forward in the gut. The contents of this area have a higher bicarbonate concentrate than does the rumen, less chloride and an almost constant temperature and pH (pH 6.6-7.8). The secondary fermentation which occurs therein represents a significant contribution to overall energy production and its importance increases when fodder is of low nutritive value.

The bacteria which have been found in fodder from the caecum/colon are the same as those from the rumen so that the genera *Bacteroides, Butyrivibrio, Fusobacterium, Selenomonas* and *Micrococcus* are all represented. There are some 10^8 cellulolytic bacteria/g of contents which, depending on the quality of fodder, may digest between 5-30% of the plant cell wall material ingested each day. Volatile fatty acid production occurs, of course, with a high acetate content (8-15% of total daily production) some butyrate and a higher incidence of branched chain fatty acids than arises from the rumen. Methane production also occurs and represents about 12% of daily output; most is excreted via the lungs. Proteolytic, deaminase and urease activity all approximate that of the rumen so that protein metabolism occurs and the ammonia concentration is always quite high. Some bacterial protein synthesis also occurs, but is of no benefit to the host since the material cannot be reclaimed.

It is most noteworthy that no protozoans have been isolated from the large intestine and while there are, as yet, no reports of a fungal population complete absence would be surprising.

Finally, the large intestine is a very efficient absorptive organ. Almost all (90%) of the water which enters with the ingesta is absorbed together with volatile fatty acids, sodium, chloride, methane and ammonia. The structure and function of the large intestine of ruminants has been reviewed by Ulyatt, Dellow, Reid and Bauchop (1975).

Intermediary Metabolism in Ruminants

The nutrients available for metabolism are, first, those that arise from microbial activity in the fore-stomachs - acetate, propionate, butyrate, higher fatty acids and protein, fat, carbohydrate and nucleic acids derived from the cellular constituents of rumen microorganisms. Second, there are those nutrients which escape degradation in the rumen, such as small amounts of starch and glucose, some protein, amino acids and long chain fatty acids from dietary lipids.

The volatile fatty acids (acetate, propionate and butyrate) are absorbed directly from the fore-stomachs and utilised immediately. Acetate passes through the rumen epithelium and liver unchanged (Hungate, 1966; Annison, Hill and Lewis, 1957) but is oxidised (via acetyl CoA) by tissue to produce energy (Jarrett and Potter, 1950); acetate may supply up to 50% of the carbon expired as CO_2 (Davis, Brown, Staubus and Nelson, 1960). Acetate introduced to the tricarboxylic acid cycle can, of course, form oxaloacetate and thus glucose by gluconeogenesis, but there is no evidence that this occurs in ruminants. The importance of acetate as a source of energy is emphasised by the low blood sugar levels of ruminants; up to half that of humans in the adult non-pregnant animal and about one third that of humans in the pregnant animal.

As with other tissues, the mammary gland uses acetate for energy production but between 17% and 45% of that which enters the gland is used in the synthesis of fatty acids (Annison, Linzell, Fazakerly and Nichols, 1967), a facility which is not shared by non-ruminants. Again unlike other animals, ruminants use acetate to produce body fat stores. According to Leng (1985) the synthesis of palmitate from acetyl CoA proceeds as follows:

$$8 \text{ acetyl CoA} + 14\text{NADPH} + 14\text{H}^+ + 7\text{ATP} + \text{H}_2\text{O}$$
$$\rightarrow \text{palmitic acid} + 8 \text{ CoA} + 14 \text{ NADP} + 7\text{ADP}$$

the NADPH is formed, in part, via the phosphogluconate pathway in which 1 mole of glucose-6-phosphate is oxidised to CO_2:

$$G\text{-}6\text{-}P + 12NADP + 7H_2O \rightarrow 6CO_2 + 12NADPH + 12H^+ + Pi$$

The overall reaction is thus;

$$24 \text{ acetyl CoA} + 42NADPH \rightarrow 3 \text{ palmitic acid} + 42NADP$$
$$3.5 \text{ glucose} \rightarrow 21CO_2 + 42NADPH$$
$$0.5 \text{ glucose} \rightarrow 1 \text{ glycerol}.$$

(One mole of glycerol phosphate is, of course, required to esterify three moles of palmitic acid).

Since four moles of glucose is oxidised to synthesise one mole of tripalmitin it is obvious that ruminants still have a requirement for, and can utilise, glucose. In fact, as Leng (1985) points out, early studies in Australia examining the role of glucose as a priming substrate for the tricarbarboxylic acid cycle (i.e. providing oxaloacetate) used acetate clearance rate as a measure of glucose sufficiency. Acetate clearance is most rapid on those diets likely to have a high glucogenic capacity.

Propionate is the most important precursor of glucose in ruminants and production may proceed directly (via succinate and phosphoenolpyruvate) or indirectly following conversion of propionate to lactate (and thence via oxaloacetate, succinate and phosphoenolpyruvate). Both synthetic pathways occur, with direct conversion taking place only in the liver and indirect conversion initiated in ruminal epithelium (propionate to lactate) and completed in the liver (Pennington and Sutherland, 1956; Leng and Annison, 1963).

Leng, Steele and Luick (1967) argue that if these two pathways are the only method by which propionate carbon is converted to glucose then the extent to which each individual route contributes can be estimated from the ratio of incorporation of C-1 to that of C-2, C-3 (of propionate) into glucose. The reaction sequences may be summarised as:

(1) propionate + CO_2 → succinate → lactate
(2) lactate + CO_2 → oxaloacetate → succinate (-CO_2) → glucose and phosphoenolpyruvate
(3) propionate + CO_2 → succinate (-CO_2) → phosphoenolpyruvate and glucose
(1 and 2 represent the indirect and 3, the direct pathways)

In reactions (1) and (3) propionate undergoes successive carboxylation and decarboxylation with a symmetrical four carbon compound as an intermediate; half the original carbon atoms at C-1 of propionate are thus substituted on conversion to lactate or glucose. Conversion of lactate to glucose (reaction (2)) causes further loss of carbon from the C-1 position of the original propionate. In the equilibration of oxaloacetate and succinate and subsequent decarboxylation to form phosphoenolpyruvate, half of the C-1 carbons are removed. The C-2, C-3 carbons of propionate are not exchanged in any of these reactions so that the dynamics may be followed using propionate with a ^{14}C label in the C-1 position or C-2, C-3 (Leng, et al., 1967). The foregoing makes it obvious that if the direct pathway (reaction (3)) is the only one operating then the C-1 incorporation in glucose will be half that of the C-2, C-3 incorporation. If, however, glucose is synthesised entirely by the indirect pathway (reactions 1 and 2) in which a further 50% of the remaining carbon in the C-1 position is lost, then the incorporation from that position can be only one quarter of that of C-2, C-3 carbons.

Leng et al., (1967) using appropriately labelled propionate infused into the rumen of sheep fed on a defined diet found that the incorporation of C-1 into glucose was one third that of C-2, C-3 incorporation, indicating that about 70% of the propionate converted to glucose was first converted to lactate. Weigand, Young and McGilliard (1972), however, concluded that the indirect route accounted for no more that 5% of the propionate converted to glucose. Diet may have had considerable influence in creating the difference since Leng et al. (1967) fed their animals on lucerne alone, whereas Weigand et al. (1972) fed theirs on a concentrated mixture of corn (65%), oats (21%), soya-bean meal (12%) together with lucerne grass or hay *ad libitum*.

While most propionate is utilised in the production of glucose, some may be used in the tricarboxylic cycle or the glycolytic sequence and it is regarded as being a strong possibility that propionate carbon may be converted to glycerol without prior conversion to glucose (Leng et al., 1967).

Most of the butyrate which is absorbed from the rumen is metabolised to 3-hydroxybutyrate and acetoacetate in the epithelial lining of the rumen (Pennington, 1952). Conversion occurs by two pathways: a deacylase reaction; and via the β-hydroxy-β-methyl glutaryl CoA (HMG-CoA) route (Baird, Hibbit and Lee, 1970 which appears to be the major pathway of physiological ketone body formation. Any butyrate not metabolised in the ruminal epithelium is converted to acetyl CoA in the liver (Leng and Annison, 1963).

The ketones (3-hydroxybutyrate and acetoacetate) are utilised for energy production in ruminants exactly as they are in other animals. In addition, however, 3-hydroxybutyrate can be incorporated into fatty acids in the milk of ruminants (Palmquist, Davis, Brown and Sachan, 1969) and appears to supply the initial four carbons of the fatty acids (Dimmick, McCarthy and Patton, 1970). Acetate and 3-hydroxybutyrate thus seem to play an interdependent role in fatty acid synthesis in the ruminant mammary gland.

Metabolism of the other nutrients does not differ substantially from that which occurs in non-ruminants.

Opportunities for Enhanced Production by Manipulation of Ruminal Processes

The rumen is a complex ecosystem in which the survival of a particular strain of bacterium, protozoa or fungus may be determined by the activity and metabolic bye-products of other organisms. Perhaps the most noteworthy of these inter-relationships concerns the lactate fermenting strain of *Selenomonas ruminantium* and the consequences attached to their relative absence, in certain circumstances. Ruminal production of lactate is fairly low in animals receiving a fibrous diet (pasture, hay, silage, etc.) and that which is produced is converted to propionate by *S. ruminantium* with the ability to ferment lactate. In contrast, large amounts of lactate are produced in the rumens of cattle/sheep etc., receiving grain or concentrate diets rich in soluble carbohydrates and, provided the change from fibrous to concentrate diet is undertaken over several days, conversion of lactate to propionate proceeds so that ruminal conditions are maintained within normal limits. The gradual introduction of the concentrate diet allows time for the population of lactate fermenting *S. ruminantium* to expand in response to the increase in the concentration of their substrate. If the change in diet is undertaken too rapidly, lactate accumulates in the rumen contents, exceeds the neutralising capacity of the buffer systems (mainly bicarbonate) and causes a fatal acidosis.

It is, however, unlikely that the presence of lactate fulfils all of the nutritional requirements of lactate fermenting strains of *S. ruminantium*, since other ruminal bacteria require branched chain fatty acids and vitamins to sustain growth and reproduction. The complexity of factors required to satisfy the growth requirement of some ruminal bacteria is well illustrated by those defined for *Methanomicrobium mobile*. A defined medium for the culture of *M. mobile* required acetate, isobutyrate, isovalerate, 2-methylbutyrate, tryptophan, pyridoxine, thiamine, biotin, vitamin B_{12}, p-aminobenzoic acid and an, as yet, unidentified growth factor found in ruminal fluid (Tanner and Wolfe, 1988).

The expansion of the population of lactate fermenting *S. ruminantium* is an example of the manipulation/adaptation of ruminal processes facilitating increased livestock production. The increased propionate production which is engendered by the expansion increases the availability of glucose which is critical in increasing growth rate and milk production, in dairy cattle (Leng, 1985). High energy concentrate feeding is, certainly, a feature of dairy farming, the world over.

The feeding of high levels of concentrates is almost always associated with reduced access to roughage, a combination which allows the pH of the rumen to fall. So long as the pH remains above 5, animals remain in apparent good health, but the ruminal production of acetate is reduced and the production of propionate and butyrate may also suffer. Since the relative amounts of ruminal acetate and propionate influence both the efficiency of fattening and milk composition, control of ruminal pH may influence livestock production (Esdale and Salter, 1972).

The addition of sodium bicarbonate to the diet raised the ruminal pH and increased the acetate : propionate ratio from 1.1 to 2.8 (Esdale and Salter, 1972). Similarly, the inclusion of 5% sodium bentonite in the ration increased the acetate : propionate ratio (Bringe and Schultz, 1969) but the mechanism by which this was achieved remains in doubt. Magnesium salts have also been used for their buffering effect in the rumen and, in common with sodium and potassium bicarbonate, have increased the fat content of milk.

A family of chemicals known as polyether ionophores alters the pattern of volatile fatty acid production when administered to ruminants, as do the glycopeptides actaplanin and avoparcin. In addition some of these compounds act as deaminase inhibitors and reduce microbial protein synthesis, so that the quantity of protein available to the host is increased to a significant degree. Such compounds are widely used to increase meat and milk production and some may even increase wool production (Graham and Edwards, 1985).

It is believed that monensin, a fairly typical ionophore growth promotant and antibiotic, enhances the supply of substrate to succinate and propionate, producing bacteria not susceptible to its action while, at the same time, reducing the supply of methanogenic bacteria by removing those bacterial species producing acetate, butyrate, formate and hydrogen (McGregor, 1983). Antibiotics such as monensin and lasalocid are also used as coccidiostats in the poultry industry. It might, therefore, be expected that the administration of such compounds to ruminants would cause reductions in the levels of ruminal protozoa and a consequent diminution in nitrogen recycling, in the rumen. The evidence for such changes is, however, very variable and defaunation (removal of all protozoa) does not appear to be an important part of the activity of either compound.

Defaunation of the rumen can be accomplished easily, at very low cost. With appropriate management ruminants can be maintained free of protozoa for prolonged periods. When protozoa are removed, the bacterial population of the rumen increases as does that of normal ruminal fungi. These increases are accompanied by enhanced digestibility of fodder roughage and significant increments in growth rates in cattle and sheep and wool production in the latter (Leng, 1985). Both reactions are indicative of an increase in the availability of protein from the protozoa-free rumen and this has been confirmed by the estimation of the flow of amino-acids to the abomasa of defaunated sheep. Re-establishment of the protozoan population, in the rumen of such sheep, always reduced the amino-acid flow.

High ruminal populations of fungi are always associated with a fibrous diet and in some situations the population may be increased by the addition of sulphur or methionine to the diet. Both additives increased feed intake and digestibility in animals so treated (Bauchop, 1985). The direct installation of fungal cultures into the rumens of roughage fed animals has also increased feed digestibility and coincidentally increased the ruminal population of cellulolytic bacteria. Significant increases in fibre digestibility have been recorded in roughage fed cattle supplemented with purified enzymes of fungal origin. The enzyme mixture possessed cellulose, protease, amylase and pectinase activity (Ralston, Church and Oldfield, 1962).

Because the rumen is such a complex ecosystem, manipulation of one or more factors to increase fodder digestibility and thus, livestock production, has to be approached cautiously. But, the number of factors available for manipulation encourages further research, since the net result of any particular set of variations cannot be predicted with any certainty. As the feeding of grain-based concentrate diets makes ruminants potential competitors for nutrients well suited to consumption by humans, future research aimed at increasing production from animals maintained on roughage diets would be most valuable. It would also obviate the circumstance in which a low acetate : propionate ratio, leading to high glucose production together with maximum milk yield, creates a demand for milk fatty-acids which cannot be satisfied from the diet of dairy cows; mobilisation of body-fat to make good the deficit then creates an acetonaemia leading to inappetance and decreased milk production. The condition can be controlled by the addition of fat to dairy cow rations but this, again, is a redirection of a potential human nutrient.

In order to obtain the maximum production of meat, milk, wool and hair from roughage-fed animals the following criteria must be satisfied:

(a) the plant varieties used as fodder must supply energy, protein, fat and inorganic requirements at the appropriate levels;
(b) moisture content of the diet should not, of itself, limit intake;
(c) ingesta must be rapidly digested and removed from the rumen; and
(d) metabolic efficiency must be maintained at the highest possible level.

The ionophore antibiotics promote propionate production and metabolic efficiency in the rumen by eliminating susceptible bacterial species. Administration of these antibiotics is associated with decreased feed intake (Graham and Edwards, 1985) which may be a direct consequence of delayed ruminal emptying because of the reduced bacterial population. If this is so then the use of such antibiotics will be limited to those situations in which the cost of fodder is a significant factor in livestock production.

Removal of ruminal protozoa is, by contrast, associated with increases in both bacterial and fungal populations and animal production (Bird and Leng, 1978). The latter arises from enhanced fodder digestibility accompanied by increased fodder intake (Bird, 1985), presumably the direct result of accelerated ruminal emptying. Defaunated sheep may be maintained protozoa-free for many months (Bird, 1985) because re-faunation can be accomplished only by direct transfer of viable organisms from one animal to another. It should thus be possible to have all the ruminants on a farm property rendered protozoa free and maintained in such a state, at very little cost. When defaunated lambs have been shown to produce 37% more wool than their faunated counterparts on the same diet (Bird and Leng, 1984), the economic advantage engendered by defaunation is quite obvious.

Ruminal fungi, *in vitro* and in the absence of bacteria, are able to degrade almost as much fibre as a normal mixed population (Akin and Rigsby, 1987). The two groups are, however, complementary in that they degrade different tissues and there is no suggestion that ruminants should, or could, be maintained in normal health and production when devoid of either. Even limiting the bacterial population, as occurs when ionophore antibiotics are administered, reduces feed intake.

Enhancement of ruminal digestion could be obtained if the digestive activity of bacteria and fungi could be increased. It may be possible to identify strains of normal ruminal flora more active than others and to alter ruminal conditions to provide for selective growth. Alternatively, mutagenesis and selective screening for mutants with augmented activity is a simple approach and has achieved success with the production of high yielding cellulose mutants of the fungus *Trichoderma reesei,* from soil (Bauchop, 1985). Another approach worthy of exploration is that of transforming bacteria and/or fungi in such a way that the transformants have an enhanced ability to produce normal digestive enzymes, or possess heterologous genes allowing the production of enzymes responsible for furthering the normal metabolic process and/or possess enzymes responsible for an alternative metabolic chain. Transformed *Saccharomyces* capable of one-step conversion of starch to alcohol, due to the secretion of normal flucoamylase and mouse amylase, already exist (Kim, Park and Matton, 1988). It should thus be possible to produce transformants capable of one-set ratios. Elimination of species producing only acetate, butyrate or succinate should enhance production (as occurs when ionophore antibiotics are administered) while, at the same time, increasing feed intake because of more rapid ruminal emptying. Acetate production would ensure that milk-fat synthesis was maintained for dairy production and, of course, the feeding of off-spring.

The main constituent of wool is the protein keratin and the production of wool in relation to absorbed amino acids is about 12 g/100g (Doyle and Egan,1983). Because cystine and methionine make up 9-13% of keratin, very significant responses in wool growth occur when 1-2 g/day of each is provided as ruminal by-pass supplement. By-pass is required because the amino acids are rapidly degraded by ruminal bacteria (Cottle, 1988 a, b). Transformed bacteria capable of synthesising high levels of either or both of the amino acids, which would become available to the host during subsequent gastric digestion, could make a worthwhile contribution to wool production. The manipulation of ruminal microbiota offers a fertile field for future research aimed at increasing the production of meat, milk and fibre by animals maintained on roughage diets.

Conclusion

The harvesting, processing, distribution and sale of products derived from ruminant animals provides employment in industries as diverse as shearing, butchery and fellmongery on one hand to advertising, modelling and sale of high fashion garments and accessories, on the other. As a consequence the ruminant animal industry stimulates fundamental and applied research in many fields and fosters technological advances in a myriad of manufacturing processes.

The breeding and selection of genetically superior animals started an industry which was well established before John Macarthur first imported merino sheep into Australia, almost 200 years ago. The industry flourishes today and has been the catalyst in the development of such techniques as artificial insemination, multiple ovulation and ovum/embryo transplantation.

Commercial herds and flocks vary from a few animals (cattle, sheep, goats etc) on small land-holdings in Europe and Asia to several thousand animals occupying large tracts of land in north and south America, Australia and South Africa. The sheep population of Australia has been as high as 190,000,000 when the human population was less than 10,000,000 and cattle stations are still measured in hundreds of square kilometres in the north of the continent. Ruminant animal management, whether intensive or extensive, supports those directly responsible for their welfare together with the industrial infra-structure associated with the wool/hair, leather, meat and dairy industries.

Indeed, humanity, through its domestication of ruminants, has harnessed a series of production systems so complex that they are, today, not capable of detailed and complete explanation, yet are amenable to modification to meet a wide variety of environmental, climatic and consumer demands. This exploitation of anaerobiosis may well have helped to shape our civilization and has, without doubt, contributed to the quality and enjoyment of life.

References

Akin, D.E. and Rigsby, L.L. (1987) 'Mixed fungal populations and lignocellulosic tissue degradation in the bovine rumen' *Applied Environmental Microbiology* 53, 1987-1995

Annison E.F., Hill, K.J.and Lewis, D. (1957). 'Absorption of volatile fatty acids from the rumen of the sheep' *Biochemical Journal* 66, 592-599

Annison, E.F., Linzell J.L., Fazakerly S.and Nichols, B.W. (1967) 'The oxidation and utilization of palmitate, stearate, oleate and acetate by the mammary gland of the fed goat in relation to their overall metabolism and the role of plasma lipids and neutral lipids in milk fat synthesis' *Biochemical Journal* 102, 637-647

Baird, G.D., Hibbitt, K.G. and Lee, J. (1970) 'Enzymes involved in acetoacetate formation in various tissues.' *Biochemical Journal* 117, 703-709

Bauchop, T. (1979) 'Rumen anaerobic fungi of cattle and sheep.' *Applied Environmental Microbiology* 38, 148-158

Bauchop, T. (1985) 'Rumen fungi - their roles and potential for manipulation.' In R.B. Cumming (ed.) *Recent Advances in Animal Nutrition In Australia* (University of New England, Armidale)

Bauchop, T. and Mountford, D.O. (1981) 'Cellulose fermentation by a rumen anaerobic fungus in both the absence and presence of rumen methanogens.' *Applied Environmental Microbiology* **42**, 1103-1110

Bird, S.H. (1985) 'Rumen protozoa - potential for manipulation.' In R.B. Cumming (ed.) *Recent Advances in Animal Nutrition In Australia* (University of New England, Armidale)

Bird, S.H. and Leng, R.A. (1978) 'The effects of defaunation of the rumen on the growth of cattle on low protein high energy diets.' *British Journal of Nutrition* **40**, 163-167

Bird, S.H. and Leng, R.A. (1984) 'Further studies on the effects of the presence or absence of protozoa in the rumen on liveweight gain and wool growth in sheep.' *British Journal of Nutrition* **52**, 607-611

Bringe, A.N.and Schultz. L.H., (1969) 'Effects of roughage type or added bentonite in maintaining fat test.' *Journal of Dairy Science* **52**, 465-471

Bryant, M.P. (1963) 'Symposium on microbial digestion in ruminants: identification of groups of anaerobic bacteria active in the rumen.' *Journal of Animal Science* **22**, 801-813

Clarke, R.T.J. (1968) 'The microbiology of "pre-gastric" fermentation.' *Australian Journal of Science* **31**, 141-146

Coleman, G.A. (1975) 'The interrelationship between ruminal ciliate protozoa and bacteria.' In McDonald I.W. and Warner A.C.I. (eds.) *Digestion and Metabolism in the Ruminant.* (University of New England, Armidale)

Coleman, G.S. (1980) 'Rumen ciliate protozoa.' *Advances in Parasitology* **13**,

Cottle, D.J. (1988a) 'Effects of defaunation of the rumen and supplementation with amino-acids on the wool production of housed Saxon merinos 1. Lupins and extruded lupins' *Australian Journal of Experimental Agriculture* **28**, 173-178

Cottle, D.J. (1988B) 'Effects of defaunation of the rumen and supplementation with amino-acids on the wool production of housed Saxon merinos 2. Methionine and protected methionine'*Australian Journal of Experimental Agriculture* **28**, 179-185

Davis, C.L., Brown, R.E., Staubus, J.R. and Nelson, W.O. (1960) 'Availability and metabolism of various substrates in ruminants 1. Absorption and metabolism of acetate.' *Journal of Dairy Science* **45**, 231-237

Demeyer, D.I. and Van Nevel, C.J. (1975) 'Methanogenesis, and integrated part of carbohydrate fermentation, and its control.' In McDonald, I.W. and Warner, A.C.I. (eds.) *Digestion and Metabolism in the Ruminant.* (University of New England, Armidale)

Dimmick, P.S., McCarthy, R.D. and Patton, S. (1970) In Phillipson A.T.(ed.), *Physiology of Digestion and Metabolism in the Ruminant* (Oriel Press)

Doyle, P.T. and Egan, J.K. (1983) 'The utilisation of nitrogen and sulphur by weaner and mature merino sheep.' *Australian Journal of Agricultural Research* **34**, 433-439

Eadie, J.M. and Hobson, P.N. (1962) 'Effect of the presence or absence of rumen ciliate protozoa on the total ruminal bacterial count in lambs.' *Nature (London)* **193**,503-505

Esdale, W.J. and Salter. L.D. (1972) 'Manipulation of ruminal fermentation iv. Effect of altering ruminal pH on volatile fatty acid production.' *Journal of Dairy Science* **55**, 964-970

Graham, C.A. and Edwards, S.R. (1985) 'Feed additives in ruminant nutrition.' In Cumming, R.B. (ed.) *Recent Advances in Animal Nutrition In Australia* (University of New England, Armidale)

Halliwell, G. (1957) 'Cellulolysis by rumen microorganisms.' *Journal of General Microbiology* **17**, 153-165

Hungate, R.E. (1966) *The Rumen and its Microbes* (Academic Press, New York)

Ibrahim, E.A. and Ingalls, J.R. (1972) 'Microbial protein biosynthesis in the rumen' *Journal of Dairy Science* **55**, 971-978

Jarratt, I.G. and Potter, B.J. (1950) 'The metabolism of acetate by sheep after injection of phlorhizin.' *Australian Journal of Experimental Biological and Medical Sciences* **28**, 595-579

Kim, K., Park, C.S. and Matton, J.R. (1988) 'High-efficiency, one-step starch utilisation by transformed *Saccaromyces* cells which secrete both yeast glucoamylase and mouse glucoamylase.' *Applied Environmental Microbiology* **54**, 966-971

Krogh, N. (1959) 'Studies on the alteration in the rumen fluid of sheep especially concerning the microbial composition when readily available carbohydrates are added to the food.' *Acta Veterinaria Scandinavia* **1**, 74-97

Leng, R.A. (1985) 'Efficiency of feed utilisation by ruminants.' In R.B. Cumming (ed.) *Recent Advances in Animal Nutrition In Australia* (University of New England, Armidale

Leng, R.A., and Annison, E.F. (1963) 'Metabolism of acetate, propionate and butyrate by sheep liver slices.' *Biochemical Journal* **86**, 319-327

Leng, R.A., Steele, J.W., and Luick, J.R. (1967) 'Contribution of propionate to glucose synthesis in sheep.' *Biochemical Journal* **103**, 785-790

Mackie, R.I., Gilchrist, F.M.C., Roberts, A.H., Hannah, P.E. and Schwartz, H.M. (1978) 'Microbiological and chemical changes in the rumen during stepwise adaptation of sheep to high concentrate diets.' *Journal of Agricultural Science* **90**, 241-254

McGregor, R.C. (1983) 'Growth promoters and their importance in ruminant livestock production.' In Haresign W. (ed.) *Recent Advances in Animal Nutrition* (Butterworths, London)

Nolan, J.V. (1975) 'Quantitative models of nitrogen metabolism in sheep.' In McDonald I.W. and Warner A.C.I. (eds.) *Digestion and Metabolism in the Ruminant* (University of New England, Armidale)

Orpin, C.G. (1975) 'Studies on the rumen flagellate *Neocallimistax frontalis*.' *Journal of General Microbiology* **91**, 249-262

Orpin, C.G. (1976) 'Studies on the rumen flagellate *Sphaeromonas communis*.' *Journal of General Microbiology* **94**, 270-280

Orpin, C.G. (1977) 'The rumen flagellate *Pironomonas communis*: its life history and invasion of plant material in the rumen.' *Journal of General Microbiology* **99**, 107-117

Orpin, C.G. and Munn, E.A. (1986) *Neocallimastix protriciarum* sp. Nov., a new member of the Neocallimasticaceae inhabiting the rumen of sheep.' *Transactions of the British Mycological Society* **86**, 178-181

Oxford A.E. (1958) 'Rumen microorganisms and their products.' *New Zealand Science Review* **10**, 38-44

Palmquist, D.L., Davis, C.L., Brown, R.E. and Sachan, D.S. (1969) 'Availability and metabolism of various substrates in ruminants v. entry rate into the body and incorporation into milk fat of D(-) 3-hydroxybutyrate.' *Journal of Dairy Science* **52**, 633-638

Pennington, R.J. (1952) 'The metabolism of short-chain fatty acids in the sheep 1. Fatty acid utilisation and ketone body production by rumen epithelium and other tissues.' *Biochemical Journal* **51**, 251-258

Pennington, R.J. and Sutherland, T.M. (1956) 'The metabolism of short chain fatty acids in sheep 4. The pathway of propionate metabolism in the rumen epithelial tissue.' *Biochemical Journal* **c3**, 618-628

Purser, D.B. and Beuchler, S.M. (1966) 'Amino acid composition of rumen organisms.' *Journal of Dairy Science* **49**, 81-84

Ralston, A.T., Church, D.C. and Oldfield, J.E. (1962) 'Effects of the addition of bentonite to high-grain dairy rations which depress milk fat percentage.' *Journal of Animal Science* **21**, 306-312

Smith, R.H. (1975) 'Nitrogen metabolism in the rumen and composition and nutritive value of nitrogen compounds entering the duodenum.' In McDonald, I.W. and Warner, A.C.I. (eds.) *Digestion and Metabolism in the Ruminant* (University of New England, Armidale)

Stirton, R.A. (1959) *Time, Life and Man* (Wiley, New York)

Tanner, R.S. and Wolfe, R.S. (1988) 'Nutritional requirements of *Methanomicrobium mobile*.' *Applied Environmental Microbiology* **54**, 625-628

Ulyatt, M.J., Dellow, B.D., Reid, C.S. and Bauchop, T. (1975) 'Structure and function of the large intestine of ruminants.' In McDonald I.W. and Warner A.C.I. (eds.) *Digestion and Metabolism in the Ruminant* (University of New England, Armidale)

Weigand, E., Young, J.W. and McGilliard, A.D. (1972) 'Extent of propionate metabolism during absorption from the bovine ruminoreticulum.' *Biochemical Journal* **126**, 201-209

Wolin, M.J. (1975) 'Interaction between bacterial species of the rumen.' In McDonald I.W. and Warner A.C.I. (eds.) *Digestion and Metabolism in the Ruminant* (University of New England, Armidale)

Chapter 14

Anoxibiosis in Living Metazoa: an Overview

C.S. Hammen

The Recognition of Anoxibiosis as a Subject of Inquiry in Physiology and Biochemistry.

Anoxibiosis in the metazoans is one of the two or three major areas of contemporary research in the comparative physiology of adaptations. The others are adjustment of osmotic balance with variable salinity, and perhaps gain aand loss of heat by terrestrial animals. The recognition of anoxibiosis as a subject distinct from oxygen utilisation and transport is relatively recent. Ernest Baldwin (1964) regarded animals as having just two problems of respiration: to get adequate oxygen, and to get rid of carbon dioxide. At that time, many animals were known to survive temporary anoxia, and the use of carbon dioxide in synthetic reactions was known to occur in representative species of most of the animal phyla. Most biologists took little note of these facts, regarding them as minor in the scheme of animal metabolism, if indeed they were aware of them at all. Marcel Florkin (1966) gave a summary of metabolic pathways showing lactic acid and ethanol as the sole products of anaerobic glycolysis, and showing the citric acid cycle as unidirectional. The role of carbon dioxide fixation as essential in the succinate fermentation of nematodes had been known for several years. If the leaders of comparative biochemistry were unaware of the fundamental explanation of anoxia tolerance then taking shape, it is understandable that biologists in general were similarly unprepared to grasp its significance.

When Bueding (1962) described the phosphorylation coupled with anaerobic fumarate reduction in *Ascaris,* he suggested that the same set of reactions might occur in other organisms that depend on anaerobic metabolism. The observation that labelled bicarbonate rapidly entered succinate in mantle tissue of the American oyster, *Crassostrea virginica,* was the first in a series of clues from several laboratories that favoured the presence of the *Ascaris*-type pathway in oysters (Hammen, 1969). The term 'facultative anaerobe', a misnomer applied to the oyster, since all metazoans are basically aerobes, was taken up with enthusiasm, but the details of energy-producing succinate formation in animals other than parasitic helminths accumulated more slowly. The focus remained on lactic acid, but the formation of alternate end-products, such as succinate, propionate, and alanine was well publicised in a discussion of anaerobic metabolism in a book on biochemical adaptation (Hochachka and Somero, 1973). The subject has prospered, and much solid information is now available to support earlier speculations.

The Discrepancy between Variety of Animals and Quantity of Physiological Research

Advances in the study of anoxia, as in other areas of comparative physiology, depend on experiments with a variety of species from different animals groups, using modern techniques. Often these criteria are not respected, as biochemists study the same few species in ever greater detail, while physiological zoologists examine an interesting array of animals with less rigour. On the one hand, biochemists need to learn that the animal kingdom consists of much more than 'animals' (mammals), 'bugs' (insects), and 'worms'. On the other hand, physiologists characterise enzyme activity by means of a single determination with presumed 'saturating' substrate concentration. This tells us very little about an enzyme's function, while a few repetitions with various amounts of substrate can give apparent K_m and V_{max}, from which rates at any concentration can be predicted. The properties of 'reverse' reactions, often unexamined, could be useful in constructing species-specific metabolic maps.

Most of our knowledge of anoxic metabolism concerns the molluscs, many of which are euryoxic, and the crustaceans, which are mostly stenoxic. These successful shelled forms have not replaced soft-bodied Metazoa, which are abundant in the living marine fauna, and present or dominate in several fossil faunas (Glaessner, 1984). If an 'average animal' must be named, it would not be a mollusc or crustacean, but a nereid (Buchsbaum, 1948). Many of the 'minor' phyla are virtually unknown with regard to anoxic metabolism. Some forms, such as the Pogonophora, inhabit the deep sea, and are hard to collect. Others are easily found, and seem to invite study because of clues already in the literature. For example, sponges have fumarate reductase activity greater than succinate oxidation, and therefore they probably carry on a succinate fermentation. Some marine nematodes are highly tolerant of anoxia, but little is known of the metabolism supporting this ability. Crustaceans are not noted for anoxia tolerance, but they are an extremely varied group, and only a few decapods have received the bulk of the research interest. The Cirripedia, especially, seem worthy of more intense effort. The mole crab, *Emerita talpoida,* appeared to maintain a high rate of heat output well after its oxygen consumption had fallen to a low level (Hammen, 1983). Other tantalising bits of information abound.

Evolutionary Origin of Anoxibiosis

When comparative physiology in relation to evolutionary theory was discussed by Prosser (1960), a major theme was the biochemical unity principle of Florkin, also recognisable as the common ground-plan of Baldwin. One example of this universality was 'the presence of glycolytic-fermentative enzymes, together with or without aerobic oxidative enzymes'. An important minor theme was the relative frequency of 'loss of synthetic capacity'. Together these themes explain the presence of both euryoxic and stenoxic metazoans today.

The pathway of glycolysis from glucose to pyruvate varies little throughout the animal kingdom. Reduction of pyruvate, however, leads to a variety of end products, not just L-lactate, as found in vertebrate muscle. Lactate dehydrogenase (LDH) seems ubiquitous, but in many species of molluscs, perhaps the majority, LDH activity is minor compared with the activity of octopine dehydrogenase (ODH; EC 1.5.1.11), which catalyses the reductive condensation of pyruvate with arginine to form octopine. Some LDH, notably those of molluscs, arachnids and polychaetes, produce and oxidise D-lactate only. The stereospecificity of these enzymes must have become fixed in different lines of descent very early in metazoan evolution.

Activities of ODH and the similar oxidoreductases forming alanopine and strombine were measured in 103 species in 13 phyla by Livingstone et al. (1983). (The Mollusca were over-emphasised, while some phyla were represented by only one species, and a dozen or more phyla were not examined. See comments in section 2). The general result was a complete array of 'opine' dehydrogenases in some of the lower forms, such as the Cnidaria, absence of one or more in most of the other invertebrate phyla, and complete absence in echinoderms, arthropods and chordates. These reactions all require amino acids to accomplish the re-oxidation of reduced NAD so that glycolysis can continue. The role of free amino acids is considerable in osmotic adjustment of marine invertebrates. Therefore, it is probable that the adaptive value of the opine dehydrogenases is determined by interaction between the demands of anoxia tolerance and salinity tolerance. Animals not subjected to variation in oxygen and salinity might be expected to lose these enzymes through lack of selective pressure.

More recently, still another enzyme of this type has been discovered, called 'tauropine dehydrogenase' (Sato and Gade, 1986). Formerly the sulphonic amino acid taurine was regarded as having no function other than that of osmotic adjustment. The distribution of TDH, as currently known, is limited to Brachiopoda and Archaeogastropoda, suggesting that it is a kind of relict enzyme. The present distribution of the opine enzymes will require further evaluation, with TDH taken into account. For the moment, the hypothesis of gradual loss of genes by chance still seems adequate.

The earliest Metazoa that left a useful fossil record were the cnidarians, annelids, and arthropods of the Ediacarian fauna from the Late Precambrian, about 660 to 560 million years (m.y.) before the present. Some estimates place the origin of the

Metazoa at 700 m.y., and the oxygen concentration at that time at 8 per cent of present atmospheric level. However, lack of definite information on the rate of increase of oxygen in the atmosphere and bottom water at the relevant time makes impossible any firm conclusions about the physiology of ancestral forms (Glaessner, 1984). One cannot say that they were anaerobes, but it seems agreed that the ancestral metazoans lived, compared with the present, a life of hypoxia. With the increase of oxygen came more dependence on oxidative metabolism, presumably, and more rapid and efficient locomotion, which enabled animals to avoid anoxic habitats. Lack of selection pressure to retain anoxic pathways would lead to evolution by loss of function, analogous to the evolution of parasites. The majority of modern animals, having evolved under conditions of constant oxygen abundance, are dependent on oxygen. A minority are euryoxic, capable of shifting from aerobic metabolism to anaerobic, as environmental factors influence them. This topic is discussed in greater detail by Dr B. Runnegar in an earlier chapter of this book.

Very Low Metabolic Rates

Reduction in metabolic rate has been recognised as one of the principal means of responding successfully to anoxia (Hochachka and Somero, 1973). How much reduction is regularly found, and how much is possible? The answer would vary with species, of course, but the anoxic metabolic rate of a particular animal might be a definite fraction of its aerobic rate at the same temperature. In the metabolism of any organism, a portion of the maintenance energy always appears as heat. Calorimetry appears the logical choice of method to study the quantitative aspect of anoxic metabolism.

Among the first to apply calorimetry to aquatic 'poikilotherms', Davies (1966) measured the heat production of goldfishes under various conditions of light and temperature, and Jackson (1968) studied diving turtles as they entered and endured anoxic conditions. After 3h of submergence and 2h of anoxia, the turtles had reduced their metabolic rate to 20 percent of their fully aerobic rate. About ten years later, Gnaiger (1977) and Hammen (1976) had the idea that, with sufficiently sensitive instruments, aquatic invertebrates could be studied in the same way.

Based on analysis of end products of glycolysis and the efficiencies of known pathways, various authors calculatd rates of anoxic metabolism of bivalves from 60 per cent to 5 per cent of aerobic rates. Variations in these estimates now appear due to not only differences in analytical methods and biochemical assumptions, but to other factors as well. The means used to produce anoxia, the time allowed for entering the anoxic state, and its duration are all important.

Suitable apparatus for simultaneously measuring heat output and oxygen consumption of small aquatic animals has become available only recently. Direct comparison of rates requires such apparatus. Rates of oxygen consumption, which appear in tables of data implying constancy, actually vary with age, sex and size of animals, with temperature, salinity, pressure, light, season of the year and even time of day. Similarly, rates of anoxic metabolism may be expected to vary in their own way with each set of conditions. One fruitful approach has been to measure both heat and oxygen concurrently, beginning with animals in fully aerated water, and following their transition to anoxia as oxygen concentration was reduced to low levels by the animals. This procedure has the merit of resembling a natural event that could occur in a tidepool or some benthic habitat where inward diffusion of oxygen might be less than metabolic demand.

Experiments of this type on nine species of marine invertebrates resulted in rates of heat production that initially agreed closely with caloric output calculated from the simultaneous rates of oxygen uptake, assuming carbohydrate oxidation (Hammen, 1983). In the bivalves especially, this agreement lasted only a few minutes, as respiratory rate fell rapidly with declining oxygen concentration, while heat production declined more slowly. Each species responded to declining [O_2] differently, but in general the VO_2 was one half the initial maximum rate when [O_2] was about 70% of air saturation. Heat output (Q_H) or total metabolism tended to be regulated, so that it reached one-half the aerobic value when [O_2] was near 20% of saturation. These experiments were limited to the first two hours of anoxia, and the minimum Q_H were 15 to 25 per cent of aerobic rates. Such rates would be expected when anoxia is periodic and temporary, as in the intertidal zone. Other studies, involving many hours of anoxia, have shown relative rates of 5 to 7 per cent.

Heat production depends primarily on rates of ATP hydrolysis, which in turn depends on basic cellular work, such as maintenance of concentration gradients across membranes, ciliary activity, etc. It should be possible to calculate the minimum metabolic rate necessary to preserve these functions. This would set the limit to duration of anoxia, and determine the tolerance of an euryoxic species. The potential contribution of combined calorimetry and respirometry is not clear. A convenient and moderately priced apparatus remains to be developed.

Ways to Further Progress

Metabolic Rates in Nature

Before the importance of anoxic metabolism can be assessed properly, more information on the fraction of time each species normally remains anoxic is needed. Is it a significant fraction of the lifespan, or only accidental and occasional? The impact of a given species on the metabolism of its community or ecosystem depends on its use of oxygen over days, weeks and years. Close observation of animals in nature is useful. For example, Taylor (1976) reported that the ocean clam *Arctica islandica* normally spends 1-7 days buried in sand, respiring anaerobically, before resuming pumping activity and oxygen consumption. While buried, the heart rate of

Arctica fell from 10 to 1 beat per minute. Experiments that place more emphasis on imitating natural conditions are also valuable. For example, Pamatmat (1983) pointed out that sediment-dwelling animals often display greatly elevated average rates of metabolism when they are not in sediment.

Defined Strains of Experimental Animals

In order to compare results from different laboratories, standard laboratory animals of known genetic constitution must be developed, paralleling the strains of small mammals and microorganisms so widely used. Offspring of at least six species of bivalve molluscs, from single pair matings or few known parents, have been reared to various ages, in efforts to assess or reduce genetic variation (Gall and Busack, 1986). Physiological research could benefit from use of inbred strains developed for aquaculture. Investigators, however, are still collecting animals from local habitats, or even purchasing them from fishmarkets. This makes it difficult to determine how much hereditary and environmental factors contribute to the physiological differences observed. Genetically defined mussels, crabs, polychaetes etc., reared under standard conditions, could eliminate some of the extreme variability that plagues not only research on anoxia but comparative physiology in general.

Another step toward standardisation is a compilation of standard blood values, to serve as indicator of good health in experimental animals. There are many measurements of components of the blood or haemolymph of various species in the literature, but usually the range of values expected in normal, healthy animals is unknown. If the haemolymph of *Mytilus edulis* from St George, N.S. has pH 7.65 ± 0.02 while in aerated sea water, and pH 7.24 ± 0.03 after 8 hours of air exposure (Booth, McDonald and Walsh, 1984), can we assume that the same is true in mussels from other locations?

Quantitative Relations

More and better mathematics will benefit the study of anoxic metabolism. On the horizon are encouraging signs that computer-assisted modelling techniques are coming into use. For example, the calculation of NADH/NAD ratio, in response to influx of pyruvate into dehydrogenase systems important in anoxia, required the continuous solution of five simultaneous equations (Fields, 1983). The technique of analysing physiological processes by constructing models, and finding equations to express volumes of compartments and rates of exchange between them was already common when Riggs (1963) described how analogue computers can be used in 'electrical curve-fitting'. Within a few years, numerous attempts at modelling complex systems had been made. Those of special interest are concerned with pools of metabolites and the properties of enzymes mediating their interconversion. The concentration of NADH and some glycolytic intermediates were both calculated from apparent equilibrium constants, and measured during the transition to anaerobiosis on turning off the air supply to a batch of yeast cells, and similar oscillations were shown (Betz, 1968). Any model of intermediary metabolism of a metazoan entering or leaving anoxia is likely to include such oscillations.

On the level of over-all metabolism, the material and energy balance sheets and flow diagrams of ecology could prove useful, especially when combined with specific equations. The thermodynamic background of invertebrate anoxibiosis has been worked out, and characterised as a balance of economy *contra* power in environmental and physiological phases, respectively (Gnaiger, 1983). Much of the data needed to develop predictive models is not yet available.

At present, the literature is somewhat marred by decimal errors, dimensional errors, and use of old units that are not part of the SI. The practice of presenting the Lineweaver-Burk double-reciprocal plot of enzyme activity is still common, although this method has the serious defect that a least-squares line cannot be fitted to unweighted data (Riggs, 1963). The growth of computer literacy will not prevent errors, of course, but it will perhaps encourage more attempts at modelling and prediction in precise mathematical terms.

Intracellular Events

The goal of reduction of a biological problem is to describe in mathematical terms the underlying cellular and molecular events that support observable phenomena. Thus it is useful to be able to look inside cells and see molecules. Results of work with the method of nuclear magnetic resonance have begun to appear (Ellington, 1983). These show that intracellular pH and relative levels of arginine phosphate changed markedly in whelk heart under anoxia while ATP and related adenylates changed very little. I look forward to at least another decade of progress in research on anoxia in Metazoa, during which some of these suggestions may be adopted, and other new approaches may be even more fruitful.

References

Baldwin, E. (1964) *An Introduction to Comparative Biochemistry* (Cambridge University Press, Cambridge)

Betz, A. (1968) 'Oscillatory Control of Glycolysis as a Model for Biological Timing Processes'. In *Quantitative Biology of Metabolism* (A. Locker, ed.) (Springer-Verlag, New York)

Booth, C.E., McDonald, D.G., and Walsh, P.J. (1984) 'Acid-Base Balance in the Sea Mussel, *Mytilus edulis*. I. Effects of Hypoxia and Air-Exposure on Hemolymph Acid-Base Status'. *Marine Biology Letters* 5, 347-358.

Buchsbaum, R. (1948) *Animals Without Backbones*. 2nd Edn. (University of Chicago Press, Chicago)

Bueding, E. (1962) 'Comparative Aspects of Carbohydrate Metabolism'. *Federation Proceedings* **21**, 1039-1046

Davies, P.M.C. (1966) 'The Energy Relations of *Carassius auratus* L. II. The Effect of Food, Crowding and Darkness on Heat Production'. *Comparative Biochemistry and Physiology* **17**, 983-885

Ellington, W.R. (1983) 'Phosphorus Nuclear Magnetic Resonance Studies of Energy Metabolism in Molluscan Tissues'. *Journal of Comparative Physiology* **153**, 159-166

Fields, J.H.A. (1983) 'Alternatives to Lactic Acid: Possible Advantages'. *Journal of Experimental Zoology* **228**, 445-457

Florkin, M. (1966) *A Molecular Approach to Phylogeny* (Elsevier, Amsterdam)

Gall, G.A.E. and Busack, C.A. (eds.) (1986). *Genetics in Aquaculture II. Proceedings of the 2nd International Symposium* (Elsevier, Amsterdam)

Glaessner, M.F. (1984) *The Dawn of Animal Life* (University Press, Cambridge)

Gnaiger, E. (1977) 'Thermodynamic Consideration of Invertebrate Anoxibiosis'. In *Applications of Calorimetry in Life Sciences* (I. Lamprecht and B. Schaarschmidt, eds) (de Gruyter, Berlin)

Gnaiger, E. (1983) 'Heat Dissipation and Energetic Efficiency in Animal Anoxibiosis: Economy Contra Power'. *Journal of Experimental Zoology* **228**, 471-490

Hammen, C.S. (1969) 'Metabolism of the Oyster, *Crassostrea virginica*.' *American Zoologist* **9**, 309-318

Hammen, C.S. (1976) 'Respiratory Adaptations: Invertebrates'. *Estuarine Processes*, Vol. 1. (M. Wiley, ed.) (Academic Press, New York)

Hammen, C.S. (1983) 'Direct Calorimetry of Marine Invertebrates Entering Anoxic States'. *Journal of Experimental Zoology* **228**, 397-403

Hochachka, P.W.S. and Somero, G.N. (1973) *Strategies of Biochemical Adaptation* (W.B. Saunders, Philadelphia)

Jackson, D.C. (1968) 'Metabolic Depression and Oxygen Depletion in the Diving Turtle'. *Journal of Applied Physiology* **24**, 503-509

Livingstone, D.R., de Zwaan, A., Leopold, M. and Marteijn, E. (1983) 'Studies on the phylogenetic distribution of pyruvate oxidoreductases'. *Biochemical and Systematic Ecology* **11**, 415-425

Pamatmat, M.M. (1983) 'Measuring Aerobic and Anaerobic Metabolism of Benthic Infauna Under Natural Conditions'. *Journal of Experimental Zoology* **228**, 405-413

Prosser, C.L. (1960) 'Comparative Physiology in Relation to Evolutionary Theory'. In *Evolution after Darwin* (S. Tax, ed.) (University of Chicago Press, Chicago)

Riggs, D.S. (1963) *The Mathematical Approach to Physiological Problems* (Williams and Wilkins, Baltimore)

Sato, M. and Gade, G.(1986). 'Rhodoic acid dehydrogenase: a novel amino acid-linked dehydrogenase from muscle tissue of *Haliotis* species'. *NaturWissenschaft* **73**, 207-209

Taylor, A.C. (1976) 'Burrowing Behaviour and Anaerobiosis in the Bivalve *Arctica islandica* L.' *Journal of the Marine Biological Association of the U.K.* **56**, 95-109

Index

Page references shown in italics indicate figures, those in bold type are tables.

AAT, *see* Asparate aminotransferase
Abalone 200
Acanthamoeba castellanii 123
Acanthocephalans 146, 147, **151**, **159**
Acetoacetyl-CoA **175**
Acetoin 147, 155, 156
Acetyl-CoA 171, *173*, **175**, 260–1
   helminths *156*, 158, 160
Aconitase 158
Acritarchs 45, 57–8
Acrobeloides beutschlii 137
Acute renal failure 240, 244
Adductor muscle, molluscs 189, **195**, 197, *198*, 203, *204*, 205, 206, 209, 210, 211, *212*
Adelaide Fold Belt 45–53, *46*, *47*
Adriamycin 21
Aerobic oxidative enzymes 273
Aerobic processes
   energy production in opisthsoma 226
   role of 156–8, *156*
Ageing 3
α-glycerophosphate dehydrogenase 99, 102
Alanine 167, 188, 189, 193, 197, *198*, 204, 205, *207*, 208, 209, 212, 221, 225, 227, 228, 230, 231
   anaerobic formation 202–5
   D-alanine 168, 205
   intracellular location of formation 206–8
   L-alanine 170
Alanine aminotransferase 168, *190*, 191, 197, 206, 208
Alanine dehydrogenase 147, 177
Alanine racemase 168, 205
Alanopine 147, 200, 273
ALAT, *see* Alanine aminotransferase
Alcohol dehydrogenase 147
Alcohols, as end-products 146
Aldehyde oxidase 21
Algal blooms 187
Allopurinol 25
Allosteric effectors, annelids 170

Alloxan-induced diabetes 23
Alma emini 165
Alvinella pompeijana 116
Amino acids 54, 55, 56, **57**, 58
   anaerobic metabolism 177
   catabolism 146, 160
Aminotransferases *190*, 203, 208
Amoeba, *see Acanthamoeba castellanii*
Ammonia, ruminal 253, 255
Ammoniogenesis, molluscs 203, 204
AMP-deaminase 205
Amphipods 219
Amphitrite sp. 165
Anaerobic
   end-products **149**, **150**, **151**, 154–6
   energy production in prosoma 226
   falcultative 271
   fumerate reduction 271
   metabolism
      high power output 197–202
      locusts 232
      low power output 189–97, *190*
   sulphate reducers 110
Anaerobiosis
   coelimic fluid, role in anaerobic metabolism 176
   end products of anaerobic energy metabolism 166–7, **167**
   environmentally-dependent 166–78, 187, 188, *190*, 193, **194**, **195**, 196, 200, 201, 203, 206, *207*, 211, 213
   excretion volatile fatty acids 177
   functional, in arthropods 218, 222–3, 226
   mitochondria role in anaerobic energy metabolism 171–5, *172*, *173*, **174**, **175**
   switching period 169–70
   tissue specific differences 176
   transition period 167–9, *168*
   unresolved problems 177–8
Animal life, emergence of *40*, 43, *56*
Anisakis sp. 147

Annelids 102, 116, 117, 131, **132**, 147, 165–81, 220, 221, 273
  beta-oxidation 180
  cytochromes 180
  early phylogeny 73, 77, 81
  euryoxia in 165, 166, 179, 180
Anodonta cygnea 197, *198*, 203, *204*
Anoxia
  avoidance at cellular level 187–8
  environmental, blowfly larvae 227
  molluscs 187
  origins 186–7
Anoxibiosis 271–7
  evolutionary origin of 273–4
    'loss of synthetic capacity' 273
  very low metabolic rates 274–5
Anthracomedusa sp. 75
Antioxidant defence mechanisms 19, 20
  evolution of 24–5
Aphelenchus avenae 131, 137–8
Arachnids 225–6, 273
Archaeocyaths 84
Archaegastropod 113, 117, 273
Archiannelids 131
Arctica islandica 275–6
Arenicola marina 112, 165–78, **167**, *168*, *172*, *173*, **174**, 175, 179, 182, 205
  coelomic fluid 176
Arenicola sp. 157
ARF, *see* Acute renal failure
Arginine phosphate 277
Arthropoda 77, 147, 188, 200, 218–33, 273
  locomotion in 218
Ascaridia galli **149**, 153, **159**
Ascaris lumbricoides 148, **149**, *152*, 153, *156*, 157, **159**, 160
Ascaris sp. **98**, 102, 103, 147, 148, 154, 155, 157, 162, 271
  muscle 153
Ascaris suum **99**, 102
Ascorbate 21–2, 25
Asparaginase 203
Aspartate 166, *168*, 169, 170, 178, 181, 221
Aspartate aminotransferase 168, 206, 208
Aspartate catabolism 170
Aspartate–malate shuttle, molluscs *207*, 210
Asteriacites quinquefolius 71
Atherosclerosis 22–3
Atmosphere, early 3–4, 19, 28–9, 91, 92

ATPase 96
ATP cycle 239, *240*
  generation rate 187
  turnover rate 187
Autoxidation reactions 6, 7, 21
Autotrophism 91

Bacillus subtilis **99**, 101
Bacteria
  aerobic photosynthetic 57, 58
  anaerobic 133, 153
  anaerobic methanogenic 255, 259, 264
  anaerobic photosynthetic 57, 58
  chemoautotrophic symbiosis 112, 116, 117, 121
  endosymbiotic sulphur 113, 115, 116, 117
  filamentous epibacteria 116–17
  passive defence 112–14, *114*
  proteolytic, ruminal 254
  purple non-sulphur 97
  ruminal *252*, 253–7, 265, 266
  sulphate reducing 42, 51, 58
  sulphur oxidising 112
  symbiotic intracellular, gastropods 117
  symbiotic sulphur 112, 116–21, *118*, *120*
  thermophilic 25
Bacteriocytes 117
Bacteriodes rumincola 255
Bacteriodes succinogenes 254, 255, 259
Balanus balanoides 221
Balanus nubilis 224
Barnacles 220, 222
  D-lactate 221
  D-lactate dehydrogenase 224
  see also Balanus balanoides, *Balanus nubilis*
Beltanella sp. 75
Benthic crabs 113
Benthic organisms 187
Beta-carotenes 20
Beta-oxidation
  annelids 180
  fatty acids 146
  helminths 147, 155, 160
Biosphere 1, 2, 4
Bivalve molluscs 71, 109, 113, 117, 122, 187, 189, 192, 193, 196, 197, 199–200, 201, 203, 205, 206, 207, 208, 209, 219, 220, 221, 274, 275, 276
  fresh water 196, 200

Index 281

intertidal 102
Lucinacea 117
Mytilidae 117
Solemyidae 117
Vesicomyidae 117
Blood
 oxygen content during diving 246
 standard values 276
 see also Red blood cells
Blowfly larvae 227
Brachiopods 84, 273
Brain damage 23
Branchial heart, cephalopod 199
Brassica campestris 26
Bronsted base 3, 11
Brooksella sp. 78
Buccinum undatum **195**, 201
Bunyerichnus sp. 53, 78
Busycon contrarium **195**, 201, 206
Butyrate 92
Butyrivibrio sp. 259
Bythograea thermydron 113, 122

Caenorhabditis briggsae 131, 137
Caenorhabditis elegans 131
Callianassa californiensis 219, 222
Callianassa jamaicense 219
Callinectes sapidus 222
Calorimetry 274–5
Calyptogena magnifica 114, 115
Cambrian 4, 38, *44*, *46*, 55, 73, 76, 77, 78–9, 81, 83–4
Canyonensis sp. 78
Carabid beetle, *see Pelochila borealis*
Carbohydrate metabolism 102
Carbon dioxide fixation 119, 121, 146–7, 148–54, 161, 271
Carbon fixation 109–110
Carbon monoxide binding to haemoproteins 157
Carcinogenic agents 23
 anti-carcinogenic agents 21
Carcinus sp. 222
Cardiac arrest 240, 241
Cardisoma sp. 222
Cardium edule 186, **195**, 206
Cardium tuberculatum 201
Carotenoids 25
Cartesian diver micro-respirometry 136, 137
CAT, *see* Catalases

Catalases 3, 20, 24, 25, 26
Catch muscle, molluscs **195**, 196, 197, 201
Catecholamines 21
Caulobacter crescentus 26
Cephalopods 186, 188, 196, 198, 199, 201, 202
Cestodes 146, 147, 148, **150**, **159**, 160
Chaoborus sp. 230, 231
 larvae 230
Charnia sp. 67, *70*, 75, 76
Charniodiscus sp. *69*, 75, 76
Chelicerata 224–6
 see also Arachnida, *Xiphosura sp.*
Chelogenic cyles 39, *40*
Chemiosmotic coupling 95
Chemotrophs 91
Cherax destructor 222, 224
Chironomids 136–7, 228, 229
Chironomus anthracinus 228
Chironomus plumosus 228
Chironomus thummi 228, 229, 230
Chitin 58, 59
Chlamys varius 197
Chloragogen tissue, annelids 176
Chloroplasts 57, 59
Chlorophyll 94, 96
Chordates 273
 origins 55
Chromosomal aberrations 23
Cirripedes 221, 272
Cirrolana borealis 221
Citrate synthase 158, 199
Citric acid cycle 189, 191–2
Clams
 ocean, *see Artica islandica*
 vent clam, *see Calyptogena magnifica*
 see also Lucinoma annulata, *Solemya reidi*
Cloudina sp. 49
Cnidaria 273
 anthozoa 74, 75–6
 ediacarian 73, 74, 75, 84
 hydrozoa 74–5, 83
 medusa 67, 75
 modern 81
 scyphozoa 74–5
CoA esters **90**
CoA transferase 155, *156*, 173, 174, 175
Cockroaches 232
Coelenterates 73–4
Coelomate origins 55, 56, 58, 59

Index

Coelomic fluids 80, 81
  in anaerobic metabolism 176
Collagen 58, 59, 73, 79, 80, 160, 199
Collembola 232
Collembola sp. 232
Conchopeltis sp. 74–5
Conomedusites sp. 74–5
Copepods 219
Coproporphyrinogen 159–60
Corbicula japonica 205
Cori cycle 202, 225
Corophium arenarium 219
Corophium volutator 219
Crabs 222, 276
  horse-shoe 225
  shallow water, *see Menippe mercenaria*
  vent, *see Bythograea therydron*
Crabtree effect 161, 238
Crangon sp. 223
Crassostrea gigans 205
Crassostrea virginica 271
Cratonisation 39–41, *40*
Crayfish, *see Orconectes sp.*
Crustacea 131, **132**, 135, 141, 272
  copepoda 135
  ostracoda 131, 135
Cryptobiosis, nematodes 129, 136, 137
Cuttlefish, *see Sepia officinalis*
CuZnSOD, *see* Superoxide dismutases
Cyanide insensitive respiration 157
Cyanobacteria 39, 41, 57, 58, 59
Cyclomedusa sp. 74
Cysteine-thiol mixed disulphate 115
Cytochromes 42, 56, **57**, 58, 59, **94**, 95, **99**, 101, 131, 138, 153, 157, 158, 160, 161, 162, 192
  dependent respiration 97
  sulphide binding 111, 113, 123
Cytochrome oxidase 20, 112, 113, *114*, 115, 122, 138
  annelids 180
  inhibited by sulphide 111, 123

Decapods 219, 220, 222, 223, 272
  high sulphide environments 113, 122
Decarboxylation 147, 148, 155
Defaunation of rumen 264, 265
Dehydrogenase systems 276
Dehydrogenation 92
Desmethylmenaquinone 99
Desmodora cazca 140

Desulphomaculum sp. 133
Desulphovibrio sp. 133
Detoxification of sulphide 113
Diatoms 187
Dickinsonia brachina 76
Dickinsonia sp. 67, *68*, 69, 73, 76, 77, 81, 82, 83
Digenea 146, 147, **151**, **159**
Dihydroorotate dehydrogenase 21
Dinoflagellates 187
Dioxygen
  atmosphere 2–3
  history 3–4
  multi-stage reduction in the cell 2, 13
  see also Oxygen
Diptera 78
Dismutases 11–12
  see also Superoxide dismutases
Diving animals 245–8
  fermentation pathways 239
  marine mammals 238, 244–8
  metabolic rate of voluntary diving seals 246
  internal SCUBA tank 245–6
  strategies during diving 247–8
Dorylaimus sp. 137
DNA repair mechanisms 19, 20
  evolution of 30
DNA strand scission 23, 30
Drosophila sp. 78

Earth
  crust 1, 3
  history 28
  mantle 3, 39
ECF, *see* Extracellular fluid
Echinoderms 188, 200, 273
Echiura 115
Ectoparasitic polychaetes, *see Spinther sp.*
Ediacarian 38, 52–3, 55, 59, 60, 67, 78, 79
  fauna 66, 67, 68, *69*, 70, *71*, 72, 73, 76–8, 81, 84, 273
  fossils 52
Ediacaria sp. 72, 75
Eels 212
Eisenia foetida 178, 179
Electrode potential 6
Electrons
  acceptors 88, 92, 93, 94, 95, 102, 171
  sinks 154, 160, 253, 255
  transfer in *Ascaris sp.* 102, 103, 171

Index   283

transport 20, 88–105, 192, 197
  'alternative' *104*
  chain 42, 58, 153, 155, 156, 158, 159, *190*
  'classical' *104*
  nematodes 138
  photosynthetic 97
  prokaryotes 57
  sulphide 123
Embden–Meyerhof pathway 169
  see also Glycolysis
Emerita talpoida 272
Emphysema 22
Endoparasites 173, 179, 180
Energy transduction 91
Enoploidea 134
Enoplus brevis 136
Enoplus communis 135, 136
Environmental factors 20, 21
  anaerobiosis 187, 188, *190*, 193, **194**, **195**, 196, 200, 203, 206, *207*, 211, 213
Eoporpita sp. 74
Ernietta sp. 78
Erythrocytes 22, 115
Escherichia coli 23, 24, 25, 26, 27, 30, **98**, **99**, *100*, 101
Ethanol
  in chironomid larvae 229
  as end-product 147, 155
Eubostrichys dianeie 116
Eudorylaimus andrassy 136
Euglena sp. 26
Euilyodrilus heuscheri 136
Eukaryotes
  early 60
  endosymbiotic origin 29, 59
  first 4
  higher, radiation of 45
Euryoxia 272, 273, 274, 275
  annelids 165, 166, 179, 180
  bivalves 208
  invertebrates 169, 179
  organisms 178
Eutrophication 187
  lakes and ponds 227
Euzonus mucronata 178
Evolution
  of early life 91, 92, *100*, 101, 112
  of earth 3
Exergonic coupling 15

Exoskeletons 80–1, 218
Experimental animals, defined strains 276
Extracellular fluid 239, 241, 242, 243

Faecal pellets 78, 134
Fasciola hepatica **151**, 153, 154, 157, **158**
Fatty acids 19, 147, 154, 155, 156, 159, 162
  beta-oxidation 146
  free fatty acids 181
  peroxidation of unsaturated 22
  polyunsaturated 21
  reductases 162
  volatile fatty acids 167, 175, 176, 177, 188, 192, 196, 203, 220
  excretion 177
  ruminal 250, *251*, *252*, 254, 257, 259, 262, 263, 264
Fe/MnSOD, *see* Superoxide dismutases
Fenton reaction 12–13
Ferritin 58
Ferrodoxins 21, 58, 59
FeSOD, *see* Superoxide dismutases
Fibrinopeptides 55
File shells 201
Fish
  sulphide tolerance 123
  sulphide toxicity 113
Flavoprotein dehydrogenase 21
Flavoproteins 95, *103*, *104*
Folic acid 25
Fossil metabolism 66
Free radicals 3, 8, 9, 11, 12
  and evolution 24
  mutagenicity 20, 23, 24, 30
  one-electron transfers 20
  reactions 20
Frost diagram 6, *8*
Fumerase 148
Fumerate, as electron acceptor, annelids 171
Fumerate reductase 88–105, 148, 153, 156, 157, 162, 171, 192, 205, 211, 272
  from *Ascaris sp.* 102, 103, 104
  from bacteria 99, 100, 101, 102
Fundulus parvipinnis 123
Fungi
  origins 58
  ruminal *252*, 258–9, 260, 264, 265, 266
Fusobacterium sp. 259

Galactose oxidase 21

GAPDH, *see* Glyceraldehyde 3-phosphate dehydrogenase
Gasterophilus sp. 227
Gastropods 187, 189, 193, 196, 197, 200, 201, 206, 207
 bacteria in 117
 marine 196
 terrestial 102, 196
Gastrotricha 113, 131, **132**, 135, 136, 141, 143
Gecarcinus sp. 222
Geothermal activity 110
Gibbs function 6, 7, 8, 12, 14, 16, **89**, **90**, 92, 95
Glaciation 38, 41, 42, 45, 48, *50*, 51, 52, 59, 60, 84
Glaessnerina sp. 76
Globin 54, 55, 60
 lamprey 55
Ginkgo biloba 26
Glucose, coelomic fluid of annelids 176
Glutamate dehydrogenase 147, 203, 205
Glutamate oxaloacetate transaminase 210
Glutamate pyruvate transaminase 191, 210
 in *Limulus sp.* 225
Glutaminase 203
Glutathione-protein mixed disulphides 115
Glyceraldehyde 3-phosphate dehydrogenase 92, 148, 158, 169, 188, 191, **198**, 212
Glycerol phosphate 233
 blowfly larvae 227
 shuttle 197, 198, 199
 spiders 226
Glycerol phosphate dehydrogenase 198
Glycogen 189, *190*, 196, 197, 200, 202, 206, *207*, 208, 220, 221, 228, 229, 238, 239, 248
Glycogen phosphorylase 168
Glycolysis 91, 92–3, 146, 147, 148, *152*, 160, 161, 167, 176, 179, 188, 198–9, 201, 202, *207*, 211, 258, 271, 173, 274
 in arthropods 221, 223, 225, 226, 229, 231, 232
 in hypoxia animals 238, 239, 247
Glycolytic flux 199, 201, 202, 211, 212, 213, 225, 238, 248
Glyconeogenesis 155, 156, 162, 202

Gnathostomulida 113, 131, 134, 135, 136, 141, 142, 143
Goldfish 212, 239, 274
GOT, *see* Glutamate oxaloacetate transaminase
GPDH, *see* Glycerol phosphate dehydrogenase
GPT, *see* Glutamate pyruvate transaminase
Granulocytes 22
Grasshoppers 232
Grenville Cycle 43–5, *44*, 45, 49

Haemocyanin 225
Haemoglobin 14, 21, 54, 55, 131, 138, 228–9
 diving mammals 245
 oesophageal, nematodes 136
 sulphate binding 113, 115, 120, 121, 122
Haemolymph 197, 199, 200, 201, 203, 220, 221, 222, 223, 225, 226, 227, 228, 276
Haemonchus contortus 155
Haemoproteins 21, 94
Haliotus dissus hannai 200
Helix aspersa 205
Helix pomatia 187, 197
Helminth
 evolution pathways 160–2
 metabolic regulations 161
 mitochondria 148, *152*, 153, 155, 157, 158, 160, 161, 162
Herbivory, adaptation to 250–1, *251*
Heterotrophs 60, 89, 90
 anaerobic 19
Hexokinase 161
Hexose fermentation **93**, 94
Hill's equation 130
Hirudines 165
Hirudo medicinales 178
Hirudo sp. 178–9
Homolactic fermentors 147, 161
Hyaluronic acid 23
Hydrogen peroxide 6, 7, 8, 9, 10, 12, 14, 20, 22, 24, 25, 29
Hydrogen sulphide 132–43
 metazoan adaptations 109–24
 neurotoxin 111–12
Hydrogen translocation 208–10, *209*, 211–13

Hydroperoxidase 24
Hydroperoxy radical 3, 7
Hydroquinones 21
Hydroxylases 159
Hydroxyl radical 3, 6, 7, 12–13
Hydroxylysine 58
Hydroxyprolin 58, 160, 162
Hymenolepis diminuta 147, 148, **150**, 153, 157, **159**
Hymenolepis microstomum 148, **150**
Hydrothermal vents 109–24
Hypothermia 244
Hypoxia 186, 187, 188, 201, 238–48, **242**
   cell functions during 241–3, *242*, 243–4
   channel arrest as defence against 238–48
   diving response 244–8
   environmental 201, 218, 219–20, 221, 227–31
   functional 201, 231–2
Hydroxyl radical 20, 24, 25
HySOD, *see* Superoxide dismutases

ICF, *see* Intracellular fluid
Iminodicarboxylic acid, *see* Opine
Indoleamine dioxygenase 21
Insects 227–32
   flight 218
Intracellular fluid 239, 241, 242, 243, 244
Invertebrates, euryoxia in 169, 179
Ion flux across cell membrane 239, 240
Ionizing radiation 19, 20
Iron formations
   banded, Greenland 4
   geological *48*, 49, 65, 66
Iron-sulphur proteins 95
Irving-Scholander diving response 245
Ischaemia 23, 25
   cerebral 241
   liver 240
   myocardial 240, 242
Isocitrate dehydrogenase 158, 171, 192
Isopods 219–20, 221

Jellyfish, *see* Cnidarian medusae, *Ediacaria sp.*

Lactate dehydrogenase 92, 148, 153, 200, 202, 273
   arthropods 224
   D-form in barnacles 224
   insects 231

   isoenzymes 202, 224
   Limulus sp. 225
Lactate fermenters, ruminal 254, *256*
Lactobacillus sp. 25–6
Large intestine, ruminal 259–60
Leech, *see Hirudo medicinalis*
Life, early forms 94
Limnoria lignorum 219, 220
Limulus sp. 224, 225
Lineweaver-Burk double-reciprocal plot 277
Lipoxygenase, in plants 157
Lithosphere 4
Litomosoides carinii 147, **149**
Littorina sp. 186
Liver fluke, *see Fasciola hepatica*
Lockeia amygdaloides 71
Locusts 232
Loligo vulgaris 202
Lucinacea 117
Lucinoma annulata 117
Lugworm, *see Arenicola marina*
Lumbricus rubellis 178, 179
Lumbricus sp. 178, 179
Lumbricus terrestris 178, 179
Lumbricus variegatus 178
Lungfish 239
Lung fluke, *see Paragonimus westermanis*
Lycopod wood 73
Lymnaea stagnalis 197
Lysosomes 22

Macrophages 21
   see also Phagocytosis
Malate aspartate shuttle *190*, 197–8, 199, 208, 209, 210, 211
Malate dehydrogenase 148, 152, 169, 170, 191, 192, 198, 208, 210, 211
Malate dismutation 148
Malic enzyme 148, 171, 192, 207, 209, 211
Mammals **98**, 240–1
   brain 241
   marine diving 238, 244–8
Manganese ions 26
Marisa cornaurietis 197
Marella sp. 77
Marywadea sp. 76
Mawsonites sp. 75
MDH, *see* Malate dehydrogenase
Medullary thick ascending limb 240–1
Meiobenthos, *see* Meiofauna

Meiofauna 129–43, **132**
Membrane functions, balance with metabolism 241–3
Menaquinone 99
Menippe mercenaria 113, 220–21
Mercenaria mercenaria 113, 208
Meretrix lamarckii 205
Mesodiplogaster lheritieri 137
Mesogloea 80, 81
Metabolism
  acidosis 188, 201
  arrest as a defence against hypoxia 238–48, *242*
    intervention strategy 240–1
    origins 238–40
Metal-dioxygen adducts 12, 14, 15, 16
Metazoa
  adaptations to hydrogen sulphide 109–24
  ATP production 123
  monophyly of 78, 79
  origins 53, 58, 60, 274
  oxygen and evolution 79, *82*
  radiation 65, 67, 78, 80–4
  relationship to ediacarian forms 67
  sulphide resistant mitochondria 122–3
Methanomicrobium mobile 263
Methemoglobin 21
  sulphide binding 111, 115
Methionine 94
2-Methylbutyrate 155, *156*
Methylmalonyl-CoA isomerase 193
Methylmalonyl-CoA mutase 155, 173, 175
Methylmalonyl-Coa racemase 155, 173, 174, 193
2-Methylvalerate 155, *156*
Microbial degradation of proteins 110
3-Mercaptopicolinic acid **174**, 197
Micrococcus lactilyticus, see Veillonella alcalescens
Micrococcus sp. 259
Midges, *see* Chironomids
Minor phyla 272
Mitochondria 57, *242*
  anaerobic energy metabolism 171–5, *172, 173,* **174, 175**
  ancestors of 97
  Ascaris sp. muscle 102, *103*
  chelonids 229
  helminths 148, *158*, 153, 155, 157, 158, 160, 161, 162

locusts' leg muscle 232
metabolism, annelids *168*, 169, 171–5, *172, 173,* **174, 175**
molluscs 188, *190*, 191, 193, 197, 199, 203, 204, 205, 206, *207*, 208, *209*, 210, 211
nematodes 138
spiders' leg muscle 226
sulphide oxidation by 117, 118, 119
sulphide resistance in metazoa 122–3
MnSOD, *see* Superoxide dismutases
Modiolus demissus 202, 208
Mole crab, *see Emerita talpoida*
Molecular clock, evolutionary 38, 54–8, *55*
Molluscs 81, 84, **98**, 115, 117, 173, 186–213, 272, 273
  foot **195**, 206, 209
  gill 197, 200
  heart 196
  mantle **195**, 196, 198, 199, 202, 205, *209*
  ventricle 187, 189, 206, 209, 211
  see also Bivalves, Gastropods
Moniezia expansa 148, **150**, 157
Moniliformis moniliformis 147, **151**
Mosquito larvae 231
mTAL, *see* Medullary thick ascending limb
Multicellularity, and adaptive radiation 65, 79–84
  evolution 60
Mushroom 26
Mussels, *see Mytilis edulis*
Mustard plant, *see Brassica campestris*
Myoglobin 55, 199, 246
Mysids 219
Mytilidae 117
Mytilis edulis 186, 189, 192, 193, **194**, *198*, 203, 205, 209, 210, *212*, 276
Mytilis sp. 136, 157, 169, 170

NADH reoxidation 211–13
NADH dehydrogenase 99
NADH/NAD$^+$ ratio 200, 276
[NAD$^+$]/[NADH] couple 153
NADH-quinone oxidoreductase 171
NADH-quinone reductase 192
Naiadids 137
Nassa mutabilis 201
Nassarius coronatus 201
Nautilus sp. 199
Nematodes **98**, 102, 113, 116, 129–43, **132**, *139*, 146–62

cryptobiosis 129, 136, 137
crystalloid inclusions 140
cytochromes 131, 138
cytology and ultrastructure 138–41
  electron transport 138
  marine 272
  mitochondria 138
  oxygen deprivation 135–6
  parasitic 59, 271–2
  salt-marsh 135
  symbiotic prokaryotes 138, 143
Nephtys hombergii 178
Nereids 178, 272
Nereis sp. 178
Nereis virens 178
Nippostrongylus brasiliensis 102, 104, 148, **149**, 153, 157, **159**
2-Nitropropane dioxygenase 21
Nucleoside diphosphate kinase 153
Nucleotide sequencing 79, 83
Nuphar luteum 26

Octocorallia sp. 76
Octopine 147, 190, 200, 201, 202, 213
Octopine dehydrogenase 201, 273
Octopine dehydrogenase isoenzymes 202
Octopus sp. 199
ODH, *see* Octopine dehydrogenase
Oligochaetes 136, 229
  marine 116
Opine 147, 188, 189, 191, 193, 200, 211, 212, 273
  crustacean 221
Orconectes limosus 220, 223, 224
Ostracoda 131, 135
Otocelis sp. 138
Ouabain 241, 243
Ovatoscutum sp. 71, 74
Owenia sp. 165
Oxidative detoxification 160
Oxidative phosphorylation 95, 99, 157, 230
Oxidases
  alternative 157
  non-respiratory chain 158
Oxidised red beds 4
2-Oxoglutarate dehydrogenase 158, 171, 173, 174, 192
Oxygen
  binding capacity 199
  carriers 13–16
  concentrations 187
  conformers 157
  consumption in helminths 157
  debt 158, **159**
  diffusion through body wall 80, 199
  divalent reduction 8
  electron acceptor 2, 104
  electronegativity 1
  'fitness' 2
  generated by photodissociation 39
  in atmosphere 1, 2, 4
  isotopes 5
  molecular structure 6–9, 7
  partial pressure in water 130, 131, 136, 157, 179–80, 181
  production
    abiotic 65
    biotic 65
  singlet 7, 9, 10, 20
  sinks 65–6
  tension 157, 160–1, 187
  tetravalent reduction 8, 20
  toxicity 1, 2, 3, 10, 19, 20, 23, 24, 80
  triplet 9, 10
  univalent reduction 9, 10, 12, 20
  uptake from atmosphere 188, 199
Oxygenases 159
Oyster, American, *see Crassostrea virginica*
Ozone 6, 7, 10, 20, 21, 24

PAL, *see* Present Atmospheric Level
Paragonimus westermanii **98**, **99**, 102
Paraplectonema sp. 137
Paraquat 21, 23, 24, 25
Parascaris equorum 148, **149**
Parasitic helminths 103, **104**, 138, 188, 271–2
Paravancorina sp. 76
Pasteur effect 158, **159**, 161, 193, 211, 238–9, 244, 247, 248
Patella vulgata 186
Pecten maximus 197
Pectinidae 186, 196
Pelochila borealis 227
Pelodera punctata 131, 137
Pennatulacea 76
Pentosan fermenters, ruminal 254
Pentose-phosphate pathway 160
PEP, *see* Phosphoenolpyruvate

PEPCK, *see* Phosphoenolpyruvate
    carboxykinase
Peroxides 3, 20, 24, 26
Peroxide 3, 7, 11
    see also hydrogen peroxide
Phagacytosis 21, 22
    macrophages 3
Phallodrilus leukodermatus 116
Phallodrilus planus 116
Phanerozoic 66, 71
Phosphagens 91, 166–7, **167**, 178, 179,
    181, 193, **194**, 221, 222, 223, 231
    Limulus sp. 225
    mosquito larvae 231
    spider 226
Phosphoarginine 91, 193, 200, 201, 211,
    221, 222, 223, 230, 232
    in locusts 232
Phosphocreatine 91
Phosphoenolpyruvate 93, 103, 169, 170,
    196
Phosphoenolpyruvate branchpoint 169,
    170, 196
Phosphoenolpyruvate carboxykinase 148,
    153, 169, 196
Phosphofructokinase 161
Phosphoglycerate kinase 161
Phospholombricine 91
Phosphopyruvate carboxylase 148
Phosphorylase 161
Phosphorylation 88, **90**, 91, 95, 97, 99,
    102, 271
    site 1 of ADP 153, 155, 156
    substrate linked 92, 166
Phosphotaurocyamine 91, 167, 170, 176
Photobacterium leiognathi 26
Photophosphorylation 96
Photosynthesis 4, 5, 19, 20, 28, 29, 91,
    109, 187
    algal 41
    bacteria 92, 96
    early 95, 97
    organisms 90
    prokaryotes (energy transfer) 42
Phototrophism 58
Phytoplankton 45, 59, 60
Placopecten magellanicus **194**, **195**, 196,
    197, 201, 202
Plankters 45, 58, 59, 60
Platelet aggregation 22
Platyhelminths 102, 162

Plectodiscus sp. 74
Pleurotus sp. 26
Pogonophora 115, 119, 272
Poikilotherms, aquatics 274
Polarity 1–2
Polychaetes 112, 116, 165, 178, 273, 276
Polymorphonuclear leukocytes 22
Polyunsaturated fatty acids 21
Pony fish 26
Porphyrins 94, 95, 96, 159
    ring 14
Porpita sp. 74
Praecambridium sp. 76
Precambrian 38, *40*, 43, 45, *46*, 49, 51, 55,
    65, 66, 67, 72, 73, 75, 76, 78–9, 80,
    84
    late 273
Present Atmospheric Level 66, 81, 83
PRH, *see* Pyruvate oxidoreductases
Proline hydroxylase 160
Prokaryotes
    fossil 41
    membrane 95
Prokaryotic ancestors 26–7, 29
Propanol, as end-product 155
Propionibacterium shermanii 27
Propionyl-CoA 155, *156*
Propionyl-CoA carboxylase 155, 175, 193
Proteinase inhibitor 22
Protein catabolism 203
Protein disulphide bridges 111, 115
Proterozoic 41–3, *44*
Proteus mirabilis 25
Protista 57, 59
Protochordates 55
Protons
    intracellular accumulation of 95
    motive force 88
    pumping 95
Protoporphyrin 159–60
Protozoa **98**, 102, 129, 133
    entodiniomorphs 257
    holotrichs 257
    ruminal 252, 257–8, 260, 265
Pseudocoelomates 116
Pseudomonas maltophilia 26
Pteridinium sp. 78
Pterin aldehyde 25
PUFA, *see* Polyunsaturated fatty acids
Pulmonata 187, 197
Pyrogallol 21

Pyruvate dehydrogenase 147, 155, 171, 192, 209, 229
Pyruvate kinase 148, 161, 169, 196
  phosphorylation of (annelids) 170
Pyruvate oxidase 224
Pyruvate oxidoreductase 198, 200–1, 211–13
Pyruvate reductase 188, 198, 199, *207*, 210, 211
Pyruvate transaminase 147

Quinone 95, 99, 100, 171
  tanning of trematode eggs 159
Quinone-fumerate reductase 192
Quinone oxidase 157, 158

Rangea sp. 75
Rat, kidney 240
RBC, *see* Red blood cells
RDZ, *see* Redox Discontinuity Zone
Red blood cells 245, 246
Redox balance 147, 148, 153, 154, 160–1, 188, 191, 198, 205, 206, 208–10, 230, 252
Redox Discontinuity Zone 133, 134, 138
Redox potential 92, *93*, **94**, 133–4, *139*, 171
Redox reactions 2, 6, 15
Reperfusion 23, 25
Respiration 4
Respiratory chain 171
  end-products 102
  proteins 80, 84
Retronectidae 135
Rhizostomites sp. 75
Rhodanase 123–4
Rhodoquinone **94**, 99, 103, *104*, 157
Riftia pachyptila 113, 114, 115, 119, *120*, 121, 122
Rotifera 131, 136
Rumen
  activity 252–3
  ecology, exploitation 263–6
  end-products fermentation *252*, 253
  fermentation 250, 252, 253
  microorganisms 250, 253–9
Ruminants 250–67, *251*
  evolution 250–1, *251*
  gastric digestion *251*
  intermediary metabolism 260–2
  microorganisms 155
Ruminococcus flavifaciens 254, 255

Sabatieria teterula 140, **142**
Sabelliditids 53, 55
Saccharomyces cereviseae **99**
Saccharomyces sp. 266
Salmonella typhimurium 23, 24, 25
Saprophagous insects 227
Scallop 186, 188, 196, 197, 200, 201
Scapharca inequivalvis 186, 206
Schistosoma mansoni 147, **151**
Sclerotisation 80
Scoloplos armiger 178, 180, 181
Scrobicularia plana 186
Seal, grey 245
Sea-pen, *see Charnia sp.*
Sediment-dwelling animals 276
Segmented worm, *see Dickinsonia sp.*
Selenomonas ruminantium 254, 255, 256, 263
Selenomonas sp. 259
Sepia officinalis 197, 198, 201, 202
Shipworm, *see Teredo navalis*
Shrimps 219, 223
  see also *Crangon sp.*
Sialic rocks 39
Silification of fossils 43
Skinnera sp. 78
Slugs 187
Snails, freshwater 187
SOD, *see* Superoxide dismutases
Solar radiation 187
Solemya reidi 117, *118*, 119, 121, 122
Solemyidae 117
Solenofilomorpohas funilis 138
Solenofilomorphidae 135
Sorbitol, in blowfly larvae 227
Sphaerolaimus sp. 140
Spiders 225, 226
Spinther sp. 73
Spleen, red blood cell reservoir 245–6
Sponges 272
Spriggina sp. 69, 74, 76, 77, 83
Squalene 159
Starch/sugars fermenters, ruminal 254
Starfish, *see Asteriacites quinquefolius*
Stenoxia 186, 272, 273
Steroids 21–2
Steroid synthesis 159
Stroke 240
Stromatolites 39, 42, 43
Strombine 167, *168*, 169, 176, *190*, 200, 201, 213, 273

Strombus luhuanus **195**, 201
Strongyluris brevicaudata 147
Stylommatophora 187, 196
Squid, *see Symphlectotenthis onalaniensis*
Succinate, coelomic fluid, annelids 176
Succinate decarboxylation, rumen
  microorganisms 255, *256*
Succinate dehydrogenase 97, 101, 102,
  103, 104, 105, 153, 155, 199
  in *Nippostrongylus brasiliensis* 104
Succinate-propionate cycle, annelids *173*,
  174
Succinate-quinone oxidoreductase 171
Succinate-ubiquinone reductase 102
Succinic thiokinase 155
Succinyl-CoA 94, 155, 160, 193
Succinyl-CoA synthetase 173, 174, 175
Sulphate
  reduction, microbial 110
  terminal electron acceptor 110
Sulphide
  anion 111
  biome 133–43
  binding proteins 112, 113, 115–16
  detoxification 113, 119, 121, 124, 138,
    140
  effects on proteins 112
  layer, meiofauna 112, 113, 114
  metabolism, metazoa 116–21, 121–4
  oxidase enzxyme, mitochondrial 123
  oxidation 109–24
  production 110
  protection mechanisms 112–16, *114*
  resistant enzymes 113
  surface impermeability 113
  symbiosis, bivalves 117, *118*
  toxicity 110, 111, 112, 113, 115
Sulphite oxidase 123
Superoxide 3, 6, 7, 8, 9, 11–12
Superoxide dismutases 3, 20, 24, 25–30, 80
  copper-zinc containing (CuZnSOD) 26,
    28, 29, 30
  iron and manganese containing (FE/
    MnSOD) 26, 28, 29–30
  iron containing (FESOD) 26–7, 29
  manganese and iron containing
    (HySOD) 27, 30
  manganese containing (MnSOD) 27, 30
Superoxide radical 20, 21, 22, 27
  toxicity 23, 30
Surface trails 79

Surface/volume ratios 80, 81
Symbionts
  cellulose-digesting, ruminal 250, *252*,
    253–6, *256*, 258, 259, 264
  endozoic bacteria-like 138
Symphlectoteuthis oualaniensis 199
Synthetic reactions, oxygen linked 158–60,
  159

Taenia taeniaeformis 148, **150**, **159**
Tanais chevreuxi 219
Tapes philippinarum 205
Tarantulas 226
Tardigrada 131, 137
Tateana sp. 74
Taurine 200, 273
Tauropine dehydrogenase 273
TCA cycle, *see* Tricarboxylic acid cycle
TDH, *see* Tauropine dehydrogenase
Tectonic processes 38–9
Tellina planata 197
Teredo navalis 219, 220
Terschellingia longicaudata *140*, **141**
Thalassinids 219
Thiobios 109, 133–43, **142** .
Thiol/disulphide exchange reaction 115
Thiols 110, 115
Thisulphate 117, *118*, 119, 122
Tissue thickness 81, *82*, 83
Tobacco smoke 21, 22
Tobrilus gracilis 137, 140
Tocopherols 21, 22, 25
Trace fossils 53, 59, 71, 83
Transamination 189, 203, 208, 211
Transhydrogenase, non-energy
  linked 148, 192
Transition metal ions 20
Translocases 148, 162
Transphosphorylation 193, 200, 201
Trematodes **98**
Tribrachidium sp. 78
Tricarboxylic acid cycle 80, 91, 97, 105,
  148, 158, 160, 171, 173, 260, 261,
  262
Trichoderma reesei 266
Trilobites 84
Trilobozoa 78
  Albumares sp. 78
  Skinnera sp. 78
  . *Tribrachidium sp.* 78
Trophosome 119, *120*, 121, 122

Trout 212
Trytophan dioxygenase 21
Tube worm, *see Riftia pactyptila*
Tubifex sp. 165, 178, 179, 180, 229
Tubificids 137, 229
Tubificoides benedii 117
Turbellaria 113, 131, **132**, 135, 141, 142, 143
  acoel 138
  rhabocoel 136–7
Turtles, diving 239, 274
Trypanosoma cruzi 102
Tyrasotaenia? 53

Ubiquinone 100, 103, 157, 171
Uca pugnax 222
Upogebia pugettensis 219, 222
Urey effect 28

Variation, intraspecific 189
Vaska's compound 14
Veillonella alcalescens 97, **98**, 99, 153, 255
Velella sp. 74
Vendia sp. 76
'Vendozoa' 77
Venus gallina 187
Venus sp. 196

Vertebrates 189, 200, 210
  ectothermy 238, 239
  endothermy 243, 244, 248
  muscle 273
  thermogenesis 243
Vesicomyidae 117
Virgularia sp. 76
Vitamin C 20
Vitamin E 20

Water
  dissociation of water 19
  parent solvent 2
  photolysis 4, 28, 38–9, 65
  radiolysis 3
Waterbugs 232
Water lily, *see Nuphar luteum*
Weathering 4
Weddell seals 245–8
Whelk 201, 206, 277
Wolinella succinogines **98**, **99**, 101

Xanthine oxidase 21, 25
Xenobiotics 21
Xiphosura sp. 224–5

Yeasts **98**, 99

DATE DUE

Demco, Inc. 38-293